D0760911

Biological Physics of the Developing Embryo

During development, cells and tissues undergo dynamic changes in pattern and form that employ a wider range of physical mechanisms than at any other time during an organism's life. *Biological Physics of the Developing Embryo* presents a framework within which physics can be used to analyze these biological phenomena.

Written to be accessible to both biologists and physicists, major stages and components of biological development are introduced and then analyzed from the viewpoint of physics. The presentation of physical models requires no mathematics beyond basic calculus. Physical concepts introduced include diffusion, viscosity and elasticity, adhesion, dynamical systems, electrical potential, percolation, fractals, reaction–diffusion systems, and cellular automata.

With full-color figures throughout, this comprehensive textbook teaches biophysics by application to developmental biology and is suitable for graduate and upper-undergraduate courses in physics and biology.

GABOR FORGACS is George H. Vineyard Professor of Biological Physics at the University of Missouri, Columbia. He received his Ph.D. in condensed matter physics from the Roland Eötvös University in Budapest. He made contributions to the physics of phase transitions, surface and interfacial phenomena and to statistical mechanics before moving to biological physics, where he has studied the biomechanical properties of living materials and has modeled early developmental phenomena. His recent research on constructing models of living structures of prescribed geometry using automated printing technology has been the topic of numerous articles in the international press.

Professor Forgacs has held positions at the Central Research Institute for Physics, Budapest, at the French Atomic Energy Agency, Saclay, and at Clarkson University, Potsdam. He has been a Fulbright Fellow at the Institute of Biophysics of the Budapest Medical University and has organized several meetings on the frontiers between physics and biology at the Les Houches Center for Physics. He has also served as advisor to several federal agencies of the USA on the promotion of interdisciplinary research, in particular at the interface of physics and biology. He is a member of a number of professional associations, such as The Biophysical Society, The American Society for Cell Biology, and The American Physical Society.

STUART A. NEWMAN is Professor of Cell Biology and Anatomy at New York Medical College, Valhalla, New York. He received an A.B. from Columbia University and a Ph.D. in Chemical Physics from the University of Chicago. He has contributed to several scientific fields, including developmental pattern formation and morphogenesis, cell differentiation, the theory of biochemical networks, protein folding and assembly, and mechanisms of morphological evolution. He has also written on the philosophy, cultural background and social implications of biological research.

Professor Newman has been an INSERM Fellow at the Pasteur Institute, Paris, and a Fogarty Senior International Fellow at Monash University, Australia. He is a co-editor (with Brian K. Hall) of *Cartilage: Molecular Aspects* (CRC Press, 1991) and (with Gerd B. Müller) of *Origination of Organismal Form: Beyond the Gene in Developmental and Evolutionary Biology* (MIT Press, 2003). He has testified before US Congressional committees on cloning, stem cells, and the patenting of organisms and has served as a consultant to the US National Institutes of Health on both technical and societal issues.

Biological Physics of the Developing Embryo

Gabor Forgacs
University of Missouri
and
Stuart A. Newman
New York Medical College

CAMBRIDGE
UNIVERSITY PRESS

CAMBRIDGE UNIVERSITY PRESS
Cambridge, New York, Melbourne, Madrid, Cape Town, Singapore, São Paulo

Cambridge University Press
The Edinburgh Building, Cambridge CB2 2RU, UK

Published in the United States of America by Cambridge University Press, New York

www.cambridge.org
Information on this title: www.cambridge.org/9780521783378

First published 2005

Printed in the United Kingdom at the University Press, Cambridge

A catalog record for this publication is available from the British Library

Library of Congress Cataloging in Publication data

ISBN-13 978-0-521-78337-8 hardback
ISBN-10 0-521-78337-2 hardback

Contents

Acknowledgments

The writing of this text, addressed simultaneously to biologists and physicists, presented us with many challenges. Without the help of colleagues in both fields the book would still be on the drawing board. Of the many who advised us, made constructive remarks, and provided suggestions on the presentation of complex issues, we wish to thank particularly Mark Alber, Daniel Ben-Avraham, Andras Czirók, Scott Gilbert, James Glazier, Tilmann Glimm, Michel Grandbois, George Hentschel, Kunihiko Kaneko, Ioan Kosztin, Roeland Merks, Gerd Müller, Vidyanand Nanjundiah, Adrian Neagu, Olivier Pourquié, Diego Rasskin-Gutman and Isaac Salazar-Ciudad. Commentary from students was indispensable; in this regard we received invaluable help from Richard Jamison, an undergraduate at Clemson University, and Yvonne Solbrekken, an undergraduate at the University of Missouri, Columbia, who read most of the chapters.

We thank the members of our laboratories for their patience with us during the last five years. Their capabilities and independence have made it possible for us to pursue our research programs while writing this book. Gabor Forgacs was on the faculty of Clarkson University, Potsdam, NY, when this project was initiated, and some of the writing was done while he was a visiting scholar at the Institute for Advanced Study of the Collegium Budapest. Stuart Newman benefited from study visits to the Indian Institute of Science, Bangalore, the Konrad Lorenz Institute, Vienna, and the University of Tokyo-Komaba, in the course of this work.

In a cross-disciplinary text such as this one, graphic materials are an essential element. Sue Seif, an experienced medical illustrator, was, like us, new to the world of textbook writing. Our interactions with her in the design of the figures in many instances deepened our understanding of the material presented here. Any reader who accompanies us across this difficult terrain will appreciate the freshness and clarity of Sue's visual imagination.

Harry Frisch introduced the authors to one another more than a quarter century ago and thought that we had things to teach each other. Malcolm Steinberg, a valued colleague of both of us, showed the way to an integration of biological and physical ideas. Judith Plesset, our program officer at the National Science Foundation, was instrumental in fostering our scientific collaboration during much of the intervening period, when many of the ideas in this book were gestated. We are grateful to each of them and for the support of our families.

Introduction: Biology and physics

Physics deals with natural phenomena and their explanations. Biological systems are part of nature and as such should obey the laws of physics. However correct this statement may be, it is of limited value when the question is how physics can help unravel the complexity of life.

Physicists are intellectual idealists, drawing on a tradition that extends back more than 2000 years to Plato. They try to model the systems they study in terms of a minimal number of "relevant" features. What is relevant depends on the question of interest and is typically arrived at by intuition. This approach is justified (or abandoned) after the fact, by comparing the results obtained using the model system with experiments performed on the "real" system. As an example, consider the trajectory of the Earth around the Sun. Its precise details can be derived from Newton's law of gravity, in which the two extensive bodies are each reduced to a point particle characterized by a single quantity, its mass. If one is interested in the pattern of earthquakes, however, the point-particle description is totally inadequate and knowledge of the Earth's inner structure is needed.

An idealized approach to living systems has several pitfalls – something recognized by Plato's student Aristotle, perhaps the first to attempt a scientific analysis of living systems. In the first place, intuition helps little in determining what is relevant. The functions of an organism's many components, and the interactions among them in its overall economy, are complex and highly integrated. Organisms and their cells may act in a goal-directed fashion, but how the various parts and pathways serve these goals is often obscure. And because of the enormous degree of evolutionary refinement behind every modern-day organism, eliminating some features to produce a simplified model risks throwing the baby out with the bath water.

Analyzing the role of any component of a living system is made all the more difficult by the fact that whereas many cellular and organismal features are functional adaptations resulting from natural selection, some of these may no longer serve the same function in the modern-day organism. Still others are "side effects" or are characteristic ("generic") properties of all such material systems. To a major extent, therefore, living systems have to be treated "as is", with complexity as a fundamental and irreducible property.

One property of a living organism that sets it apart from other physical systems is its ability and drive to reproduce. When physics is used to understand biological systems it must be kept in mind that many of the physical processes taking place in the body will be organized to serve this goal, and all others must at least be consistent with it. The notion of goal-directed behavior is totally irrelevant for the inanimate world.

Physics and biology differ not only in their objects of study but also in their methods. Physics seeks to discover universal laws, valid everywhere in time and space (e.g., Newton's laws of motion, the laws of thermodynamics). Theory expresses these general laws in mathematical form and provides "models" for complex processes in terms of simpler ones (e.g., the Ising model for phase transitions, diffusion-limited aggregation models of crystal growth). Biology also seeks general principles. However, these are recognized either to be mechanisms or modes of organization limited to broadly defined classes of organisms (eukaryotes vs. prokaryotes; animals vs. plants) or to be molecular commonalities reflecting a shared evolutionary history (the use of DNA as hereditary material; the use of phospholipids to define cell boundaries). Biological "laws," where they exist (e.g., the promotion of evolution by natural selection, the promotion of development by differential gene expression) are rarely formulated in mathematical terms. (See Nanjundiah, 2005, for a discussion of the role of mathematics in biology.) There are very few general laws in biology and the ones that exist are much less exact than in physics.

This is not to say that there are no good models for biological processes, but only that they have a different function from models in physics. Biologists can study subcellular systems, such as protein synthesis or microtubule assembly, in a test tube, and cellular interactions, such as those producing heartbeats and skeletons, in culture dishes. Both types of experimental set-up – for cell-free systems and for living tissues outside the body – have been referred to as *in vitro*. It is always acknowledged, however, that, unlike in physics, the fundamental process is the *in vivo* version in its full complexity, not the abstracted version. There is always the hope that the experimentally accumulated knowledge of biological systems will lead eventually to the establishment of fundamental organizing principles such as those expressed in physical laws. It is however possible that the multileveled and evolutionarily established nature of cells and organisms will continue to defeat this hope.

Physicists and biologists also look at the same things in different ways. For a physicist, DNA may simply be a long polymer with interesting elastic properties. For the biologist, DNA is the carrier of genetic information. The sequence of bases, irrelevant to the physicist's concerns, becomes of central importance to the biologist studying how this information is stored in the molecule and how it is processed to produce specific RNAs and proteins. Because biologists must pay attention to the goal-directed aspects of living systems, the properties studied are always considered in relation to possible contributions toward the major goal of reproduction and subsidiary goals such as locomotion toward nutrients and increase in size and complexity.

Because biological systems are also physical systems, phenomena first identified in the nonliving world can provide models for biological processes too. In many instances, in fact, we may assume that complexity and integration in living organisms have evolved in the context of forms and functions that originally emerged by straightforward physical means. In the following chapters we will introduce physical mechanisms that may underlie and guide a variety of the processes of early animal development. In certain cases simple physical properties and driving forces are the determining factors in a developmental episode. In other cases developmental causality may be multifactorial, which is to say that evolution has recruited physicochemical properties of cells and tissues on many levels. An appreciation of the connection between physics and biology and the utility of biological physics for the life scientist will ultimately depend on the recognition of both the "simple" and the multifactorial physical determination of biological phenomena. When we come to consider the evolution of developmental mechanisms we will discuss scenarios in which simple physical determination of a biological feature appears to have been transformed into multifactorial determination over time.

The role and importance of physics in the study of biological systems at various levels of complexity (the operation of molecular motors, the architectural organization of the cell, the biomechanical properties of tissues, and so forth) is being recognized to an increasing extent by biologists. The objective of the book is to present a framework within which physics can be used to analyze biological phenomena on multiple scales. In order to bring coherence to this attempt we concentrate on one corner of the living world – early embryonic development. Our choice of this domain is not entirely arbitrary. During development, cells and tissues undergo changes in pattern and form in a highly dynamic fashion, using a wider range of physical processes than at any other time during the organism's life cycle.

Physics has often been used to understand properties of fully formed organisms. The mechanics of locomotion in a vertebrate animal, for example, involves the suspension and change in orientation of rigid bodies (bones) connected by elastic elements (ligaments, tendons, muscles). The ability of some of the elastic elements (the muscles) to generate their own contractile forces distinguishes musculoskeletal systems from most nonliving mechanical systems – hence "biomechanics." A developing embryo, in contrast, is much less rigid: rather than simply changing the orientation of its parts, it continuously undergoes remodeling in shape and form. Embryonic cells can slip past one another or be embedded in pliable, semi-solid matrices. Thus, the physical processes acting in an early organism are predominantly those characterizing the behavior of viscoelastic "soft matter" (a term coined by the physicist Pierre-Gilles de Gennes; de Gennes, 1992), rather than the more rigid body systems typical of adult organisms.

Another reason to concentrate on early development is that it is here that the role of physics in constraining and influencing the outcomes of biological processes is particularly obvious. Early developmental phenomena such as blastula formation and gastrulation are examples of *morphogenesis*, the set of mechanisms that create complex biological forms out of simpler structures. While each episode of developmental change is typically accompanied by changes in the expression of certain genes, it is clear that gene products – RNA and protein molecules – must act in a specific physical context in order to produce three-dimensional forms and patterns. The laws of physics establish that not every structure is possible and that programs of gene expression can only produce shapes and forms of organisms and organs within defined limits.

The development of the embryo is followed from the fertilized egg to the establishment of body plan and organ forms. We close the developmental circle by discussing fertilization, the coming together of two specialized products of development – the egg and sperm – whose interaction employs certain physical processes (such as electrical phenomena) in a fashion distinct from other developmental events.

We conclude with a discussion of how developmental systems were likely to have originated from the physical properties of the first multicellular forms. The topology and complexity of gene regulatory networks may have had independent evolutionary histories from their associated biological forms. We therefore also review in this section computational models that test such possibilities.

Major stages of the developmental process and the major participating cellular and molecular components are introduced in terms familiar to students of biology, and sufficient background is provided to make these descriptions accessible to non-biologists. These developmental episodes are then analyzed from the viewpoint of physics (to the extent allowed by our present knowledge). No preparation beyond that of introductory calculus and physics courses will be needed for an understanding of the physics presented. Physical quantities and concepts will be introduced mostly as needed for the analysis of each biological process or phenomenon. Complex notions that are important but not essential for comprehending the main ideas are collected in boxes in the text and a few worked examples are included in the early chapters. We avoid presenting the basics of cell and molecular biology (for which many excellent sources already exist), beyond what is absolutely needed for understanding the developmental phenomena discussed. Each chapter concludes with a "Perspective," which briefly summarizes its major points and, where relevant, their relation to the material of the preceding chapters.

We have drawn on an abundant and growing literature of physical models of biological processes, including phenomena of early development. The choice of models reflects our attempt to introduce the biology student to the spirit of the physical approach in the most straightforward fashion and to help the physics student appreciate the range of biological phenomena susceptible to this approach. We continually emphasize the constraints associated with any realistic application of physical models to biological systems. In line with this, we focus on the biology-motivated formulation of quantitative models, rather than the solution of the resulting mathematical equations.

We have attempted, as far as possible, to make each chapter self-contained (with ample cross-referencing). Moreover, because biological development employs a wide range of physical processes at multiple spatial and temporal scales, we have made a point, wherever relevant, of introducing novel physical concepts and models for each new biological topic addressed.

Finally, biological physics is a relatively young discipline. With the constant improvement of experimental and computational techniques, the possibility of studying complex biological processes in a rigorous and detailed fashion has emerged. To be capable in this endeavor one has to be versatile. Biologists and medical researchers today and in the future will increasingly use sophisticated investigative techniques invented by physicists and engineers (e.g., atomic force microscopy, magnetic resonance, neutron scattering, confocal microscopy). Physicists will be called on to characterize systems of increasing complexity, of which living systems are the ultimate category. There is no way in which anyone can be an expert in all aspects of this enterprise. What will be required of the scientist of tomorrow is the ability to speak the language of other disciplines. The present book attempts to help the reader to become at least bilingual.

Chapter 1

The cell: fundamental unit of developmental systems

For the biologist the cell is the basic unit of life. Its functions may depend on physics and chemistry but it is the functions themselves – DNA replication, the transcription and processing of RNA molecules, the synthesis of proteins, lipids, and polysaccharides and their building blocks, protein modification and secretion, the selective transport of molecules across bounding membranes, the extraction of energy from nutrients, cell locomotion and division – that occupy the attention of the life scientist (Fig. 1.1). These functions have no direct counterparts in the nonliving world.

For the physicist the cell represents a complex material system made up of numerous subsystems (e.g., organelles, such as mitochondria, vesicles, nucleus, endoplasmic reticulum, etc.), interacting through discrete but interconnected biochemical modules (e.g., glycolysis, the Krebs cycle, signaling pathways, etc.) embedded in a partly organized, partly liquid medium (cytoplasm) surrounded by a lipid-based membrane. Tissues are even more complex physically – they are made up of cells bound to one another by direct adhesive interactions or via still another medium (which may be fluid or solid) known as the "extracellular matrix." These components all have their own physical characteristics (elasticity, viscosity, etc.), which eventually contribute to those of the cell itself and to the tissues they comprise. To decipher the working of even an isolated cell by physical methods is clearly a daunting task.

The eukaryotic cell

The types of cells discussed in this book, those with true nuclei ("eukaryotic"), came into being at least a billion years ago through an evolutionary process that brought together previously evolved "prokaryotic" living units. Among these were "eubacteria" and "archaebacteria," organisms of simpler structure in which information specifying the sequence of proteins was inherited on DNA present as naked strands in the cytosol, rather than in the highly organized DNA–protein complexes known as "chromatin," found in the nuclei of eukaryotes.

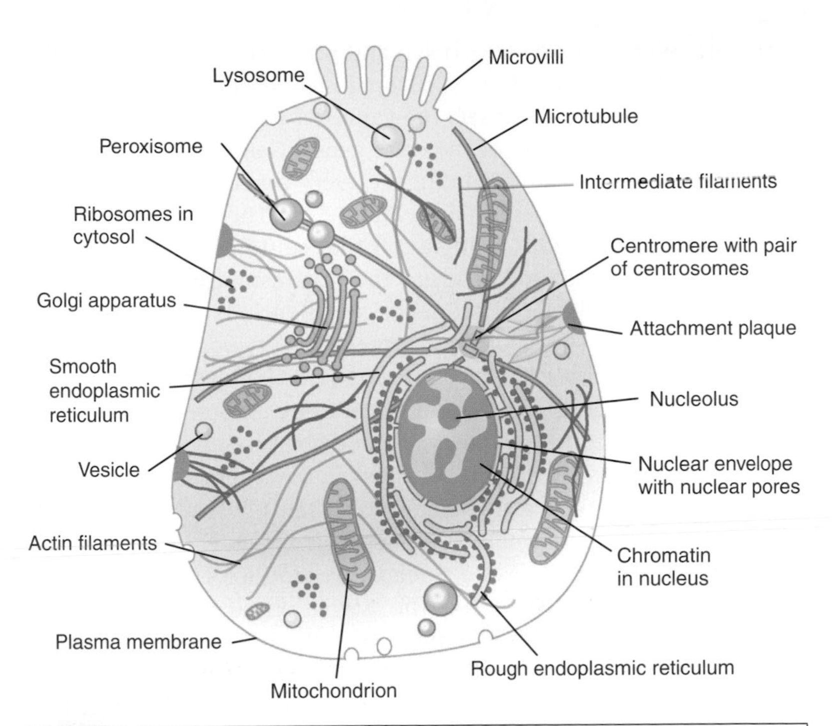

Fig. 1.1 Schematic representation of an animal cell. Most of the features shown are characteristic of all eukaryotic cells, including those of protists (e.g., protozoa and cellular slime molds), fungi, and plants. Animal cells lack the rigid cell walls found in fungi and plants, but contain specializations in their plasma membranes that permit them to bind to other cells or to extracellular matrices, hydrated materials containing proteins and polysaccharides that surround or adjoin one or more surfaces of certain cell types. These specializations are depicted in Fig. 4.1.

Modern eukaryotic cells, whether free-living or part of multicellular organisms ("metazoa"), have evolved far beyond the ancient communities of prokaryotes in which they originated. Cellular systems with well-integrated subsystems are at a premium in evolution, as are organisms with flexible responses to environmental changes. The result of millions of years of natural selection for such properties is that a change in any one of a cell's subsystems (by genetic mutation or environmental perturbation) will have repercussions in other subsystems that tend either to restore the original functional state or to bring the subsystem in question to another appropriate state. The physicist's aim of isolating relevant variables to account for the system's behavior would seem to be all but impossible under these circumstances.

Despite these complications, no one questions that all cellular processes are subject to physical laws. For a cell to function properly, the values of its physical parameters must be such that the governing physical laws serve the survival and reproduction of the cell. If the osmotic pressure is too high inside the cell, it may lyse. If the voltage across the membrane is not appropriate, a voltage-gated ion channel will not function. If the viscosity of the cytoplasm is too large, diffusion across it slows down and processes that are normally coordinated by diffusing signals become incoherent.

One way in which a cell sets the values of its physical parameters is by regulated expression of the cell's genes. Genes specify proteins, which themselves control or influence every aspect of the cell's life, including the values of its physical parameters. The embryo begins as a fertilized egg, or zygote, a single cell that contains the full complement of genes needed to construct the organism. The construction of the organism will necessitate changes in various physical parameters at particular stages of development and typically this will be accomplished by new gene expression. These changes will be set off by signals that may originate from outside the embryo or by interactions between the different parts of the embryo itself. As cells reach their definitive differentiated states later in development their appropriate functioning will require stability of their physical properties. If these change in an unfavorable direction then other signals are generated and transmitted to the nucleus to produce further proteins that reverse or otherwise correct the altered physical state.

In the remainder of this chapter we will consider certain transport and mechanical properties that are essential to an individual cell's existence. Because these properties have physical analogues at the multicellular level they will also figure in later discussions. We will begin with the "default" physical description, the simplest characterization of what might be taking place in the interior of the cell. We will then see how this compares with measurements in real biological systems. Where there is a disparity between the theoretical and actual behaviors we will examine ways in which the physical description can be modified to accommodate biological reality.

Diffusion

Free diffusion

The interior of the zygote, like most other cells, represents a crowded aqueous environment (Ellis and Minton, 2003) with many thousands of molecules in constant motion in the presence of complex structures such as organelles. Metabolism, the transformation of small molecules within the zygote, requires nutrients to be brought in and waste molecules to be carried out. In the course of intracellular signaling, information is passed on from one molecule to another and finally delivered to various destinations. How do biological molecules move inside the cell? Unless they are bound to a surface these molecules typically bump into each other and collide with water molecules and thus are constantly changing their direction of motion. Under such conditions how can they reach their destinations and carry out their tasks with high fidelity?

Motion requires energy. In general it can be active or passive. Of the two, only active motion has a preferred direction. Molecular motors moving along cytoskeletal filaments use the chemical energy of adenosine triphosphate (ATP) and transform it into kinetic energy. They shuttle from the plus to the minus end of microtubules ("plus-directed motor") or in the opposite direction ("minus-directed motor").

As an example of passive motion we may consider a molecule inside the cell, kicked around by other molecules, just as a stationary billiard ball would be by a moving ball, and triggering a cascade of collisions. In the case of a billiard ball, motion is generated in the first place by a cue; cytoplasmic molecules are not set in motion by any such external device. Rather, they have a constant supply of energy that causes them to be in constant motion. The energy needed for this motion is provided by the environment and is referred to as "thermal energy." In fact, any object will exchange kinetic energy with its environment, and the energy content of the environment is directly proportional to its absolute temperature T (measured in kelvins, K). For everyday objects (e.g., a billiard ball weighing 500 g) at typical temperatures (e.g., room temperature, 298 K = 25 °C) the effect of thermal energy transfer on the object's motion is negligible. A cytoplasmic protein weighing on the order of 10^{-19} g, however, is subject to extensive buffeting by the thermal energy of its environment.

For a particle allowed to move only along the x axis the kinetic energy E_{kin} imparted by thermal motion is given by $E_{\text{kin}} = m\langle v_x^2\rangle/2 = k_B T/2$, where m is the mass of the particle, v_x is its velocity, and k_B is Boltzmann's constant. The symbol $\langle\ \rangle$ denotes an average over an ensemble of identical particles (Fig. 1.2). Averaging is necessary since one is dealing with a distribution of velocities rather than a single well-defined velocity. To understand better the meaning of $\langle\ \rangle$, imagine following the motion of a particle fueled exclusively by thermal energy, for a specified time t (at which it is found at some point x_t) and measuring its velocity at this moment. If the particle is part of a liquid or gas it moves in the presence of obstacles (i.e., other particles) and thus its motion is irregular. Therefore, repeating this experiment N times will typically yield N different values for v_x^2. The average over these N values (i.e., the ensemble average) gives us the interpretation of $\langle v_x^2\rangle$ (Fig. 1.2). For a G-actin molecule at 37 °C, thermal energy would provide a velocity $\langle v_x^2\rangle^{1/2} = 7.8$ m/s. (For a billiard ball at room temperature the velocity would be a billion times smaller.) In the absence of obstacles, this velocity would allow an actin monomer to traverse a typical cell of 10 μm diameter in about 1 microsecond.

The interior of a cell represents a crowded environment. A molecule starting its journey at the cell membrane would not get very far before bumping into other molecules. Collisions render the motion random or diffusive. When discussing diffusion (referred to as Brownian motion in the case of a single particle), a reasonable question to ask is, what distance would a molecule cover on the average in a given time? For one-dimensional motion the answer is (see, for example, Berg, 1993 or Rudnick and Gaspari, 2004)

$$\langle x_t^2\rangle = 2Dt, \tag{1.1}$$

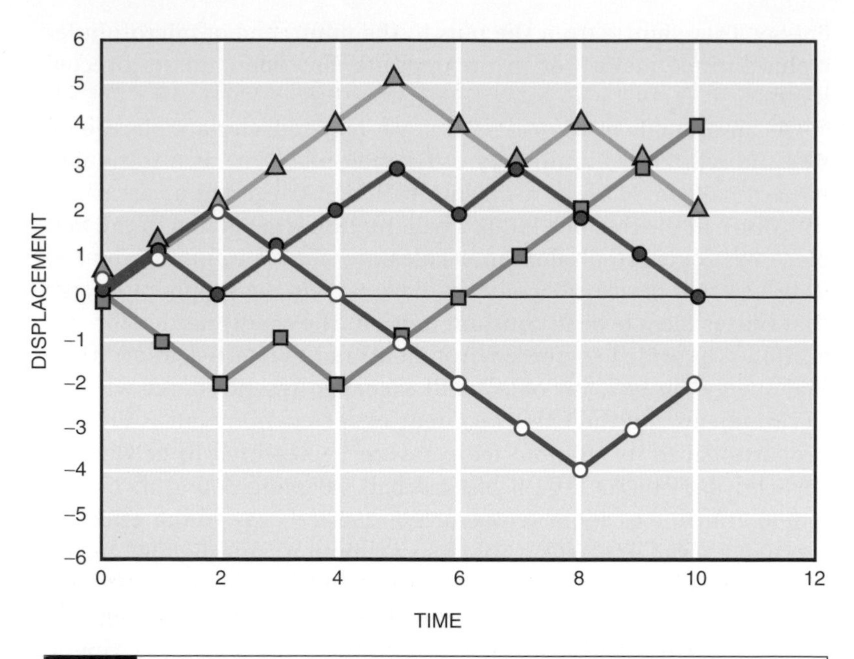

Fig. 1.2 The meaning of the ensemble average $\langle\ \rangle$. The figure can be interpreted as showing either the diffusive trajectories of four random walkers in one dimension, allowed to make discrete steps of unit length in either direction with equal probability along the vertical axis, or four different trajectories of the same walker. The trajectories are shown up to 10 time steps. The walkers make one step in one (discrete) unit of time. (Displacement and time are measured in arbitrary units.) The average distance after 10 time steps is $\langle x(t=10)\rangle = \frac{1}{4}(4+2+0-2) = 1$, the average squared displacement is $\langle x^2(t=10)\rangle = \frac{1}{4}(16+4+0+4) = 6$. The mean first passage time of arrival at $x=2$ is $\langle T_1(x=2)\rangle = \frac{1}{4}(2+2+4+8) = 4$. These values are independent of how the trajectories are interpreted. Note the difference between $\langle T_1\rangle$ and t. The latter is a fixed quantity (10 in the figure), whereas the former is a statistical quantity.

assuming that at $t = 0$ the particle was at $x = 0$. (Diffusion in three dimensions can be decomposed into one-dimensional diffusions along the main coordinate axis, each giving the same contribution to the three-dimensional analogue of Eq. 1.1. Thus, when x^2 above is replaced by $r^2 = x^2 + y^2 + z^2$, r being the length of the radius vector, the factor 2 on the right-hand side of Eq. 1.1 changes to 6.) Here D is the diffusion coefficient, whose value depends on the molecule's mass, shape and on properties of the medium in which it diffuses (temperature and viscosity). The average is calculated as above for the case of v. The diffusion coefficient of a G-actin molecule in water at 37°C is approximately 10^2 μm^2/s, which allows its average displacement to span the 10 μm distance across a typical cell in about 1 second. Comparing with the earlier result, obtained assuming unimpeded translocation with the thermal velocity, we see that collisions slow down the motion a million-fold.

There are several remarks to be made about Eq. 1.1. The most striking observation is that distance is not proportional to time: there is no well-defined velocity in the sense of the distance traversed by a single particle per unit time. A "diffusion velocity" of sorts could be defined by $v_D = \langle x_t^2 \rangle^{1/2}/t = (2D/t)^{1/2}$. This "velocity" is large for small t and gradually diminishes with time. Because of the statistical nature of these properties, Eq. 1.1 does not tell us where we will find the molecule at time t. On average it will be a distance $(2Dt)^{1/2}$ from the origin (in three dimensions the average particle will be located at the surface of a sphere of radius $(6Dt)^{1/2}$). But it is also possible that the molecule has reached this distance earlier than t (or will reach it later, see Fig. 1.2). We may be interested to know the time $\langle T_1 \rangle$ at which a given molecule will, on average, arrive at a well-defined target site (e.g., the cell nucleus) for the first time. This "first passage time" is in many instances a more appropriate or useful quantity than $\langle x_t^2 \rangle$ (Redner, 2001). For diffusion in one dimension, the first passage time for a particle to arrive at a site a distance L from the origin is $\langle T_1 \rangle = L^2/(2D)$ (Shafrir *et al.*, 2000). Even though this expression resembles the expression $t = \langle L^2 \rangle/(2D)$ (Eq. 1.1), owing to the meaning of averaging, $\langle T_1 \rangle$ and t are entirely different quantities (see also Fig. 1.2). In particular, t denotes real time measured by a clock, whereas $\langle T_1 \rangle$ cannot be measured, only calculated.

Equation 1.1 relates to the Brownian motion of a single molecule. For any practical purpose, molecules in a cell are represented by their concentration and one needs to deal with the simultaneous random motion of many particles. Brownian motion in this case ensures that even if the molecules are initially confined to a small region of space, they will eventually spread out symmetrically (in the absence of any force) towards regions of lower concentration (see Fig. 1.3). In one

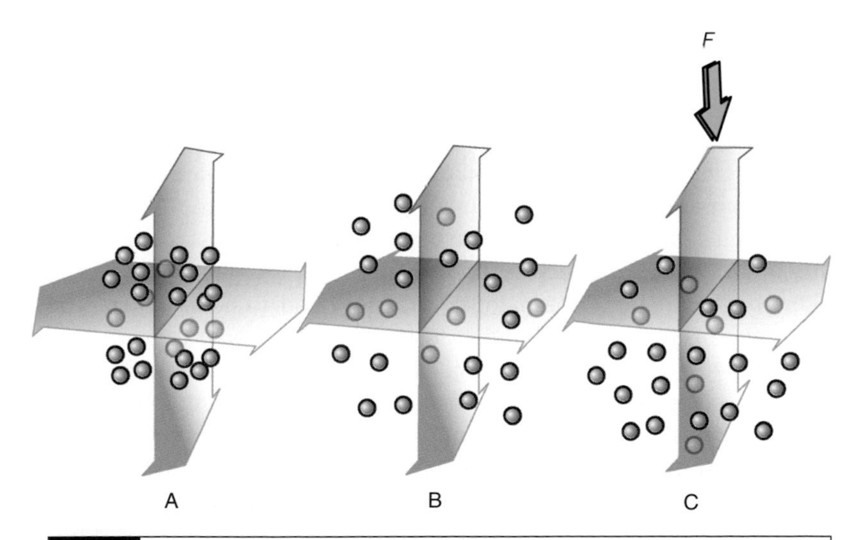

Fig. 1.3 The simultaneous Brownian motion of many particles. Particles confined initially to a small region of space (A) diffuse symmetrically outward in the absence of forces (B) or, when an external force is present, preferentially in the direction of the force (C).

dimension, if all the particles are initially at the origin with concentration c_0 then after time t their concentration at x is

$$c(x, t) = \frac{c_0 e^{-x^2/4Dt}}{(4\pi Dt)^{1/2}}.$$

(The mathematically more sophisticated reader will recognize this expression as being the solution of the one-dimensional diffusion equation

$$\frac{\partial c}{\partial t} = D\frac{\partial^2 c}{\partial x^2}.$$

For the meaning of the symbol ∂ see Box 1.1 below.)

Diffusion inside the cell

So far we have assumed that the diffusing particles execute their motion in a homogeneous liquid environment, free of any forces or obstacles other than those due to random collisions with other particles. The inside of the cell is far from being homogeneous: it represents a crowded environment (Ellis and Minton, 2003) with a myriad of organelles. The cytoskeleton, the interconnected network of cytoplasmic protein fibers (actin, microtubules, and intermediate filaments), is a dynamical structure in fertilized eggs, changing markedly in organization in different regions and at different stages (Capco and McGaughey, 1986; see Fig. 1.1). These objects and structures are likely to hinder the motion of any molecule (Ellis and Minton, 2003). It is not surprising, therefore, that simple diffusion is increasingly seen to be not an entirely adequate mechanism for extended intracellular transport (Ellis, 2001; Hall and Minton, 2003; Medalia *et al.*, 2002); at best it can act on a submicrometer scale (Goulian and Simon, 2000; Shav-Tal *et al.*, 2004).

Modern experimental techniques make it possible to follow the motion of individual molecules inside the cell. Goulian and Simon (2000), for example, tracked single proteins for close to 250 milliseconds in the cytoplasm and nucleoplasm of mammalian cells. They found that even on this small time scale the observed motion cannot be modeled by simple diffusion with a unique diffusion constant. To explain the experimental findings a broad distribution of diffusion coefficients, or "anomalous" diffusion, had to be assumed. The latter notion refers to fitting data using the more general expression $\langle x_t^2 \rangle = 2Dt^{\alpha}$, with α different from 1 (cf. Eq. 1.1). (Anomalous diffusion typically takes place in heterogeneous materials with complex structure. In special cases the exponent α can be determined theoretically. For a review on the subject see Ben-Avraham and Havlin, 2000). Other authors (reviewed by Agutter and Wheatley, 2000) claim outright that diffusion is an incorrect description of intracellular molecular transport. Such views are based on the supposition that the intracellular milieu represents a "dynamical gel" rather than a fluid medium (Pollack, 2001). Even though the major component of the cell interior is water, intracellular filaments and organelles render its architecture highly structured. The transport of molecules is more likely to take place along structural elements rather than in the intervening cytoplasm or nucleoplasm (von Hippel and Berg, 1989; Kabata *et al.*, 1993; Agutter and Wheatley, 2000; Kalodimos *et al.*, 2004).

Shafrir and coworkers (2000) presented a model of intracellular transport in which the relevant structural elements are the filaments of the cytoskeleton. Since diffusion is assumed to take place along these linear elements it is effectively one-dimensional. Using numerical techniques, the authors demonstrated that for realistic filament densities and diffusion coefficients this constrained transport mechanism is no slower than free diffusion. In comparison with free diffusion, the proposed mechanism has however distinct advantages. The filaments provide guiding tracks and thus transport becomes more focused. Because the cytoskeleton avoids organelles, movement is less hindered. As many of the cell's proteins are bound to the cytoskeletal mesh (Janmey, 1998; Forgacs *et al.*, 2004), the network may provide sites of concentrated enzyme activity for metabolic transformation during transport. Diffusive transport along cytoskeletal components should not be confused with molecular motor-driven motion. In the former case no extra source of energy is needed (other than thermal energy), whereas in the latter case a constant supply of energy, provided by ATP hydrolysis, is required.

In summary, contrary to common notions simple diffusion across the crowded intracellular environment cannot be the principal mechanism for distributing molecules within the cell. Classical diffusion as described earlier may be relevant on small length scales, especially for small molecules such as Ca^{2+} or cyclic AMP (estimated to be around 20 nm by Agutter and Wheatley, 2000). For larger distances, various transport mechanisms utilizing cellular structural components or elements (e.g., cytoskeletal filaments, DNA) take over. While extended free diffusion inside the cell is unlikely, conditions in the extracellular space are more favorable for it (Lander *et al.*, 2002). As we will see in Chapter 7, in combination with biochemical processes diffusion is an important mechanism in setting up molecular gradients that give rise to specific tissue patterns in the embryo.

Diffusion in the presence of external forces

Although free diffusion is not a realistic large-scale transport process within the cell, it does occur on a scale that is larger than most biomolecules (but small relative to the dimensions of the cell). This will be sufficient to transport molecules through pores in the plasma membrane or nuclear envelope or short distances in the cytoplasm. In many cases this diffusion will be subject to external forces, which can accelerate or reverse the direction of passive flow. The presence of external forces will modify the diffusive process discussed so far and, significantly, permit insight into another basic physical parameter of cellular and embryonic materials – the viscosity.

Let us consider particles with concentration $c(x, t)$, which in addition to random collisions experience a constant attractive or repulsive force F in the positive x direction (say due to electrostatic interactions). When $F = 0$, there will be a net diffusion current, $j_D(x, t) = -D\,\partial c(x, t)/\partial x$, along the concentration gradient, although for each individual particle in the flow $\langle x \rangle = 0$. (The minus sign indicates that the diffusion current is directed from higher concentration to lower concentration.) The diffusion current is zero for uniform concentration. For $F \neq 0$ there will be an additional current. Force causes acceleration (equal to force/mass), but as a result of the numerous collisions experienced by the diffusing particles they quickly reach a terminal velocity called the drift velocity, v_d. Thus, even for constant c the motion is biased in the direction of v_d (see Fig. 1.3). (The total particle current is now $j_T(x, t) = -D\,\partial c(x, t)/\partial x + v_d c(x, t)$.) The drift velocity v_d is defined as F/f, f being the friction coefficient, a characteristic property of the medium in which the motion takes place. If F acts opposite to the diffusion current, it may reverse the direction of motion: flow may proceed against the concentration gradient.

A specific case of diffusion in the presence of an external force is of particular interest to biologists. If a diffusing molecule is electrically charged then its motion will also be influenced by any electrical potential difference in its environment (e.g., a membrane potential). The overall mass transport of a collection of such molecules will be due to the combination of the concentration gradient and the electrical gradient, which is termed the electrochemical gradient. (See Chapter 9 for a description of the role of electrochemical gradients during fertilization).

The term "diffusion" turns up in a number of contexts in cell and developmental biology and it is important to understand how its different uses relate to the concepts described above. A cell biologist will use "facilitated diffusion" to refer to free diffusion under conditions in which specific channels permit the selective passage through a barrier (usually a membrane) of molecules for which this would not otherwise be possible. Molecules with certain shapes, for example, can be facilitated in their diffusion by pores with a complementary structure. Such facilitation, of course, is not capable of causing mass transport against a concentration gradient. For more details on diffusion in biological systems the reader may consult the excellent book by Berg (1993).

"Diffusion" of cells and chemotaxis

Locomoting cells, in the absence of any chemical gradient, typically execute random, amoeboid motion without preferred direction. As will be seen later, to interpret some aspects of such motion it is useful to introduce an effective diffusion coefficient. It is important, however, to keep in mind that the randomness here is not due to thermal fluctuations but is the result of inherent cellular motion powered by metabolic energy. If such "diffusive" motion of cells such as slime mold amoebae or bacteria occurs in the presence of a chemical gradient, the gradient can be considered as the source of an external force that biases the direction of cell flow. This phenomenon is called "chemotaxis" and, while it is a unique property of living systems, the source of the biased motion can be interpreted in physical terms.

Osmosis

A phenomenon closely related to diffusion, with important biological implications, is *osmosis*, the selective movement of molecules across semi-permeable membranes. An example of such a membrane is the lipid bilayer surrounding the eukaryotic cell. It is permeable to water but not to numerous organic and inorganic molecules needed for the cell's survival. (These must be transported through special pores composed of proteins embedded in the lipid bilayer, as discussed above. Here we will ignore this facilitated transport.)

Consider Fig. 1.4, which shows a container with two compartments (L and R) separated by a semi-permeable membrane, permissive for the solvent (e.g., water), but restrictive for the solute (e.g., sugar). The movable walls in L and R act as pistons: if they are attached to appropriate gauges the pressure inside the two compartments can be measured. At equilibrium one finds that the two pressures are not the same: $p_L > p_R$. The reason for this is as follows. The pressure is due to the bombardment of the container walls by molecules executing thermal motion. Since the solvent can freely diffuse across the membrane, it will do so until the average number of collisions per unit time of its molecules with the movable container walls (i.e., the partial

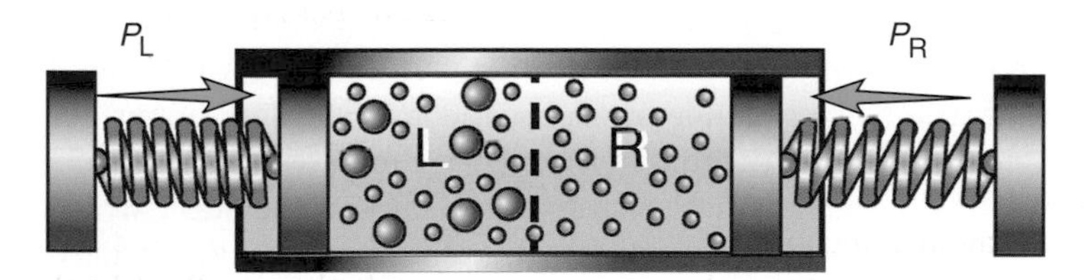

Fig. 1.4 Physical origin of osmotic pressure. The two compartments, L and R, are separated by a semi-permeable wall, represented by the broken line. The two walls at the ends of the compartments are movable (and can be thought of as the stretchable membranes of "cells" L and R) and attached to springs, which measure the pressure in L and R. The larger, pink particles (the solute) cannot pass through the wall in the middle, but the smaller blue particles (the solvent) can. At equilibrium the pressure exerted by the solvent on the movable walls in L and R is the same. If the two compartments contain the larger particles at different concentrations (in the figure their concentration in R is zero), the pressures they exert on the movable walls are not equal: their difference is called the osmotic pressure. In particular, if the two springs are made of the same material, the one attached to L will be more compressed, corresponding to the membrane of "cell" L being more stretched.

pressure due to the solvent) is the same in the two compartments. The extra pressure in L, $p_{os} = p_L - p_R$ (acting, in particular, on the semi-permeable membrane) is the osmotic pressure and is due to the imbalance in the concentration of the solute in L and R.

The biological significance of osmosis is obvious. Osmotic pressure is clearly an important determinant of cell shape. As long as the overall concentration of organic and inorganic solutes inside the cell remains higher than outside, the cell membrane is stretched (because water enters the cell in an effort to balance the concentration difference, thus increasing the volume of the cell). The membrane can tolerate increases in osmotic pressure only within limits, beyond which it bursts. To avoid this happening, cells have evolved active transport mechanisms: they are able to pump molecules across the lipid bilayer through channels.

Viscosity

Viscosity of cytoplasm

There are numerous biological transport mechanisms other than those discussed so far. For instance, microscopic observation of the interior of certain cells has revealed the phenomenon of "cytoplasmic streaming." Streaming is an example of convection. Unlike diffusion, which can take place in a stationary medium, convection is always associated with the bulk movement of matter (e.g., flowing water, flowing blood). Cytoplasmic streaming is seen in certain regions of locomoting cells such as amoeba, where it contributes to the cell's reshaping and movement, as well as in axons – the extended processes of nerve cells – where it is employed to move molecules and vesicles to the axon's end or terminus. Molecules or organelles present in the streaming cytoplasm are transported just as an unpowered boat would be in a river. Convective flow is maintained by a pressure difference (as in blood flow), whereas diffusion current is due to the difference in concentration. Yet another transport mechanism, less important in biological systems, is conduction, which is not associated with any net mass transport. A typical example is heat conductance, which is possible due to the collisions between atoms performing localized thermal motion around their equilibrium positions.

When there is a relative velocity difference between a liquid and a body immersed in the liquid (either because the body moves through the liquid or because the liquid moves past the body), the body experiences a drag force. The type of drag to which organelles and cytoskeletal fibers moving through the cell's cytoplasm are subject is *viscous drag*.

An ideal gas (whose molecules collide elastically with each other but do not otherwise interact) will flow without generating any internal resistance. For any other fluid, including cytoplasm, interactions among the molecular constituents, collectively leading to internal

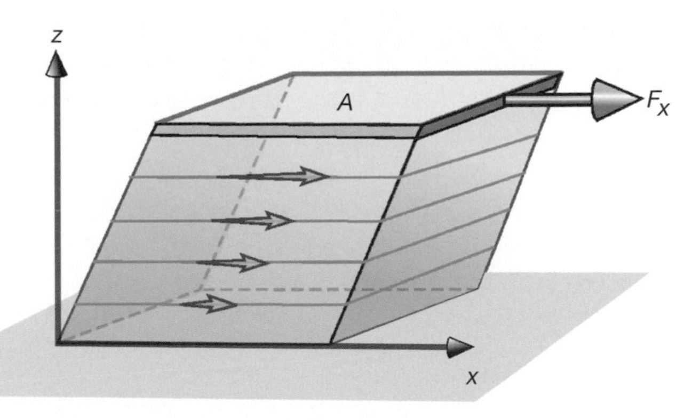

Fig. 1.5 Illustration of the phenomena described by Eqs. B1.1a and 1.8. For Eq. B1.1a consider the figure as showing a plate of area *A* pulled through a liquid in the *x* direction with a shearing force F_x. As a result of internal friction, as the plate moves the fluid particles will also be displaced. The horizontal blue arrows show schematically the magnitude of the fluid's z component of velocity in the vicinity of the plate. Only the fluid below the plate is shown. For Eq. 1.8 the object shown in the figure is to be considered as an originally rectangular solid block with its lower surface immobilized. The shearing force F_x now acts along the top surface of the block. The horizontal blue arrows denote the magnitude of the block's displacement in the *x* direction as a function of z.

friction, will slow down the flow and contribute to the bulk property known as *viscosity*. It is intuitively obvious that diffusion and friction in a liquid cannot be independent of each other: the stronger the friction the slower the diffusion. In addition, the higher the temperature, the more thermal energy the molecule has and the more intense its diffusion. Indeed, under very general conditions we have $D = kT/f$, which is known as the Einstein–Smoluchowski relation (Berg, 1993). Since $F = fv$, the faster an object moves, the stronger the friction it experiences.

To illustrate the molecular basis of friction in liquids, imagine pulling a plate (or any object) of area *A* through a liquid with a constant force F_x in the *x* direction (see Fig. 1.5). For it to move, the plate has to displace the liquid molecules it encounters. The state of the liquid is thus perturbed. This perturbation in the state of the liquid is called *shear* and leads to friction, i.e., viscous drag, acting on the plate (Howard, 2001). A measure of the shear, or rather the rate of shearing, is the modification of the liquid's velocity in the vicinity of the plate. For forces that are not too strong, the shear rate is proportional to F_x. The proportionality constant between F_x and the shear rate is the viscosity η, obviously a property of the liquid (for the precise definition of viscosity see Box 1.1). The customary unit of η is the Pa s (pascal second). The viscosity of water is 0.001 Pa s. The viscosity of the cytoplasm varies strongly with cell type (for an overview, see Valberg and Feldman, 1987) and even with location within the cell (Bausch *et al.*, 1999; Yamada *et al.*, 2000; Tseng *et al.*, 2002).

Box 1.1 | Definition of viscosity

Consider Fig. 1.5, where a force of magnitude F_x is applied along a layer of a liquid in the x direction. For simplicity, we first assume that the velocity of the resulting flow v_x depends only on a single variable, z, and increases linearly in the z direction, and that the liquid is at rest in the plane $z = 0$ (no-slip boundary condition). This is illustrated by the blue arrows, which represent the magnitude of the velocity as a function of z, and trace a straight line. Under these conditions, the defining equation for the viscocity η is

$$F_x = \eta A \frac{\Delta v_x(z)}{\Delta z}. \tag{B1.1a}$$

Here A is the area of the liquid layer acted upon by F_x and Δv_x and Δz are small changes in the corresponding quantities. Note that because of the assumed linearity of v_x on z, the ratio $\Delta v_x/\Delta z$ is constant and gives the slope of the line traced by the arrowheads.

More generally, the velocity profile is not linear (e.g., for flow in a pipe it is parabolic), in which case $\Delta v_x/\Delta z$ itself depends on z and should be evaluated in the limit when the changes in both v_x and z become infinitesimal. This procedure defines the *derivative* of v_x with respect to z, dv_x/dz (or for a function $f(x)$, df/dx). A derivative thus represents the rate of change of one quantity with respect to another. In the future, when it will create no confusion, derivatives will be written as ratios of finite differences as in the above equation.

Even more generally v_x might depend on other variables, in which case the notation for the ordinary derivative, d, is replaced by the notation for a partial derivative, ∂. Thus, in the most general case, Eq. (B1.1a) becomes

$$F_x = \eta A \left. \frac{\partial v_x(x, y, z, t)}{\partial z} \right|_{z=h}, \tag{B1.1b}$$

where

$$\frac{\partial v_x}{\partial z} = \frac{\partial}{\partial z}\left(\frac{\partial x}{\partial t}\right) = \frac{\partial}{\partial t}\left(\frac{\partial x}{\partial z}\right)$$

(the order of multiple differentiation can be interchanged) is the shear rate expressed in terms of the liquid's velocity profile, the derivative being evaluated at the location of the plate in Fig. 1.5 (assumed to be at height h); $\partial x/\partial z$ represents the shear.

The shear rate caused by the moving object is difficult to calculate (it is usually found from measurement); for this we need the velocity of the perturbed liquid as a function of position, which requires the solution of the complicated Navier–Stokes equation (this is discussed in more detail in Chapter 8). Such calculations can be carried out for simple cases. Thus, for a sphere of radius r moving in a liquid of viscosity η, with constant velocity v, the frictional drag is given by the Stokes formula (see, for example, Hobbie, 1997)

$$F = 6\pi \eta r v. \tag{1.2}$$

As a consequence, for a sphere of radius r the friction coefficient (defined earlier by $F = fv$) and the diffusion coefficient (related to f through the Einstein–Smoluchowski relationship $D = kT/f$, see above) are given respectively by $f_{\text{sphere}} = 6\pi\eta r$ and $D_{\text{sphere}} = kT/(6\pi\eta r)$. These expressions imply that f and D are strongly shape-dependent. For example, for a long cylindrical molecule of length L and diameter d the friction coefficient depends on whether the motion is lengthways (parallel to L) or sideways (perpendicular to L). In the limit of large aspect ratio, $L/d \gg 1$, the corresponding friction coefficients are (Berg, 1993; Howard, 2001)

$$f_{\|} = \frac{4\pi\eta d}{\ln\left(\frac{2L}{d}\right) - \frac{1}{2}}, \qquad f_{\perp} = \frac{8\pi\eta d}{\ln\left(\frac{2L}{d}\right) + \frac{1}{2}}. \tag{1.3}$$

Viscous transport of cells

Everyone is familiar with two extreme behaviors of moving bodies: there are “inertial” objects that when set in motion tend to remain in motion (as described by Newton’s first law) and “frictional” objects that won’t move unless you continue to push them along. Since in this book we will often be interested in the motion of individual cells in the embryo, we will again switch scales from molecular to the cellular (as we did for diffusion) and see what our analysis of viscosity can tell us about cell motion. Here we will use one of the favorite tools of physicists, “dimensional analysis,” to show the relative contribution of inertial and viscous behaviors to the movement of cells. Dimensional analysis allows us to compare the magnitudes of the different factors contributing to a complex process after having rendered them nondimensional.

According to Newton’s second law, mass times acceleration = the sum of all forces acting on a body. Let us assume that a cell moving through a tissue experiences other forces, collectively denoted by F, along with the frictional forces. According to Newton,

$$m\frac{\mathrm{d}^2 x}{\mathrm{d}t^2} = F - f\frac{\mathrm{d}x}{\mathrm{d}t}. \tag{1.4}$$

Here m is the mass of the cell, x and t denote distance and time respectively, and in the expression for the frictional force the velocity is denoted as the derivative of distance. The minus sign in the last term expresses the fact that the direction of the frictional force is opposite to the direction of motion.

Let us introduce nondimensional (hence, unit-less) quantities $s = x/L$, $\tau = t/T$. Here L and T are some typical values of the distance and time. In terms of these parameters (and with a slight rearrangement of the various factors), Eq. 1.4 becomes

$$\frac{m}{fT}\frac{\mathrm{d}^2 s}{\mathrm{d}\tau^2} = \frac{T}{fL}F - \frac{\mathrm{d}s}{\mathrm{d}\tau}. \tag{1.5}$$

Each term in Eq. 1.5 is now dimensionless and we can thus compare their magnitudes (note that $d^2s/d\tau^2$ and $ds/d\tau$ are themselves dimensionless). We can take L to be the typical linear size of a cell, so that $m = \rho L^3$, ρ being the density. Thus the coefficient of the dimensionless acceleration is $\rho v L^2/f$, where we have introduced the typical velocity $v = L/T$. Since $f/L \approx \eta$ (see above in connection with Eq. 1.2), the ratio of the inertial term (proportional to the mass) and the frictional term contains the expression $\rho v L/\eta$. We can now plug in known values of these factors. The typical size L of a cell is of order 10 micrometers and the cellular density is of order that of water. A characteristic time T could be identified with the early-embryo cell cycle time. For the sea urchin embryo, which we will discuss in Chapters 4 and 5, T is about an hour ($\approx 10^3$ s), thus $v = L/T \approx 10^{-2}$ µm/s ($1\ \mu\text{m} = 10^{-6}$ m). Using these values we obtain $\rho V L/\eta \approx 10^{-7}$, a very small number (remember, this result is independent of the units of measurement). This analysis shows that when dealing with physical motion in the early embryo, inertial effects can safely be neglected; see also the example in Box 1.2.

Box 1.2 Inertial versus frictional motion; coasting of a bacterium

Let us consider the motion of a bacterium in the viscous intracellular environment. A bacterium is propelled into motion by a rotary motor in its tail. The typical speed of such motion is 25 µm/s. We now ask the question, how long will the bacterium coast once its rotary motor stops working?

The bacterium will keep moving due to its inertia. To see how far this inertial motion will take it, we have to solve the appropriate equation of motion, which states that mass m times acceleration equals the sum of all forces acting on the bacterium. Once the motor is turned off the only force acting is the viscous force, fv. Thus

$$m\frac{dv}{dt} = -fv \tag{B1.2a}$$

The solution of this equation is $v(t) = v(0)e^{-t/\tau}$, where $\tau = m/f$ and $v(0)$ is the speed of the bacterium at the moment when its motor turned off. Approximating the bacterium by a spherical particle of radius $r = 1$ µm and density ρ that of water, we obtain $m = (4/3)\pi r^3 \rho \approx 4 \times 10^{-15}$ kg. Using Stoke's law (see Eq. 1.2) we have $f = 6\pi\eta r \approx 20$ nN s/m. The total distance the bacterium coasts is speed times time and is approximately $v(0)\tau = v(0)m/f \approx 5\text{pm} = 5 \times 10^{-12}$m, a minuscule distance even on the scale of the bacterium. (Since the speed varies with time, the mathematically accurate way of obtaining the distance is to integrate the speed with respect to time from zero to infinity, which would lead to a value of the same order of magnitude.) This example illustrates that inertial effects indeed can be neglected when the motion of cells is considered. The ratio of inertial forces to viscous forces is known as the *Reynolds number*. For a fascinating discussion of "life at low Reynolds number" see Purcell (1977).

Elasticity and viscoelasticity

A viscous material will readily change its shape (deform) when a force is applied to it. The following experiment showed that cells are not constructed entirely of viscous materials. Ligand-coated magnetic beads were attached to transmembrane proteins on the surfaces of cells linked to the cytoskeleton. These beads, and thus the cytoplasm, were subjected to a twisting force by an applied magnetic field. It was found that wild-type cells (i.e., genetically normal cells) exhibited higher *stiffness* and greater *stiffening* response to applied stress than cells that were genetically deficient in the cytoskeletal protein vimentin or wild-type cells in which vimentin filaments were chemically disrupted (Wang and Stamenovic, 2000). The properties of stiffness and stiffening measured in these experiments were directly related to the *elastic modulus* (or Young's modulus) of the cytoplasm.

Cells and tissues can easily be deformed by external forces and to some extent without sustaining any damage, because they have elastic properties (just push with your finger on your stomach). The prototype elastic device is the spring. The force needed to compress or extend a spring is

$$F = k\Delta x, \tag{1.6}$$

where k is the spring constant or stiffness and Δx is the deviation of the spring from its equilibrium length. Equation 1.6 is Hooke's law, which expresses the fact that for elastic bodies the deformation force is proportional to the elongation (or in more general terms to the magnitude of the deformation caused). Hooke's law often is written in a slightly different form, namely

$$\frac{F}{A} = E\frac{\Delta L}{L}. \tag{1.7}$$

Equation 1.7 states that if an elastic linear body (a rod, for example) of original length L and cross-sectional area A is extended by ΔL, the stress (F/A) needed to achieve this is proportional to the strain (i.e., the relative deformation, $\Delta L/L$). The parameter E is the Young's modulus of the rod's material; its unit is the pascal, Pa. Even though Eqs. 1.6 and 1.7 are defined for a linear body, one can define an elastic modulus for any biological material. Its value would be determined, for example, by simply stretching a piece of such material with known force in some direction and measuring the original length and the deformation in the same direction. The cross-sectional area then is measured perpendicular to the direction of the force. (For nonisotropic materials, the stiffness varies with direction). Any material if stretched or compressed with sufficiently moderate force (in the "Hookean regime") will obey Eq. 1.6. Davidson *et al.* (1999) listed the stiffness of a number of cells and tissues. Comparing Eqs. 1.6 and 1.7 one can define an effective spring constant for any material in terms of its Young's modulus using the relation $k = EA/L$.

Another type of deformation to which cells and tissues are often exposed to is shear. It has been discussed in connection with viscosity (Box 1), but it can also be defined for any elastic material. If the upper face of an elastic rectangular body is moved with a force F (acting parallel to that face) relative to its lower face (see Fig. 1.5) then

$$\frac{F}{A} = G\frac{\Delta x}{\Delta z}. \tag{1.8}$$

Equation (1.8) expresses the fact that in the elastic regime the shear stress F/A is proportional to the shear; the constant of proportionality, G, is the shear modulus in Pa. Note that in the case of viscous liquids shear stress can be maintained only if it causes the shear to vary in time (Eq. B1.1a, $v_x \approx \Delta x/\Delta t$). This is often used as the criterion distinguishing solids from liquids. (For a mathematically more formal discussion of various deformations arising in biological materials see Fung, 1993 or Howard, 2001.)

The most conspicuous property of a spring (or any perfectly elastic material) is that upon the action of a force it adapts instantaneously via deformation: as soon as F is applied the displacement Δx in Eq. 1.6 is established. As we have seen, in the viscous regime a shearing force determines the rate of deformation rather than the deformation itself. The prototype of such behavior is a piston moving in oil: more force needs to be applied to move the piston faster.

There exists a large class of materials, including most cells and tissues, which exhibit both elastic and viscous properties; such materials are termed *viscoelastic*. When a viscoelastic material is deformed, on a short time scale it behaves mostly as an elastic body whereas on a longer time scale it manifests viscous liquid characteristics. When a piece of tissue is compressed with a constant force, the resulting deformation (i.e., strain) shows a characteristic time dependence: the tissue first quickly shrinks in the direction of the force (just as a spring would), but the final deformation is reached through a slow flow. Alternatively, if one imposes a definite deformation on a viscoelastic material, the resulting stress varies in time until a final equilibrium state is reached.

The mathematical description of viscoelasticity is rather complicated. We will deal with it, in a somewhat simplified manner, later, where it is relevant to understanding certain developmental phenomena. (For a comprehensive discussion of viscoelasticity in biological materials, see Fung, 1993).

Perspective

Basic, "generic" physical mechanisms can provide insight into many processes that occur within and between living cells. It is essential to recognize, however, that neither cytoplasm nor multicellular aggregates are the sort of "ideal" materials that physics excels in describing. Each cellular and tissue property will be, in general, a result of many superimposed physical properties. Thus any standard physical quantity (e.g., a diffusion coefficient, an elastic modulus) will have a more restrictive meaning and the temporal and spatial range of any simple physical law will be limited. As we have seen, however, it is possible to build a certain amount of complexity into a physical representation and come closer to capturing biological reality. This can be achieved by defining "effective" physical parameters, which incorporate the complexity of the biological system, and using them in the same equations and relationships that their standard counterparts obey. Typically, the validity of such equations cannot be deduced from first principles and must be checked experimentally. The effectiveness of studying living systems using physics, therefore, will often lie in establishing analogies rather than equivalences between complex biological phenomena and well-understood processes in the inanimate world.

Chapter 2

Cleavage and blastula formation

In the previous chapter we saw how the simple physical assumption that the cell is a droplet of liquid comes into conflict with experimental evidence when the transport of molecules in its interior or the response of an individual cell to mechanical stress are considered. By adding more physics to the default concept of diffusion (external forces, viscosity, elasticity) we were able to approach the biological reality of cell behavior more closely. This analysis also had the premium of helping us to identify levels of organization (e.g., chemotaxis in a colony of bacteria or amoebae) at which physical laws that are too simple to explain individual cell behavior may nonetheless be relevant.

In this chapter we will describe the transition made by a developing embryo from the zygotic, or single-cell, stage to the multicellular aggregate known as the *blastula*. Here again the simplest physical model for both the zygote and the early multicellular embryo that arises from it is a liquid drop. As in the examples in Chapter 1 our understanding of real developing systems will be informed by an exploration of how they conform with, and how they deviate from, the basic physical picture.

The cell biology of early cleavage and blastula formation

The blastula arises by a process of sequential subdivision of the zygote, referred to as *cleavage*. Cleavage, in turn, is a variation on the process of cell division that gives rise to all cells. In cell division both the genetic material and the cytoplasm are apportioned between the two "daughter" cells. To ensure that the resulting cells are genetically identical to their progenitor, the DNA (essentially all contained in the cell's nucleus) must be replicated before division. A duplicate set of DNA molecules is thus synthesized in the nucleus, well before the cell exhibits any evidence of dividing into two; it will do this using the separated strands of the original double helix as templates.

The cell's DNA is complexed with numerous proteins, forming a collection of fibers known as *chromatin*. Although the chromatin fibers form a dense tangle when the nucleus is intact, each fiber is actually a separate structure, a *chromosome*. In the organisms that we are considering in this book ("diploid" organisms), all the cells of the embryo and the mature body, except the egg and sperm and their immediate precursors, contain two distinct versions of each chromosome, one contributed by each parent during fertilization (see Chapter 9). Thus a human cell contains 23 pairs, or 46 chromosomes.

Just before the cell divides the *nuclear envelope* (the membranes and underlying proteinaceous layer enclosing the nucleus) disassembles, while the chromosomes separately consolidate and can now be visualized in the cytoplasm as the individual structures they actually are. Since DNA synthesis has occurred by this time, there are two identical copies of each chromosome, referred to as *sister chromatids*, still attached to each other by means of a structure called a *centromere*, which contains a molecular glue (the protein cohesin). Human diploid cells at this stage contain 46 pairs of such sister chromatids.

At this point a series of changes takes place that lead to: (i) the separation of sister chromatids; (ii) the two resulting sets of chromosomes being brought to opposite ends of the cell ("mitosis"); and (iii) the cytoplasm and surrounding plasma membrane of the cell dividing into two equal portions ("cytokinesis"). Mitosis is guided by a piece of molecular machinery known as the mitotic apparatus or spindle, made up of protein filaments called microtubules and microtubule-organizing centers known as *centrosomes*, located at opposite sides of the cell and forming the poles of the spindle (Fig. 2.1). Immediately before mitosis takes place a single centrosome, located near the nucleus, separates into two, which, upon assuming their polar locations, extend long microtubules. These microtubules (the same number from each centrosome) either attach to a portion of each of the two chromosomes of the sister chromatids (the *kinetochore*) or form *asters*, star-like arrays of shorter microtubules. The spindle employs molecular motor proteins, such as microtubule-associated dynein and the BimC family of kinesin-like proteins located in the centrosomes, to exert tension on the kinetochores and to separate the sister chromatids (Nicklas and Koch, 1969; Dewar *et al.*, 2004).

Cytokinesis is regulated by another class of cytoskeletal filaments, composed of the protein actin as well as additional molecular motor molecules such as kinesin. Actin-containing microfilaments form a contractile ring beneath the cell surface and, in association with the molecular motors, cause the formation of a groove or furrow between the two incipient daughter cells that eventually pinches the cells apart. Once cell division is completed the chromosomes rearrange themselves into a ball of chromatin around which the nuclear envelope reforms.

The processes just described are collectively known as the "cell cycle," which is schematized into four discrete phases: M (mitosis, including cytokinesis), G1 (time gap 1), S (DNA synthesis), and G2

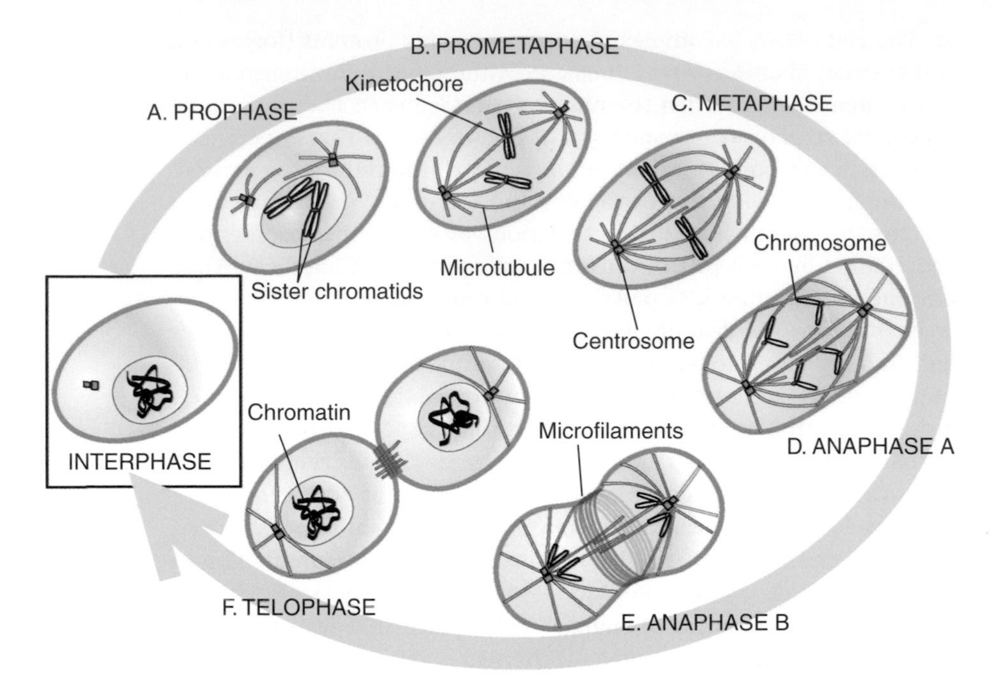

Fig. 2.1 The cell division cycle. The cell spends most of its lifetime in the interphase state, during which period its intact nucleus contains the full set of chromosomes in the form of the tangled DNA–protein complex called chromatin. If the cell divides then its DNA is replicated and its centrosome is duplicated during interphase. In the division of typical somatic cells (as pictured) the cell size also increases during interphase, but for cleavage divisions (see Fig. 2.2) cell size remains constant. Cell division is accomplished by mitosis (A–F). At prophase (A), the asters (blue microtubules) assemble at the two centrosomes (small blue boxes), which have moved away from one another. Inside the nucleus the replicated chromosomes, each consisting of two attached sister chromatids, condense into compact structures. For illustrative purposes two chromosomes are pictured here; diploid cells have two copies of each of several chromosomes. At the beginning of prometaphase (B), the nuclear envelope breaks down abruptly. The spindle then forms from the microtubules that extend from the centrosomes and attach to the kinetochores of the chromosomes. The chromosomes then begin to move toward the cell equator, defined by the location of the centrosomes, which are now at opposite poles of the cell. At metaphase (C), the chromosomes are aligned at the equator and sister chromatids are attached by microtubules to the opposite spindle poles. At anaphase A (D), the sister chromatids separate to form daughter chromosomes. By a combination of shortening of the kinetochore microtubules and further separation of the spindle poles, the daughter chromosomes move toward opposite ends of the cell. At anaphase B (E), the chromosomes are maximally separated and a microfilament-containing contractile ring begins to form around the cell equator. At telophase (F), the chromosomes decondense and a new nuclear envelope forms around each of the two complete sets. The contractile ring deepens into a furrow, which, as cytokinesis proceeds, pinches the dividing cell into two daughters.

(time gap 2). The latter three collectively form the *interphase* (Fig. 2.1). For most cells M lasts less than an hour and S one to several hours, depending on the genome size. The two gap phases G1 and G2 are periods in which various synthetic processes take place, including those in preparation for the events during the M and S phases. Since the activities in G1 and G2 are keyed to the requirements of different cell types, their durations vary widely. Cleavage-stage embryos typically utilize molecules that have been synthesized and stored during the process of egg construction or *oogenesis*, and therefore have G phases that are brief or nonexistent.

DNA synthesis, mitosis, and cytokinesis are unique events in the life of any cell, but considered in the continuity of cellular life they are periodic processes. As such, it would be natural for them to be controlled by molecular clocks, and indeed several such regulatory clocks exist in dividing cells. Molecular clocks that regulate entry into DNA synthesis and mitosis are based on temporal oscillations of the concentrations of members of the cyclin family of proteins. Such oscillations are the physical consequences of positive and negative feedback effects in dynamical systems, such as that represented by the cell's biochemistry, and will be discussed in the following chapter. The control of cytokinesis is less well understood.

In contrast with the cell division that occurs later in embryogenesis and in the tissues of growing and mature organisms, which (like the cell division of free-living cells) is typically associated with a doubling in cell mass, in cleavage a single large cell is subdivided without increase in its mass. This has the consequence that with each successive subdivision of the zygote the ratio of nuclear to cytoplasmic material increases. In the frog embryo, the "midblastula transition," a set of molecular and cell behavioral changes leading to the morphological reorganization of the embryo known as *gastrulation* (Chapter 5), is regulated by the titration of one or more cytoplasmic components resulting from this changing ratio (Newport and Kirschner, 1982a, b).

The geometry and topology of the blastula, although they differ for different types of organisms, are relatively simple (Fig. 2.2). Most typically, the end result is a ball of cells with an interior cavity (the "blastocoel"). The ball can be of constant thickness, as in the sea urchin or *Drosophila* (fruit fly) embryo, where it is a single layer of cells called the *blastoderm* ("cell skin"). In amphibians, such as the frog, the ball is of nonuniform thickness as a result of different rates of cleavage at opposite poles of the zygote. In mammals, such as the mouse and human, the outer surface of the ball consists of a layer of flat cells (the "trophoblast"), which gives rise to the extraembryonic membranes that attach to and communicate with the mother's uterus. A cluster of about 30 cells termed the "inner cell mass," which forms at one pole of the trophoblast's inner surface, gives rise to the embryonic body. In certain cases, such as in species of mollusks that develop from large eggs, the ball of cells develops with no interior cavity (Boring, 1989). This is called a "steroblastula" (solid blastula).

The routes by which the blastula takes form also vary in different groups of organisms. While cleavage can be a symmetrical process over multiple division cycles, the sizes of the cells resulting from cleavage are often unequal. The reason is that the zygote typically has within its cytoplasm, stored in a spatially nonuniform fashion, materials provided to the egg by the mother's tissues during oogenesis.

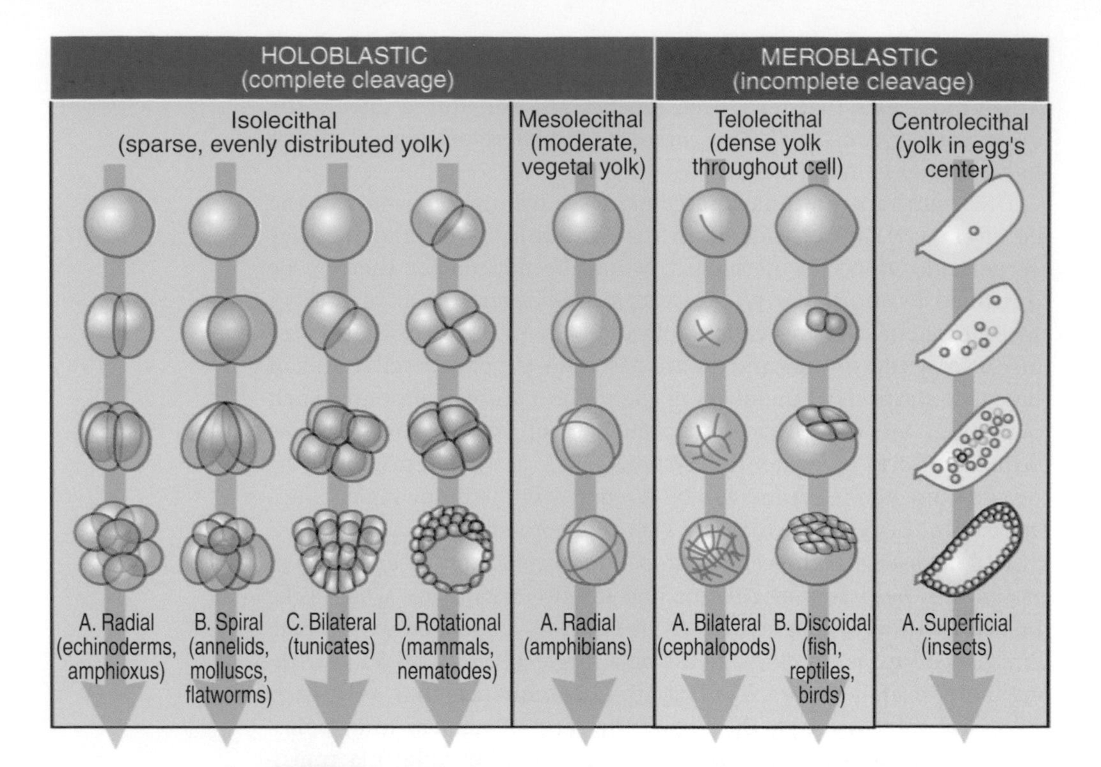

Fig. 2.2 Shapes of blastulae and patterns of cleavage in various organisms. The distribution of yolk, a dense, viscous material, is a major constraint in the pattern of cleavage. In centrolecithal meroblastic cleavage (characteristic of the embryo of the fruit-fly *Drosophila*, see Chapter 10), unlike the other types pictured, the nuclei divide in a common cytoplasm at the center of the egg during early development. This mode of cleavage is completed as the nuclei move to the egg-cell periphery, where they become separated from one another and the rest of the egg contents by the formation of surrounding membranes. (After Gilbert, 2003.)

These materials can include nutrient yolk (typically composed of proteins and lipids), messenger RNAs, and granules composed of several kinds of macromolecules. Structural and compositional asymmetries of the zygote will lead to unequal *blastomeres*, the cellular products of cleavage.

Cleavage in which the zygotic mass is completely subdivided (whether the initial blastomeres are equal or unequal) is called *holoblastic* (Fig. 2.2). If only a portion of the zygote is subdivided, cleavage is referred to as *meroblastic*. Mammals and sea urchins exhibit holoblastic cleavage, whereas cleavage in *Drosophila* embryos, where just the nuclei divide at first, and only become incorporated into separate cells after they have migrated to the inner surface of the egg, is meroblastic. Despite being vertebrates and therefore phylogenetically closely related to mammals, fish, reptiles and birds undergo meroblastic cleavage, with the yolky portion of the egg cell failing to become subdivided (Fig. 2.2).

Physical processes in the cleaving blastula

Embryonic development starts with a simple spherical or oblate fertilized egg and produces complex shapes and forms. Shape will therefore be a major concern in what follows. Although organisms use the complex machinery of DNA synthesis, mitosis, membrane biogenesis, and cytokinesis to subdivide the material of the fertilized egg and become a blastula, the simplest physical model for this subdivision process is the behavior of a liquid drop. Such a description has often been used to explain the behavior of individual cells exposed to mechanical deformations (Yoneda, 1973; Evans and Yeung, 1989) or to interpret the rheology of neutrophils during phagocytosis (van Oss *et al.*, 1975). Many biological functions such as mitosis and membrane biogenesis are based on molecular interactions that are not inherently three-dimensional and thus have no preferred geometry. Since they occur in a context in which physical forces are also active, it is reasonable to expect that in the final arrangement the physical determinants will dominate.

The shapes of simple liquid drops are determined by one of their physical properties, the *surface tension* (see below). Is this a good physical picture for the shape of a single cell, such as a fertilized egg? We will quickly conclude that it is not – the complexities of the cell's interior discussed in the previous chapter undermine its simple fluid behavior. Most importantly its bounding layer, the plasma membrane, does not have the extensibility that characterizes the surface of a uniform liquid. Nonetheless, it will be seen that when the cell's complex topography and the properties of the submembrane and extracellular layers are factored in, the physics of surface tension gives a reasonable first approximation to the cell shape. In addition, as in our discussion in Chapter 1, "generic" physical descriptions will be found to reappear at a higher level of organization, that of the multicellular aggregate.

Cells in suspension, or just prior to division, are spheroids. Eggs of most multicellular organisms have this shape (Fig. 2.2). If the egg were simply a drop of liquid, as early thinkers believed, its spheroid shape would be the straightforward consequence of surface tension. But the fact that some eggs are not spheres (see the asymmetric oblate shape of the *Drosophila* egg in Fig. 2.2) indicates that more than surface tension is determining the shape of these cells and thus, most likely, that of spherical cells as well.

This being acknowledged, can the physics of surface tension and related liquid phenomena tell us anything about how the fertilized egg divides first into two and subsequently into a cluster of cells? Indeed, theoretical analysis has been performed of "liquid drop cells" that expand in size owing to the osmotic pressure exerted by the synthesis and diffusion of macromolecules. Such objects have a tendency to break up into smaller portions of fluid, each with a higher surface-to-volume ratio than the original drop (Rashevsky, 1960).

Possibly this inherent tendency is mobilized during the division or cleavage of real cells and zygotes. But, as we will see, the basis of apparent surface tension is not straightforward in a living cell; its possible mobilization during division will be correspondingly complex.

Further, as the embryo undergoes consecutive cleavages and gradually becomes an aggregate of cells, it is reasonable to ask what physics can tell us about how the cells become arranged in such aggregates. Here we must consider the physical properties (e.g., elasticity) of the acellular materials – the hyaline layer, the zona pellucida – that variously surround the blastulae of different species. Because the blastula eventually "hatches" from such enclosing structures, we must also consider cell–cell adhesive forces, without which the divided cells would simply drift apart. (Indeed these adhesive forces are known to be present already in the prehatching blastula; see Chapter 4.) We will find that under experimentally confirmed assumptions about these adhesive forces, cells in an aggregate will easily move past one another in analogy to molecules moving randomly in a liquid. Thus, as we progress to the higher level of organization represented by the multicellular embryo, we will find that the liquid drop again becomes a relevant physical model, and we can make many strong inferences about the shapes and behaviors of cell aggregates from the physics of surface tension.

Surface tension

A liquid drop, left alone, in the absence of any force (in particular, gravity) will adopt a perfectly spherical shape. In other words, it will minimize the interfacial area with its surroundings in order to achieve a state with minimal surface or interfacial energy. (For a given volume the sphere is the shape with minimal area.) The surface or interfacial energy per unit area of a liquid is called its surface or interfacial tension and is denoted by σ. (Surface tension, strictly speaking, is the term reserved for the case when the liquid is in a vacuum but it is customarily used when the liquid is surrounded by air. The term "interfacial tension" is used when the liquid adjoins another liquid or a solid phase.)

To see surface tension in action one can prepare a small wire frame with one movable edge (of length L), as shown in Fig. 2.3 and wet the frame with a soap solution. The area of the wire frame (and thus of the liquid surface) can be increased by pulling on the movable edge with a force F by Δx in the x direction. According to the definition of σ, we have $W = F\,\Delta x = \sigma\,\Delta A$, where W is the work performed to increase the area by ΔA. Since $\Delta A = L\,\Delta x$ (Fig. 2.3), $F = \sigma L$, which provides another way of defining the surface tension: it is the force that acts within the surface of the liquid on any line of unit length. The surface of a liquid is thus under a constant tension which acts tangentially to the surface in a way as to contract the surface as much as possible. (A simple demonstration of this is to carefully place a small light coin on the surface of water in a cup; the coin will

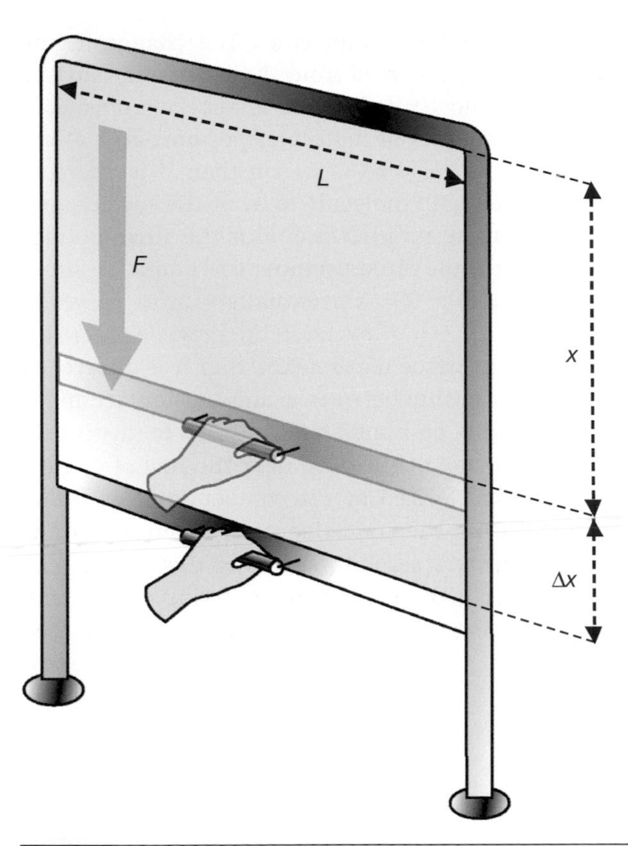

Fig. 2.3 Schematic representation of surface tension σ. The wire frame, whose original area is $A = L\,x$, carries a soap solution film. The surface tension is the energy required to increase the area of the film by one unit. The work $F\Delta x$ done by the force to increment the original area by $L\,\Delta x$ is also equal to $\sigma L\,\Delta x$. From the equality of these two expressions, it is evident that σ can also be interpreted as a force acting perpendicularly to the contact line (along the handle) between the liquid and the surrounding medium and directed into the liquid, thus opposing the effect of the external force F.

not sink. If the coin is placed on its edge, however, it will sink: the pressure it exerts on the liquid surface in this configuration exceeds that of the surface tension.)

The unit of σ is thus either J/m^2 or equivalently N/m. The surface tension of water is approximately 72 mN/m. Since σ is exclusively the property of the liquid, it does not depend on the magnitude of F applied to increase the area in Fig. 2.3. (Note that for a solid the surface or interfacial energy both depend on F and the initial area; see also below.)

The origin of surface and interfacial tension is easy to trace. When a liquid molecule is forced to leave the bulk and move to the surface, it is in an energetically less favorable state if its interaction energy with the dissimilar molecules at the surface is higher than with the molecules like itself that surround it in the liquid's interior. In the interior or bulk, the energy of a molecule in equilibrium with its surroundings is the cohesive energy E_C; at the surface it is the adhesion energy E_A (E_C and E_A refer to one molecule). The change in the molecule's energy when it is transferred from the bulk to the surface is $\Delta E = E_A - E_C$. According to its definition, the surface tension is $\sigma = N\Delta E$, where N is the number of molecules per unit area of the surface or interface. Thus, if $\Delta E = E_A - E_C < 0$ then it is energetically more favorable for the liquid molecule to be at the surface, and the liquid is driven to increase its interface with the surroundings as much as possible. Under these circumstances the liquid is said to "wet" the interface; in principle, it can eventually thin to a layer of molecular dimension.

The second definition of surface tension (i.e., that it is a force) allows one to establish a relationship between σ and the contact angle. Consider a liquid drop L placed on a solid substratum S in medium M (Fig. 2.4). The equilibrium shape of the drop near the line of contact with the substratum is characterized by the contact angle α, whose value can be determined from the condition of mechanical equilibrium (a special case of Newton's second law) along the contact line between L, S, and M. Equating the sum of forces acting on the unit length of the contact line along the x direction to zero (see Fig. 2.4), we obtain

$$\sigma_{SM} = \sigma_{LS} + \sigma_{LM} \cos\alpha, \tag{2.1}$$

where $\sigma_{SM}, \sigma_{LS}, \sigma_{LM}$ are the respective interfacial tensions between the three phases. Equation (2.1) is known as Young's equation (see for example Israelachvili, 1991). The cases $\alpha > 0$ and $\alpha = 0$ ($\cos\alpha = 1$) correspond respectively to partial and to complete wetting of the solid substratum by the liquid.

Liquid molecules are mobile whereas atoms or molecules in a solid are not. Therefore a liquid can increase its surface area by exporting molecules from the bulk while retaining the same surface density. Solids can increase their surface area only by stretching the distance between surface atoms, which results in a decrease in the surface density. As a consequence, the surface tension of a liquid drop is independent of the drop's area at any time, whereas the analogous quantity for a solid, the surface or interfacial energy per unit area, increases with area since it becomes progressively more difficult to pull atoms apart.

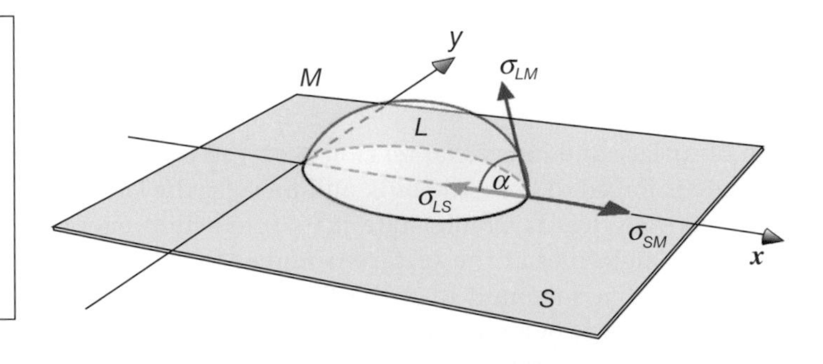

Fig. 2.4 Illustration of Young's equation. At mechanical equilibrium the shape of the liquid drop L near the contact line with the solid S and the medium M is characterized by the contact angle α. Its magnitude is determined from Newton's law (see the main text for details).

Laplace's equation

Living cells in most cases are not perfectly spherical. This does not, by itself, make surface tension or related concepts irrelevant to understanding cell shape, since liquid drops, when exposed to external forces, are typically also nonspherical. The shape of the liquid drop shown in Fig. 2.4 could be the result of gravity, for example. Under such circumstances the full equilibrium shape of an ideal incompressible liquid drop is determined by Laplace's equation (Israelachvili, 1991)

$$\sigma\left(\frac{1}{R_1}+\frac{1}{R_2}\right)+p_e=\lambda. \tag{2.2}$$

Here R_1, R_2 are the radii of curvature, $\sigma(1/R_1+1/R_2)=\sigma C$ is the *Laplace* (or curvature) *pressure*, C being the average curvature, and p_e is the pressure due to external forces. (Equation 2.2 is valid for any liquid surface, not only drops.) The radius of curvature is the radius of the circle that matches a curve or a surface at a given point. (A surface can be matched by circles along two orthogonal directions; see Fig. 2.5.) Equation 2.2 is simply the consequence of Newton's second law under the condition of incompressibility. Incompressibility enters the equation through the constant λ, a Lagrange multiplier. (Lagrange multipliers are used to incorporate specific constraints into mathematical equations. In the present case the constraint is due to the incompressible nature of the liquid drop: its volume is constant while its shape can change.) Note that for a flat surface (with infinite radii of curvature) the Laplace pressure is zero, whereas for a convex closed surface it produces an inward-directed force. (A closed surface is convex or concave according to whether it respectively bends away from or towards an enclosed observer. The average curvature at convex or concave points is respectively positive or negative, whereas at a saddle point – where one of the radii of curvature is positive and the other is negative – it can have either sign.)

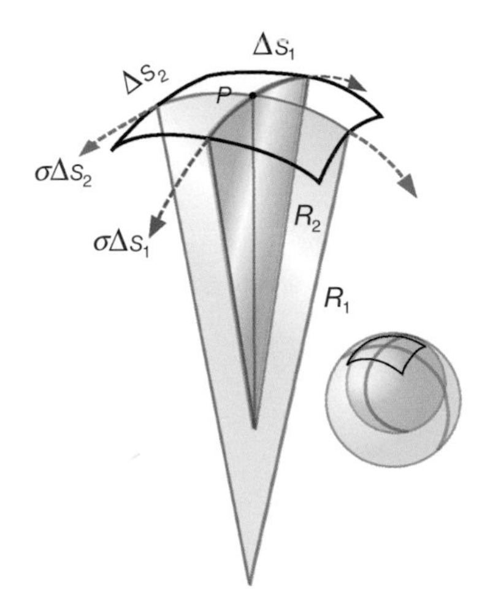

Fig. 2.5 Illustration of the radius of curvature. A point on a two-dimensional surface (*P*) can be matched with two orthogonal circles, whose radii determine the radii of curvature of the surface at that point. The figure also illustrates by the broken arrows that the surface or interfacial tension is the force perpendicular to any line of unit length within the liquid surface or interfacial area.

Box 2.1 Application of the equations of Laplace and Young

1. *Shape of a spherical drop under compression.*

Consider Fig. 2.6, which illustrates the shape change of an originally spherical liquid drop when compressed between parallel plates with a known constant force F (we ignore the force due to gravity). When Eq. 2.2 is evaluated along the flat plates, where the Laplace pressure is zero ($R_1 = R_2 =$ infinity) one can determine λ since $\lambda = p_e = F/(\pi R_3^2)$, where R_3 is the contact radius. Once the constant λ is known, applying Laplace's equation at the equator (point P in Fig. 2.6, where $p_e = 0$) yields

$$\sigma\left(\frac{1}{R_1} + \frac{1}{R_2}\right) = \frac{F}{\pi R_3^2}. \tag{B2.1a}$$

Here R_1 is the radius of the drop at its equator and R_2 is its radius of curvature in a meridianal plane, as shown in Fig. 2.6. By measuring the three radii, Eq. B2.1a provides the value of σ.

2. *Immiscible liquid drops in contact.*

Let us consider the situation shown in Fig. 2.7, which depicts contacting drops of immiscible liquids A and B embedded in a medium M. We assume that the shape of each of the three interfaces can be approximated by a part-spherical cap, with radii as indicated in Fig. 2.7, and we ask, what is the condition of equilibrium for given values of the interfacial tensions? Applying Young's equation to the vertical components of the interfacial tensions at point P in Fig. 2.7 yields the relationship between θ_{AM}, θ_{BM}, θ_{AB} and σ_{AM}, σ_{BM}, σ_{AB}:

$$\sigma_{AM} \cos\theta_{AM} + \sigma_{AB} \cos\theta_{AB} = \sigma_{BM} \cos\theta_{BM}. \tag{B2.1b}$$

Equation B2.1b provides information on the location (but not the shape) of the contact line and thus on the extent of the engulfment of drop A by drop B. When phase B just fully envelops phase A (i.e., it completely wets the interface between phase A and the medium) we have $\theta_{AM} = \theta_{BM} = 0$ and $\theta_{AB} = 180°$, and Eq. B2.1b reduces to

$$\sigma_{AB} = \sigma_{AM} - \sigma_{BM}. \tag{B2.1c}$$

This result will be used in Chapter 4 (see Fig. 4.5), in connection with the mutual envelopment of tissues.

To gain information on the radius of curvature of the contact line (i.e., R_{AB}) we combine Laplace's equation for a spherical drop $\Delta p = 2\sigma/R$, discussed above in the text, with the identity $0 = (p_A - p_M) - (p_A - p_B) - (p_B - p_M)$, where p_A, p_B, and p_M are respectively the pressures in phase A, phase B, and the medium. This results in

$$\frac{\sigma_{AB}}{R_{AB}} = \frac{\sigma_{AM}}{R_A} - \frac{\sigma_{BM}}{R_B}. \tag{B2.1d}$$

Equation 2.2 is valid locally, i.e., at each point of the surface. Whereas for an ideal liquid σ is a constant, the other quantities (except for λ) may vary along the surface. Thus in the absence of external forces, $p_e = 0$, the average curvature is the same at each point

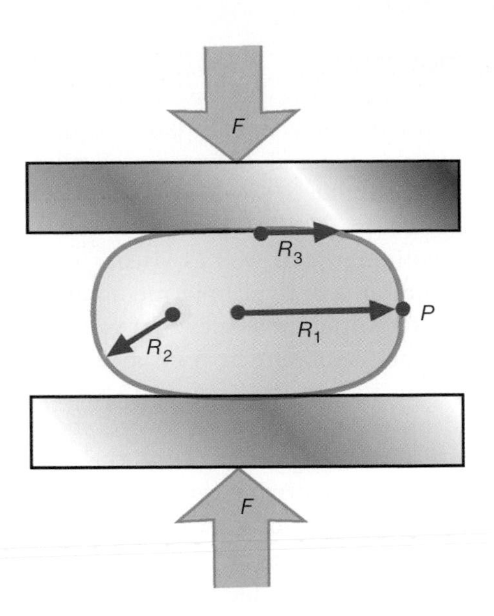

Fig. 2.6 The typical shape of an originally spherical liquid drop when compressed uniaxially, as indicated, between parallel plates. The three radii are used to establish a relationship between the compressive force and the surface tension in Example 1 in Box 2.1.

of the drop's surface. Consequently $1/R_1 + 1/R_2 = 2/R$ and the drop assumes a spherical shape with radius R, as expected. The (convex) spherical shape arises because the pressure inside the sphere, p_i, is larger than the external pressure, p_e. At equilibrium, therefore, the pressure difference across the spherical surface must be balanced by the Laplace pressure, $\Delta p = p_i - p_e = 2\sigma/R$.

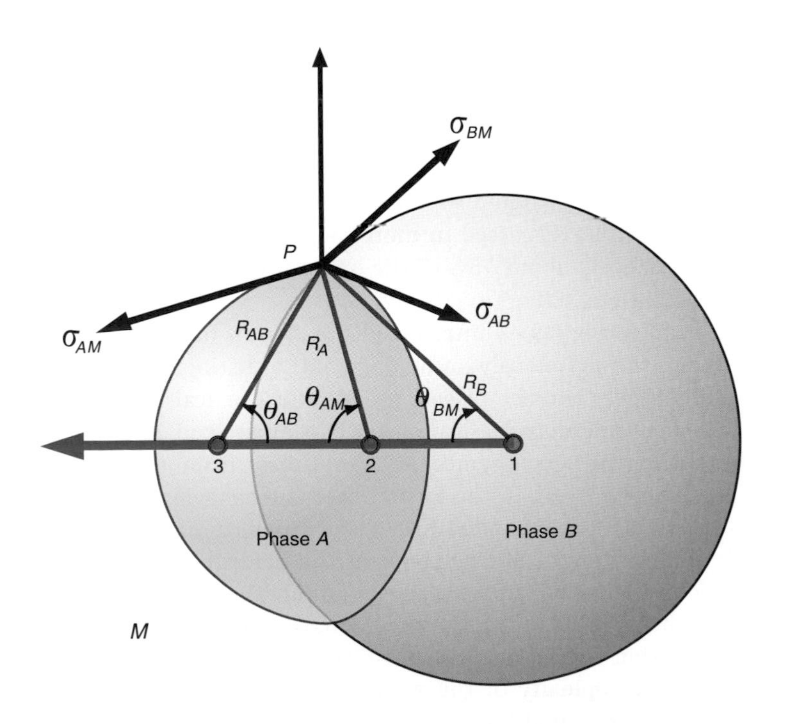

Fig. 2.7 Illustration of the system of contacting drops A and B considered in Example 2 in Box 2.1. The red arrow is drawn through the centers (1, 2) of the sphere B and the part-sphere A and meets the surface of A at 3. For further details see the text in Box 2.1.

Cortical tension

No cell, including the egg or the zygote, is purely liquid. Even intracellular water is only partly osmotically active. The water of hydration (composed of water molecules tightly bound to proteins and intracellular organelles), in particular, is osmotically inactive. And, of course, the interior of the cell is structured by fibrous proteins, with elastic as well as viscous properties. The cell is bounded by a more or less rigid lipid membrane. It has no bulk reservoir of free-floating lipid molecules that could be mobilized to increase the area of the membrane by transporting material from the interior of the cell to its surface. As a consequence, one cannot define a pure liquid surface tension, as described above, for the cell. Nevertheless, a physically analogous quantity, the *cortical tension*, turns out to be useful for characterizing certain shape changes in living cells.

The membranes of most cells can be represented by a highly convoluted two-dimensional surface decorated by spiky protrusions (microspikes, microvilli, filopodia), sheet-like extensions (lamellipodia), and invaginations (Fig. 2.8). When exposed to mechanical forces (e.g., distension or compression), therefore, a cell may mimic liquid drop behavior, its apparent surface area undergoing rapid increase or decrease by opening and closing the "wrinkles." (Specifically, the pool of invaginated membrane can be considered the analogue of the bulk reservoir of liquid molecules.) How easy it is to accomplish this area change strongly depends on the density of the cortical network of actin filaments underlying and attached to the plasma membrane (Tsai *et al.*, 1994, 1998), which must locally be separated from the membrane and resealed every time the cell surface is changed. Thus the energy needed to increase the area of a cell by one unit (its effective surface tension) has two contributions (Sheetz, 2001): (a) the membrane–cytoskeleton adhesion energy, which must be overcome to separate the membrane from the cytoskeleton; and (b) the energy needed to increase the area of the lipid membrane itself (which can be considered as a sheet of liquid within which the lipid molecules can flow freely, as will be discussed in more detail in Chapter 4; see Fig. 4.3). In a typical cell, about 75% of the apparent membrane tension is due to cytoskeletal adhesion (Sheetz, 2001).

Evans and Yeoung (1989) introduced the notion of "cortical tension" to make simultaneously an analogy and a distinction between pure liquid surfaces and the plasma membrane. Cortical tension can be measured by micropipette aspiration (Needham and Hochmuth, 1992) or pulling tethers (membrane nanotubes) from the plasma membrane (Sheetz, 2001). In the former case the behavior of a single cell is followed as it is aspirated into a micropipette narrower than the cell's diameter (Fig. 2.8). The shape of the deformed cell in equilibrium is considered as if it were purely liquid. The cortical tension, similarly to the purely liquid surface tension, is then determined from Laplace's equation applied to this shape. Here, despite the molecular and structural complexity of the actual biological system, its morphology can be accounted for by a physical law that governs a set of generic properties. Therefore, in what follows we will retain the term

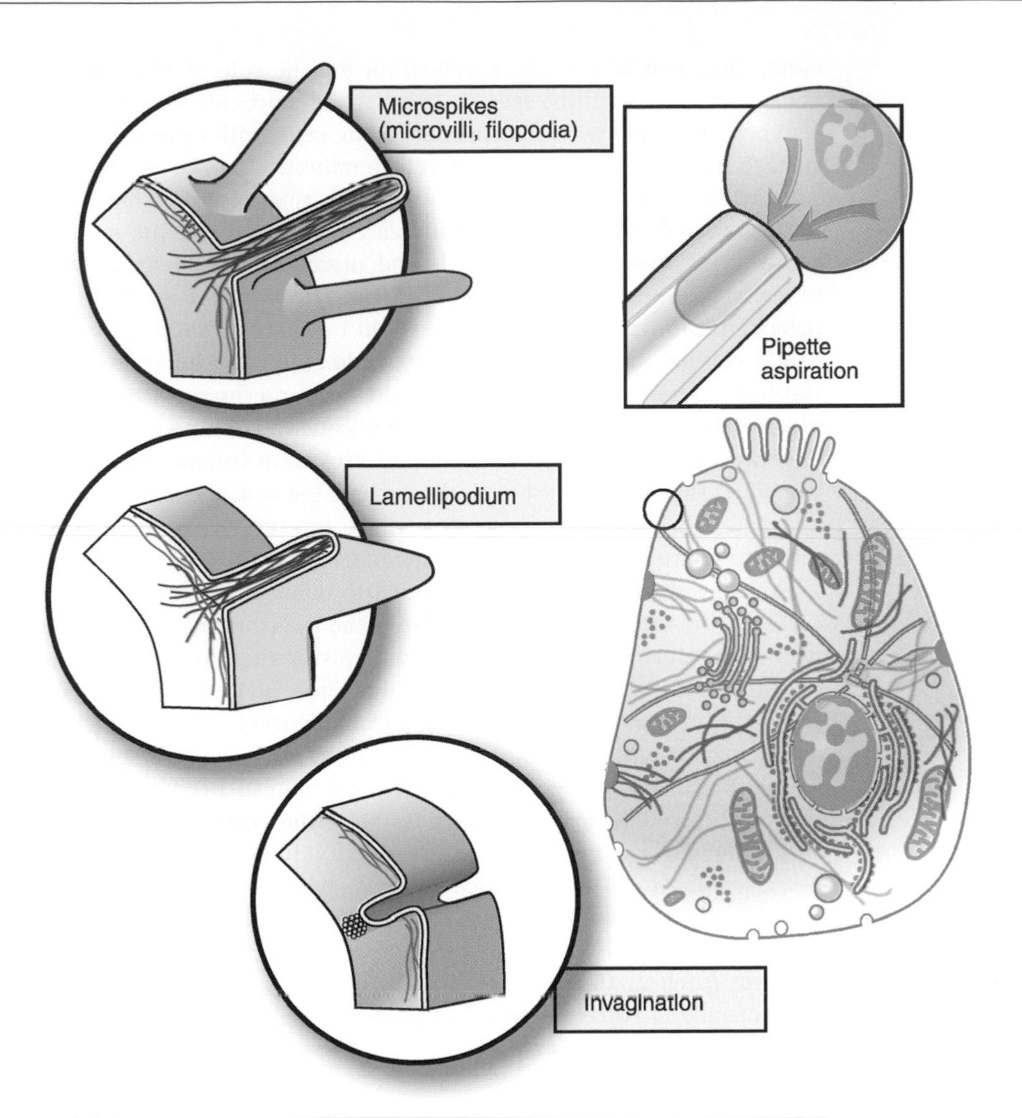

Fig. 2.8 Topographic specializations of the cell surface. The fibers underlying the membrane and extending into the microspikes and lamellipodia are typically made of actin filaments or microtubules. The figure also illustrates schematically how a small portion of the cell membrane (circled in the lower right panel) is aspirated into a micropipette to measure the cortical tension.

"cellular surface tension". It will need to be remembered, however, that its origin lies in the interaction of the cortical actin network with the cell membrane.

A true liquid is isotropic: its physical properties do not depend on orientation. In particular, the liquid surface tension is the same along the two orthogonal directions defined by the two radii of curvature in Laplace's equation, Eq. 2.2. For a viscoelastic material such as a cell, Hiramoto (1963) proposed a generalization of the Laplace pressure to $\sigma_1/R_1 + \sigma_2/R_2$. Experiments of Yamamoto and Yoneda (1983) suggested that whether a cell can be described in terms of isotropic or anisotropic surface tension depends on its position in the cell cycle. This may be a further indication for the role of the cortical actin network, as will be seen in the discussion of cleavage below.

Those familiar with the properties of cells will recognize that surface material (in the form of cell membrane) can be added and subtracted from the cell in other ways than that described above. In a growing cell, for example, the increase of the cytoplasmic mass is coordinated with the biosynthesis and insertion of new membrane in the smooth endoplastic reticulum (Fig. 1.1) and its transport and insertion into the cell surface. If this did not happen the membrane would rupture and the cell contents would spill out. Unlike the continuous, reversible exchange between the bulk and the surface in a liquid, however, such an expansion of the plasma membrane is mainly a one-way affair: cells do not reverse their growth in the normal course of events. Thus, in contrast with the cortical tension, which mobilizes and demobilizes a reserve of cell surface, the properties of the cell membrane are not a good formal analogue to surface tension.

Another way material can be added to or removed from the cell surface is by the fusion of vesicles from the interior with the plasma membrane during the process of exocytosis or their removal by endocytosis. This material exchange between the surface and the bulk could, indeed, potentially serve as an analogue of surface tension, but in fact it also is not a good one. The reason is that exocytosis and endocytosis are regulated by the physiological needs and functions of the cell rather than being responsive to mechanical stresses as are the surface tension of a pure liquid and the cortical tension described above. Although endocytosis has been observed in the cells of early-stage mammalian embryos (Fleming and Goodall, 1986), exo- and endocytosis are probably not extensive enough in most types of embryos (such as that of the sea urchin – see below) to contribute substantially to surface forces during cleavage.

Curvature energy

It is clear that the notion of constant cortical tension can give at best only a crude approximation to the shape of the cell membrane. The extent to which a convoluted cell membrane can be smoothened depends not only on the composition of the cortical actin network but also on the composition of the membrane itself. If the membrane is fully distended, no increase in the cell surface area can be achieved without pulling individual lipid molecules apart from one another. As the liquid drop model breaks down, elastic effects are bound to show up. Even before this extreme state is attained it is quite possible that the increase in the membrane area would depend on the total area. For example, stretching actin fibers could depend on the extent to which they have already been stretched. (This would, of course, violate the assumption of liquidity.) Indeed, as discussed by Needham and Hochmuth (1992) and Derganc *et al.* (2000) the energy of a neutrophil membrane W_n is well approximated by

$$W_n = \gamma A + \frac{K}{2A_0}(A - A_0)^2 . \tag{2.4}$$

Here we write the constant part of the cortical tension as γ, to distinguish it from the true liquid surface tension σ; K is the area expansivity modulus (another elastic parameter), whose units are J/m^2 (or equivalently N/m) and A_0 is the area of the free spherical neutrophil. According to the measurements of Needham and Hochmuth (1992) the membrane tension and area modulus of neutrophils are in the range 0.01 mN/m $< \gamma <$ 0.08 mN/m and 0.015 mN/m $< K <$ 1.751 mN/m.

Cells are continuously exposed to mechanical forces whose origin is in the external environment. Most of these forces are weak and cause no drastic changes in the integrity of the cell. Often they are only able to change the local curvature of the cell membrane without changing the overall surface area. For example, experiments performed with artificial vesicles made of amphipathic molecules (which contain a water-soluble, hydrophilic, part and a water-insoluble, hydrophobic, part) show that small changes in temperature (i.e., in thermal energy) may lead to marked changes in shape without any change in the surface area of the vesicle (see the review by Seifert, 1997). This would also be the dominant mode of shape change for cells with no hidden (convoluted) area. Such changes are controlled exclusively by the bending rigidity or bending stiffness (yet another elastic constant), i.e., the capacity of the membrane to resist changes in its local curvature.

According to Helfrich (1973) the curvature energy ΔW_c of a small membrane surface ΔA is given by

$$\Delta W_c = \frac{\kappa}{2}(C - C_0)^2\,\Delta A. \tag{2.5}$$

The parameters κ, C, and C_0 are the bending rigidity (a material property whose unit is the joule), the local average curvature introduced earlier, and the spontaneous curvature, respectively. The spontaneous curvature reflects the possible asymmetry of the membrane and characterizes its shape in the state of lowest curvature energy. For an inherently flat membrane the spontaneous curvature is zero, whereas for a vesicle with a preferred spherical shape of radius R it is $1/R$.

Physical models of cleavage and blastula formation

So far in this chapter we have entertained the notion that, with regard to their shapes, individual cells and in particular the zygote can be modeled as liquid drops. When we reached the plausible limits of this simple model we introduced additional parameters; this seemed to be necessary for characterizing the physical shape of cells. Following on these preliminary steps, we will now develop a quantitative description of the first cleavages and blastula formation. Although our discussion, with appropriate modification, can be applied to embryos of various species, we will concentrate mainly on the sea urchin, since experimental results in this system permit specification of most of the parameters thus far introduced. Finally, while development can proceed by relatively abrupt morphological changes, some of which will be analyzed later in this book, we will assume that the embryo is "well behaved" as a physical system before and after such changes. Thus, we consider the impact of variations in gene expression on the relevant physical parameters to be gradual. Indeed, if this were not the case, efforts to interpret morphogenesis in terms of such parameters would be largely futile.

As mentioned earlier, the initial divisions of the embryo of a multicellular organism are unusual in that they take place at constant embryo mass. At this time, in most cases, including that of the sea urchin, the embryo utilizes stored RNAs and proteins rather than synthesizing them itself. Thus, divisions are rapid (of order 1.5 hours in the sea urchin, reflecting the absence of G1 and G2 phases) and lead to smaller cells with increasing nucleo-cytoplasmic ratio. Taking this into account, we next present a simple surface-tension-based model (where "surface tension" in fact represents the cortical tension) of the early cleavage stages in the sea urchin (for comprehensive reviews, see Rappaport 1986, 1996). The results of the model will also characterize the first and later cleavage stages of other classes of organisms.

Cleavage: the White–Borisy model

The sea urchin zygote is initially spherical. Its cytoplasm has the same composition throughout the cell cortex and it thus possesses an isotropic surface tension (Yoneda, 1973). The only structure that can select a direction and thus the actual position of the cleavage plane is the mitotic spindle. Indeed, the cleavage plane is always perpendicular to this structure. If the spindle's position is physically manipulated (translated or rotated) before the furrow is fully established, the cell assembles the contractile ring so as to retain its relative orientation to the spindle. If the formation of the spindle is blocked (by the addition of chemicals such as colchicine, which disrupts microtubules), no furrow is formed. If the spindle is disrupted or even completely removed from the cell after the cleavage plane has fully been established (Hiramoto, 1968), however, the separation of the two daughter cells will progress to completion. The actual progression of the furrow thus is not dependent on the presence of the mitotic spindle, which seems to be needed only to set up the initial conditions for furrow formation. This suggests that once furrowing is under way, it is autonomous. Furthermore, the role of the asters of the mitotic spindle in the initiation of cytokinesis is well established (Rappaport, 1966): there is both a minimum and maximum separation distance of asters from the cell cortex beyond which a furrow will not be induced.

On the basis of these experimental results, White and Borisy (1983) proposed that cytokinesis is initiated by a signal emitted from the mitotic apparatus (the astral signal) and transported to the cell cortex (see Fig. 2.9). The signal polarizes the initially uniform cortex by relaxing the surface tension in a distance-dependent fashion, resulting in a gradient of the surface tension. This gradient ($\Delta\sigma/\Delta x$) can lead

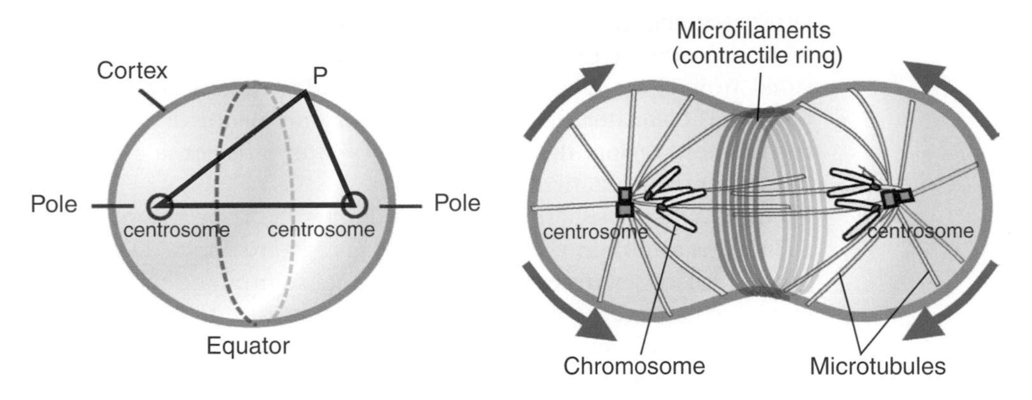

Fig. 2.9 Illustration of the physical process underlying the White–Borisy model. A putative biochemical cue, the astral signal, emanates from the centrosomes and at the cortex modifies the zygote's surface tension in a distance-dependent manner. Thus at any point *P* (left panel), the surface tension decreases in proportion to the inverse second or fourth power of the distances measured from the centrosomes. The resulting gradient in the surface tension sets up a flow of cortical elements from the pole region towards the equator (right panel), which eventually results in a contractile ring and a constricting cleavage furrow.

to shear and produce a flow in the peripheral cytoplasm, according to

$$\frac{\Delta\sigma}{\Delta x} = \eta \frac{\Delta v}{\Delta z}. \tag{2.6}$$

Here x and z measure the distance respectively along and perpendicular to the cell surface and η is the viscosity of the cortical cytoplasm. (Note that Eq. 2.6 is analogous to Eq. B1.1a in Chapter 1, since the units of $\Delta\sigma/\Delta x$ are force per area (F/A).)

The flow carries microscopic contractile (i.e., "muscular") elements from the polar region to the equator (see Fig. 2.9), where during this rearrangement a spontaneous alignment of actin and myosin filaments into an organized ring takes place. The ring is capable of exerting sustained circumferential tension of the order of 10^{-8} N (Hiramoto, 1978), while progressively constricting in the form of a groove. This moving cleavage furrow ultimately pinches the cell in half via a "purse-string" action.

The above "standard model," which is supported by a number of experiments (see the overview by He and Dembo, 1997), provides a good example of the interplay of generic and genetic processes. A biochemical signal (the astral signal) leads to a change in the cell's surface tension. This generates a physical process (cytokinesis), which eventually leads to renewed gene activity in the daughter cells.

The specific assumptions White and Borisy make about the astral signal are: (a) the stimulating activity of the asters is isotropic and depends only on the distance (but not the direction) measured from the asters; (b) the signals from the two asters combine in an additive way; (c) the magnitude of the stimulus decreases with distance r away from the centrosomes according to $1/r^2$ or $1/r^4$. This latter assumption ensures that the maximum decrease in the surface tension occurs at the poles, where the sum of the distances measured from the two asters is maximal, and progressively decreases towards the equator. Since motion occurs only in the dense narrow cortical-fluid region, the rest of the cell's interior is considered to be more watery and static.

The originally spherical egg, with radius R_0, is in a state of mechanical equilibrium because the uniform inward-directed force due to the Laplace pressure, $2\sigma/R_0$, σ being constant along the cell surface, is compensated by the elevated pressure inside the cell (due, for example, to osmosis). According to the assumption of this model, the astral signal breaks the uniformity of the surface tension and leads to its decrease. If no simultaneous change occurs in the radii of curvature then as the surface tension decreases so does the Laplace pressure, which becomes smaller than the original internal pressure. This leads to a force imbalance. Since the volume of the embryo is constant, the embryo bulges outward at the poles (where the decrease of the Laplace pressure is maximal), and flattens in the equatorial region (Fig. 2.9). White and Borisy demonstrated that mechanical equilibrium can be reestablished if the embryo displaces its surface so as to modify its radii of curvature. The length and velocity of the displacement are respectively proportional to the magnitude of the force imbalance and the viscosity of the cytoplasm. Under these circumstances, the distance between the asters is increased (in agreement with the experimental results of Hiramoto, 1958). The radii of curvature of the modified surface are then recalculated, and so is the internal pressure. The pressure, considered to be uniform since the embryo is modeled as a liquid drop, is obtained as the integrated mean of the Laplace pressures at each point of the surface. The above procedure is iterated, with the end result shown in Fig. 2.10 for an axially symmetric cleavage configuration.

In the simplest formulation of the model the astral signal leaves the surface tension forces isotropic, that is, not dependent on direction along the surface of the cell. As a consequence the furrow does not progress to completion. The reason is that, as the furrow forms, the cell surface develops "saddle points" along the equator, at which one of the radii of curvature gradually becomes negative. This provides another reason, in addition to the astral signal, for the Laplace pressure to decrease continuously. Thus, along the equator the Laplace pressure and the intracellular pressure eventually balance each other locally: an equilibrium is reached and the furrow stops elongating. This limitation of the model is removed if σ is allowed to vary with orientation, as discussed above in connection with the cortical tension. The anisotropy of σ in the present case is caused by the movement and reorientation of the contractile elements as they flow to the equator to assemble the contractile ring (White and Borisy, 1983).

Despite its success in providing a coherent description of cytokinesis in a number of cleavage patterns, the standard model has a

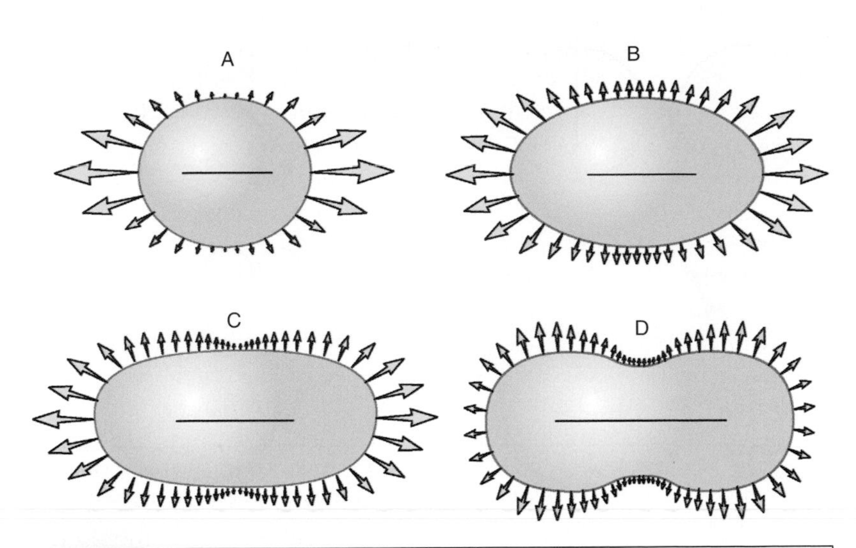

Fig. 2.10 Time evolution (A–D) of the shape of the zygote in the course of the first cleavage, as described by the White–Borisy model. The sequence of events is driven by the astral signal-induced changes in the surface tension and the subsequent changes in shape to reestablish mechanical equilibrium, as explained in the main text. The length of an arrow is proportional to the magnitude of the decrease in surface tension at a given stage of the process and a given point on the cell surface. The horizontal line indicates the distance between the centrosomes. (After White and Borisy, 1983.)

number of shortcomings (He and Dembo, 1997). The most obvious is that it does not provide a molecular basis for the astral signal. It artificially divides the cytoplasm into two spatially distinct regions, the gel-like or more viscous cortex and the watery interior fluid, but nevertheless considers the cell as a liquid drop with uniform internal pressure. Furthermore, the model does not regard chemical changes in the composition of the cytoskeleton (the polymerization and depolymerization of cytoskeletal filaments) as being an essential part of cytokinesis. In particular, it postulates that the assembly of the contractile ring and its constriction result exclusively from circumferential tension, which is contrary to experimental findings (Schroeder, 1975). Moreover, it does not provide an explanation for the origin of this tension or the mechanism of movement during furrowing. These, and several other deficiencies of the standard model, were addressed by He and Dembo (1997). Starting from the analysis of White and Borisy, these authors constructed a comprehensive, alternative, mathematical model of the first cleavage division in the sea urchin embryo. (The He–Dembo model is too complex to be discussed here, but more advanced students are encouraged to work through their paper.)

Blastula formation: the Drasdo–Forgacs model

As cleavage continues, the total number of cells in the embryo grows exponentially (Gilbert, 2003). In addition to the complex processes

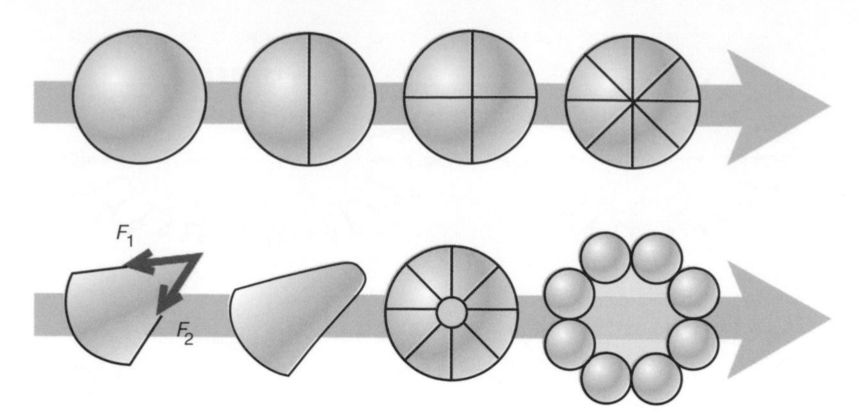

Fig. 2.11 Schematic sequence of cell divisions in holoblastic cleavage. The upper panel shows an idealized situation where the daughter cells occupy the original volume of the zygote. The individual daughter cell from the eight-cell stage shown in the lower panel has extremely sharp bends and thus high bending energy. As a consequence, it experiences forces (denoted here by F_1 and F_2) that tend to smoothen the sharp edges. Depending on the bending rigidity of the cell membrane, the rounding of each cell results in a cell configuration similar to the last diagram in the lower panel (for large bending rigidity) or the last but one (for small bending rigidity). The requirement that a change in cell shape leaves the total cell mass, or the cell volume, unchanged leads to a net displacement of the cell mass away from the center.

taking place during individual cell divisions, discussed above, cooperative effects involving groups of cells become increasingly more important. It is this cooperativity that finally gives rise to the first nontrivial morphogenetic structure, the blastula. In what follows we describe a model of blastula formation (Drasdo and Forgacs, 2000) for the case of perfect, radial, holoblastic cleavage, such as occurs in the sea urchin. The model combines a physical mechanism for cleavage at the single-cell level, with physical interactions between cells (occurring on a multicellular scale). Thus, for the first time, we will be dealing with a phenomenon involving several length scales, as is typical of most biological processes. The salient features of the model are the following.

1. Cells begin spheroidal and remain so during cleavage. Strong bends or kinks in the material of their membrane are therefore energetically disfavored. Thus, it is assumed that large changes in the local geometry (i.e., the curvature) of the membrane generate a restoring force, which tends to flatten the variation in geometry. As Fig. 2.11 illustrates, such an assumption requires that during cleavage the cell mass shifts outward, away from the center of the embryo.
2. Cells of the sea urchin embryo bind to an extracellular ("hyaline") layer. Hence the mechanical properties of the extracellular layer influence those of the blastula (Davidson *et al.*, 1999).

3. Cells in the early embryo are polar (i.e., spatially nonuniform) and, as a consequence of the inhomogeneous distribution of their adhesion molecules (see Chapter 4), form cell–cell contacts in special regions of their membrane, resulting in preferred cell configurations (Newman, 1998a; Lubarsky and Krasnow, 2003). These configurations are assumed to correspond to local minima in the (free) energy. Deviations from preferred cell shapes and configurations increase the energy and thus are unfavorable. The energy of a cell configuration contains the following contributions.
   (i) An *interaction energy* V_{ij} of neighboring cells with both attractive and repulsive contributions. Attraction is due to the adhesive bonds between cells (see Chapter 4). If the cells in isolation are spheroidal, contact between them leads to the stretching of their membranes. Furthermore, cells are composed mostly of water and therefore have small compressibility. Membrane deformations and limited compressibility give rise to repulsion. The manifestation of local physical interactions (i.e., between individual cells) at a larger scale (i.e., the blastula) is quite insensitive to the detailed shape of the corresponding interaction energy (Odell

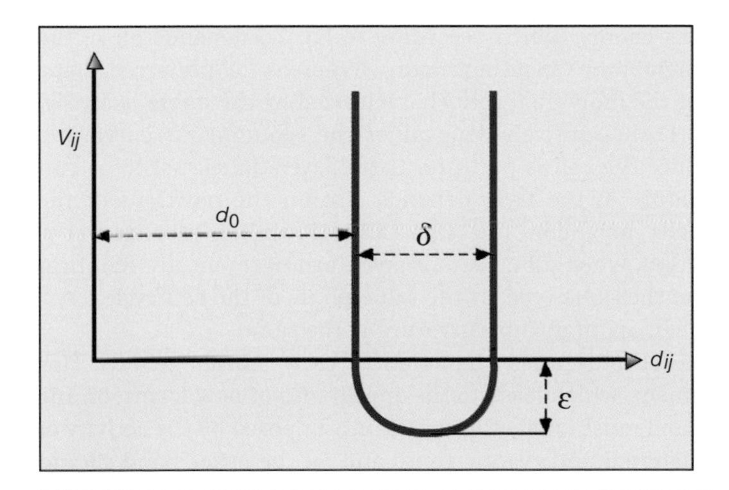

Fig. 2.12 The interaction energy V_{ij} in the contact region between two neighboring cells i and j as a function of the distance d_{ij} between the cell centers in the Drasdo–Forgacs model. The shape of V_{ij} reflects the limited compressibility of the cells and incorporates the entropic contributions of their membranes. Limited compressibility implies that d_{ij} cannot be smaller than a minimal distance d_0, at which a cell would lyse. To prevent this, V_{ij} is set to infinity at d_0. Furthermore V_{ij} contains contributions from direct cell–cell adhesion and from elastic changes resulting from any alterations in cell shape. The parameter δ is the range of interaction between neighboring cells. It determines the distance over which a cell can be stretched or compressed, as well as the interaction range of cell-adhesion molecules. The parameter ε characterizes the intensity of the elastic and direct cell–cell interactions. To ensure that cells remain in contact during blastula formation and gastrulation, the value of V_{ij} at $d_0 + \delta$ is set to infinity.

et al., 1981; Drasdo *et al.*, 1995). Therefore the combination of the attractive and repulsive energy contributions is modeled by the interaction energy V_{ij} shown in Fig. 2.12. The physical interactions responsible for the competing energy contributions have a characteristic range δ and magnitude ε. These parameters are determined by the deformability of the cell and its immediate molecular microenvironment (the "glycocalyx," see Chapter 4), the properties of the hyaline layer and the nature of the adhesion molecules that bind the cells to each other and to the hyaline layer.

(ii) A *curvature energy.* Polarized cells often form two-dimensional single-cell layers or sheets. The preferred geometries of the layer and the shape of the cells within the layer depend on the location of the cell-adhesion molecules, as shown schematically in Fig. 2.13. In analogy with polymer membranes, the preferred shape of the sheet at the position of the ith cell has a local spontaneous curvature (see Eq. 2.5). In the minimum-energy state the local average curvature equals the spontaneous curvature of the sheet. Any bending distorts the membrane, resulting in a deviation of the curvature from the spontaneous curvature and, consequently, in the increase of the membrane's bending energy, which according to Eq. 2.5 depends on κ, the membrane's bending rigidity. Whereas the preferred shape of the individual cell (characterized by the angle β_0 in Fig. 2.13A) exclusively determines the spontaneous curvature, once the cell is part of a tissue layer the actual local curvature of the layer depends also on the positions of the cells' neighbors (see Fig. 2.13B, C). In the idealized situation, when all cells in a given arrangement are identical (of the same type, at the same phase of the cell cycle, etc.), their spontaneous curvature is the same.

4. Active cell movement characterizes early morphogenesis. This movement, which leads to the appearance of new forms, on the one hand must satisfy the constraints imposed by the activity of the maternal and zygotic genes and on the other hand should proceed according to the governing physical mechanisms, which exert forces on the cells. These forces depend on the explicit form of the interaction energy and curvature energy discussed above, as well as on many other factors, and should eventually lead to configurations with minimal mechanical stresses (at least temporarily). In order to incorporate additional factors, it is assumed that, during cleavage, extracellular components provide a friction-like resistance to the displacements and orientational changes of the cells within the developing tissue sheet. The model incorporates friction and other biological and chemical processes, such as metabolism, intra- and extracellular transport, movements of the cytoplasm, and reorganization

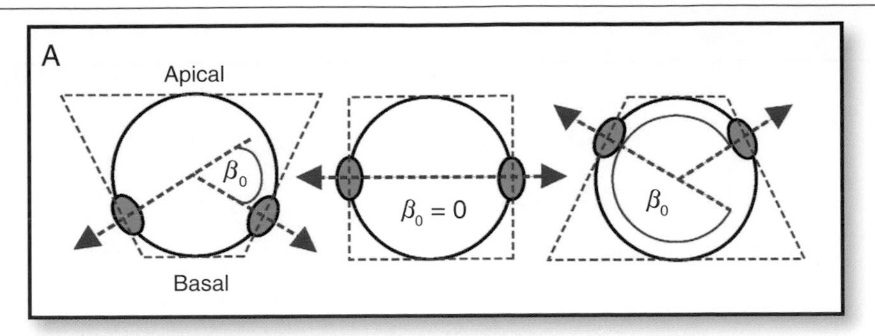

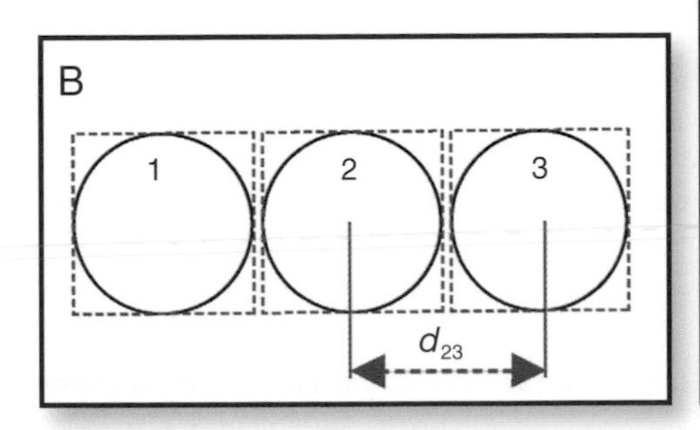

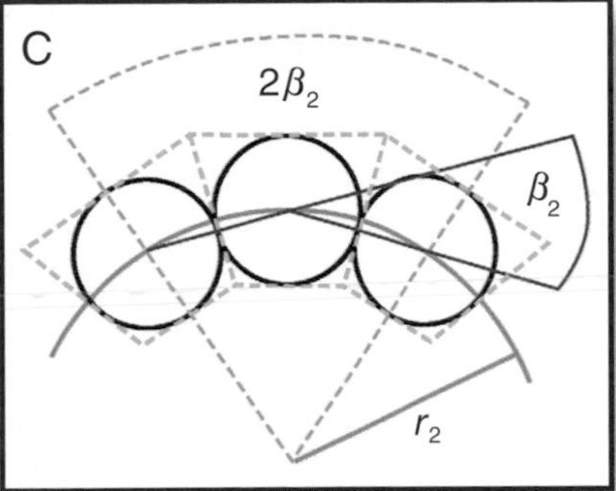

Fig. 2.13 Illustration of the curvature energy of a cellular layer. (A) The preferred individual cell shapes, denoted by the broken contours, which depend upon the location of adhesion molecules (filled areas). In this two-dimensional representation, the angle β_0 uniquely determines the preferred shape of the cell and thus the local spontaneous curvature of a cellular layer. The circles within the broken-line contours demarcate the simplified shape used to represent cells in the computer simulations of the Drasdo–Forgacs model. The optimal configuration of a cellular layer, containing only cells with preferred shapes in the left-hand panel ($\beta_0 < 180°$) or right-hand panel ($\beta_0 > 180°$) is a closed surface with the basal side, as defined here, oriented respectively towards the interior (see panel C) or the exterior. (B) The preferred cell shape when $\beta_0 = 0$ results in an optimal configuration with an open, planar, cell sheet and an equal distance between the centers of the cells. (C) A deviation from the optimal configuration shown in panel B. For a cell type with $\beta_0 = 0$ any bend (characterized by the existence of a finite local radius of curvature r and deviation angle β) increases the bending energy. Here the radius of curvature and the deviation angle are shown only for cell 2. Note that, for a cell type with $\beta_0 < 180°$ (shown in A), the configuration illustrated in panel C is optimal if $\beta_j = \beta_0$, where j denotes any cell in the sheet.

of the cytoskeleton, by imposing an additional stochastic force on the cells.

5. A cell's mobility, its geometric environment, and its interaction with its neighbors affect the observed cell-cycle time. The cell-cycle time also depends on the cell's intrinsic properties, which are incorporated by introducing the intrinsic cell-cycle time τ, a quantity that is influenced by the chemical environment (e.g., by growth and inhibitory factors). For an isolated cell not affected by physical interactions with neighboring cells, τ is the average cell-cycle time. If interactions are present, the

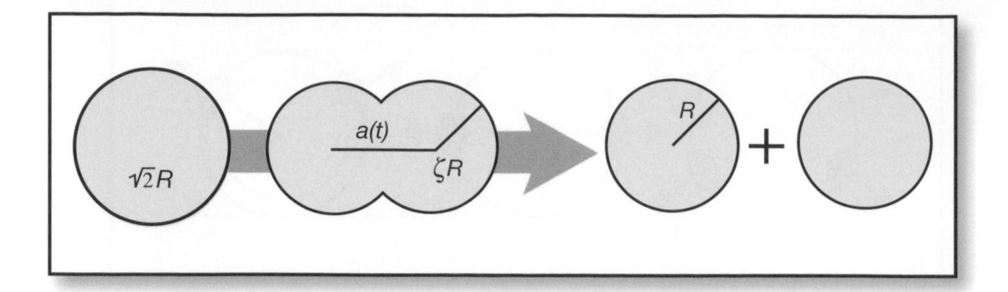

Fig. 2.14 Cell division as simulated in the Drasdo–Forgacs model. The cell is deformed by decreasing its instantaneous radius in randomly chosen small steps from the original radius, $\sqrt{2}R$. The parameter $\zeta < \sqrt{2}$ denotes the cumulative effect of these steps in an intermediate stage of cell division. As the radius decreases, the axis $a(t)$ increases to keep the total area of the cell constant during one division cycle, which is guaranteed by the dumb-bell shape. Area conservation also requires the two daughter cells to have radius R.

observed cell-cycle time is typically larger than the intrinsic cell-cycle time.

6. The described friction-limited stochastic dynamics is simulated using the Monte Carlo method (Metropolis *et al.*, 1953). The simulation chooses a cell randomly and modifies its state by (i) a shape deformation in accordance with the adopted mechanism of cell division (see Fig. 2.14), and (ii) a small random displacement. After each division a chosen cell deforms in small steps by decreases in its instantaneous radius and simultaneous increases in the distance between the centrosomes. After each modification in the state of an individual cell the total energy of the cell configuration is recalculated. If the modification increases the energy, the cell returns to its original shape with probability $P = 1 - e^{-\Delta V/F_{\mathrm{T}}}$, otherwise it stays in its new position. Here $\Delta V = V' - V$, V and V' being the total energy of the cell assembly respectively before and after the change in state of the randomly chosen single cell. The quantity F_{T} is a reference energy analogous to the thermal energy discussed in Chapter 1. (The characteristic energy scale for the motion of fluid particles is the thermal energy, whereas the characteristic energy scale for cellular motion must have its origin in the cytoskeletally driven cell-membrane fluctuations fueled by metabolic energy.) The value of F_{T} during blastula formation in the sea urchin is not known but F_{T} was estimated for chicken embryonic cells by Beysens *et al.* (2000). The above exponential form of the probability is the most typical one used in Monte Carlo simulations.

Monte Carlo simulations were performed in two dimensions, for a representative cross section of the approximately rotationally symmetric sea urchin embryo, using experimentally measured or derived

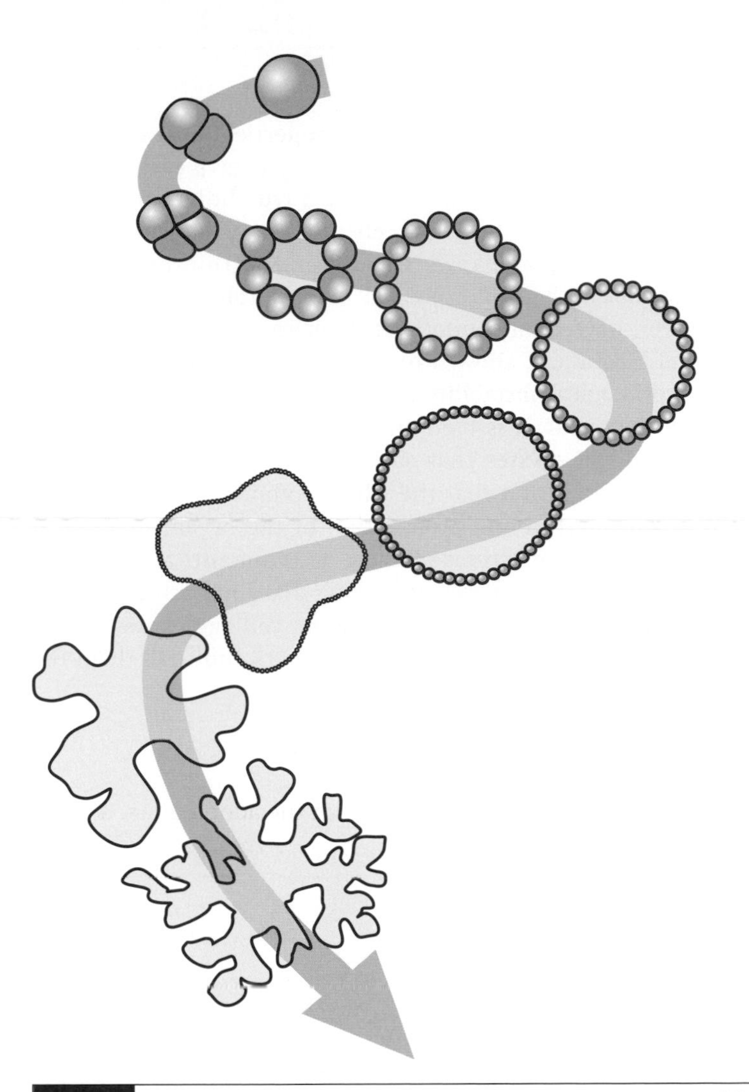

Fig. 2.15 Typical evolution of the cellular pattern in a computer simulation of the Drasdo–Forgacs model with specific, experimentally determined values of the model parameters. A dynamical instability in the shape of the blastula sets in at around the 64-cell stage, which in a spherical embryo would correspond to about 2000 cells. With further increase in cell number the folding of the blastula becomes more pronounced. Individual cells are indicated as circles up to the 64-cell stage.

values for the basic model parameters (Davidson *et al.*, 1995, 1999). The results of typical simulations are shown in Fig. 2.15. The number of cells grows exponentially and after a series of cell divisions a hollow spherical blastula forms. With further divisions the spherical symmetry disappears: the blastula becomes unstable and folds. Within the model this instability is generic and shows up over a wide range of parameter values. We will return to the question of this instability in Chapter 5 in our discussion of models of gastrulation.

Perspective

Simple models, based on realistic physical properties of cells and their interactions, can accurately reproduce many aspects of early cleavage and blastula formation. For instance, while a true liquid surface tension *per se* is inapplicable to living cells, its analogue, cortical tension, has explanatory power in this realm and, within limits set by the cell itself, obeys the same physical laws. Purely generic mechanisms, however, have their limitations. A physical process that leads to a spherical blastula, if allowed to proceed beyond a certain point leads to morphological instability and ultimately to the derailment of normal development. It is the multileveled determination of cells and embryos which ensures that each driving force is constrained by other driving forces and that the whole complex of forces is subordinated to the survival and propagation of the organism. Central to the integration of the various physical determinants are the complex biochemical signaling processes taking place in each individual cell, and across the blastula and later-forming cell aggregates, which control the values of cellular physical parameters and with them the consequences of the physical interactions.

Chapter 3

Cell states: stability, oscillation, differentiation

In the previous chapter we considered the forces leading to subdivision of the fertilized egg and the large-scale morphological consequences of successive cleavages. We followed the processes leading to the generation of the blastula, an important developmental step. In our treatment, however, the blastula was represented as a mass of identical cells. In reality, at the blastula stage not all cells of the embryo are identical: the "totipotent" zygote (with the potential to give rise to any cell type) first generates cells that are "pluripotent"– capable of giving rise to only a limited range of cell types. These cells, in turn, diversify into cells with progressively limited potency, ultimately generating all the (generally unipotent) specialized cells of the body. At the same time, in the life of each dividing cell, regardless of its level of developmental potency, there are transitions from state to state as the cell performs various functions of the cell cycle.

It is the purpose of this chapter to provide a framework for understanding how genetically identical cells can change their physical states in reliable and stable ways in time and space. We will also explore how these states, in turn, can provide a physical basis for a cell's performance of different tasks at different phases of its life cycle and for its descendants' performance of different specialized functions in different parts of the developing and adult organism.

The transition from wider to narrower developmental potency is referred to as *determination*. This stage of cell specialization generally occurs with no overt change in the appearance of cells. Instead, subtle modifications, only discernable at the molecular level, set the altered cells on new and restricted developmental pathways. A later stage of cell specialization, referred to as *differentiation*, results in cells with vastly different appearances and functional modifications – electrically excitable neurons with extended processes up to a meter long, bone and cartilage cells surrounded by solid matrices, red blood cells capable of soaking up and disbursing oxygen, and so forth. Cells have generally become determined by the end of blastula formation; successive determination increasingly narrows the fates of their progeny as development progresses. When the developing organism requires specific functions to be performed, cells will typically undergo differentiation.

Gene expression and biochemical state

Since each cell of the organism (except for the egg and sperm and their immediate precursors) contains an identical set of genes, a fundamental question of development is how the same genetic instructions can produce different types of cell. This question pertains to both determination and differentiation. Since these two kinds of cell specialization are formally similar and probably employ overlapping sets of molecular mechanisms, we will refer to both as "differentiation" in the following discussion, unless confusion would arise. Multicellular organisms solve the problem of specialization by activating only a cell-type-specific subset of genes in each cell type.

The *biochemical state* of a cell can be defined as the list of all the different types of molecules contained within it, along with their concentrations. During the cell cycle the biochemical state changes periodically with time. If two cells have the same complement of molecules at corresponding stages of the cell cycle then they can be considered to be of the same differentiated state. (The cell's biochemical state also has a spatial aspect – the concentration of a given molecule might not be uniform throughout the cell – but we will ignore this complication for the time being). Certain properties of the biochemical state are highly relevant in understanding developmental mechanisms. The state of differentiation of the cell (its *type*) can be identified with the collection of proteins that it is capable of making. This, in turn, comes down to which genes are potentially active (transcribable or "template competent"). Another important aspect of determination and differentiation that is less well understood, and will not specifically enter into our discussion, concerns the generation of different proteins from a given active gene by "alternative splicing" of the gene's RNA transcript (see Maniatis and Tasic, 2002).

The cell type is determined by the structural properties of the DNA–protein complex in the cell nucleus known as *chromatin*. Genes in the transcribable subset must be in particular physical states, and this is accomplished by variations in the packaging of different stretches of DNA by proteins. If a gene is template competent, the stretch of DNA that contains it is in a conformation referred to as "open." In cell types in which the same gene is inactive it is in a "closed" conformation (Gregory *et al.*, 2001; Cunliffe, 2003).

Although having different sets of template-competent genes defines cells as differentiated from one another, it is important to distinguish between two classes of these transcribable genes. Of the estimated 20 000–25 000 human genes (International Human Genome Sequencing Consortium, 2004), a large proportion is constituted by the "housekeeping genes," involved in functions common to all or most cell types (Lercher *et al.*, 2002). In contrast, the template competency of a relatively small number of genes – more than a hundred, but possibly fewer than a thousand – specifies the type of determined or differentiated cell; these genes are *developmentally regulated* (i.e., turned on and off in a stage- and position-dependent fashion) during embryogenesis. The generation of the approximately 250 different cell types of the adult human body is accomplished by packaging this special group of developmentally regulated genes into appropriate chromatin states in distinct cell populations. Some more familiar genes that are developmentally regulated are beta-globin in the reticulocyte (the red blood cell precursor), keratins in skin, and insulin in the pancreatic beta cell.

It is also important to recognize that template-competent genes in a differentiated cell type are not constantly being transcribed. The open chromatin state of template-competent genes is structural or conformational and is essentially irreversible when the cell remains in its natural context. The products of a special class of template-competent genes are involved in maintaining established gene expression patterns across cell generations (Brock and Fisher, 2005).

The foregoing discussion implies that the biochemical state of a cell has at least two hierarchically organized aspects – the *differentiated state*, defined by the stable set of template-competent, developmentally regulated genes it contains, and the *functional state*, defined by the (potentially varying) levels of housekeeping and developmentally regulated proteins that it produces. Differentiated cell types are among the end-products of development. Moreover, the range of functional states that a given cell can assume is limited by its state of differentiation. Significantly, though, development itself is implemented by variations in the functional states of emerging cell types at the early and intermediate developmental stages.

How physics describes the behavior of a complex system

The biochemical state of a living cell does not remain fixed. Amino acids and nucleotides are constantly being consumed to make proteins and nucleic acids. These and other small molecules are continually being synthesized from precursors or being brought into the cell from the outside. Usually, the cell is also involved in transporting functionally useful products and waste materials across its boundary. Given all this, it would be surprising if there were not wide swings in the concentrations of many of the cell's internal components. As mentioned above, however, the cell's state of differentiation provides limits to these swings.

When physicists have to deal with a complex system containing many interacting components – a "dynamical system" – they typically represent its state as a point in a multidimensional space. Let us imagine that such a system contains only two interacting components. The system then resides in a two-dimensional "state space," also called the "phase plane," defined by pairs (A, B), where A is the concentration of the first component, measured along the x axis and B is the concentration of the second component, measured along the y axis.

The first question we can ask about such a representation is: at a given instant of time, where will we find the system on the phase plane? It is important to emphasize that all the dynamical systems we will be considering are *open systems*, that is, they can exchange chemicals and energy with their environment. Indeed, these are the only kinds of system that can exhibit interesting behaviors, including the characteristics of being alive. All closed chemical systems (ones that have no interaction with their environment) eventually settle into a dead state, determined by the initial concentrations, referred to as "chemical equilibrium." Open systems that consume energy and operate far from thermodynamic equilibrium are known as *dissipative systems*. As we will see, such systems can exhibit interesting types of spatiotemporal organization of their constituents.

Since our system of interest is open, we are at liberty to prepare it in any initial condition we choose. If, like a cell, it is bounded by a membrane, then we can simply inject enough of the two components to bring the system to any desired point. But will it stay there? In fact, dynamical systems are governed by physical relationships among their components. In a simple example, if the components are two types of molecule and the system's chemistry dictates that the first is reversibly transformed into the second then there is a constraint between their concentrations A and B: when one goes up the other goes down. The laws of chemical transformation also dictate the direction of change: if we add more of the first component to the system, for instance, some of it will be transformed into the second. Thus the system will not remain in any arbitrary initial state.

The dynamical interactions in such systems are usually represented by sets of coupled differential equations. We saw in Chapter 1 that a derivative represents the rate of change of one variable as a function of another variable. (In dynamical systems the derivatives of the state variables with respect to time are of particular concern). In a differential equation the derivative of one variable (say the concentration A of the first component in the system discussed above) is equated to an expression containing other variables and possibly including that variable itself. A system of coupled differential equations represents the time dependence of each state variable expressed in this fashion.

To gain some insight into how dynamical systems are analyzed mathematically, consider, as an example of a two-component dynamical system, a chemical system governed by the following set of coupled differential equations (Leslie, 1948; Maynard Smith, 1978):

$$\begin{aligned} \frac{\mathrm{d}A}{\mathrm{d}t} &= 4A - A^2 - AB, \\ \frac{\mathrm{d}B}{\mathrm{d}t} &= B - \frac{B^2}{A}. \end{aligned} \tag{3.1}$$

(Remember the meaning of $\mathrm{d}/\mathrm{d}t$ explained in Box 1.1). Wherever the system may start out on the phase plane, Eqs. 3.1 assign an arrow to that point indicating where the system will move to next. As time progresses, the arrows thus outline a trajectory along which the system moves in time in the phase plane. The map of all possible sets of arrows (i.e., trajectories) on the phase plane is called a "vector field." At certain points in this state space the time derivatives of both variables vanish simultaneously. Any such solution is called a "fixed point" or "stationary state" or "steady state." Setting both time derivatives in Eqs. 3.1 to zero we find fixed points at (0, 0) and (2, 2).

Not all fixed points are equivalent, however. In some cases the fixed point will be stable against perturbations, i.e., if the system is displaced to a nearby point then changes in reaction rates will be induced that will return it to the original steady state (think of a marble sitting at the bottom of a round-bottomed cup). In other cases a tiny change in state will be sufficient to drive the system away from the fixed point (think of a marble at the top of an inverted round-bottomed cup). In the case of the two stationary states of Eqs. 3.1, (2, 2) is stable and (0, 0) is unstable (see Fig. 3.1). (The general mathematical technique to study the nature of a fixed point, whether it is stable or unstable, is called linear stability analysis. In the Appendix to Chapter 5 we demonstrate how this method works.)

Of course, our dynamical system may have other forms. Consider, for example,

$$\frac{dA}{dt} = (1 - A^2 - B^2)A - B,$$
$$\frac{dB}{dt} = A. \tag{3.2}$$

Here there is an unstable fixed point at (0, 0). Interestingly, the circle of unit radius, $A^2 + B^2 = 1$, surrounding the unstable solution is a stable closed orbit. Note that the time derivatives do not vanish on the circle.

Whereas any small displacement of the system state away from an unstable fixed-point solution will take it still farther away, if the system finds itself on a stable closed orbit, a circle in the case of the above equations, any small displacement will bring it back to the same circle (but not necessarily to the same point). Indeed, if the system begins in any initial state inside or outside the circle it will eventually wind its way onto it and then remain on it forever (Fig. 3.1). Such orbit is called a "limit cycle": along it the system periodically revisits each point. To prove that the unit circle is a stable limit cycle of the dynamical system defined by Eqs. 3.2 requires the application of linear stability analysis. However, it is easy to see that it is indeed a solution of Eqs. 3.2, in the sense that the equations map the circle onto itself. The trigonometric identity $1 = \cos^2 t + \sin^2 t$ implies that if A and B satisfy the equation for the unit circle, their time dependence, for the particular case in which $A(0) = 1$ and $B(0) = 0$, can be written as $A(t) = \cos t$ and $B(t) = \sin t$. Inserting these expressions into Eqs. 3.2 (noting that the derivative of $\sin t$ is $\cos t$, whereas the derivative of $\cos t$ is $-\sin t$), we arrive at the desired result.

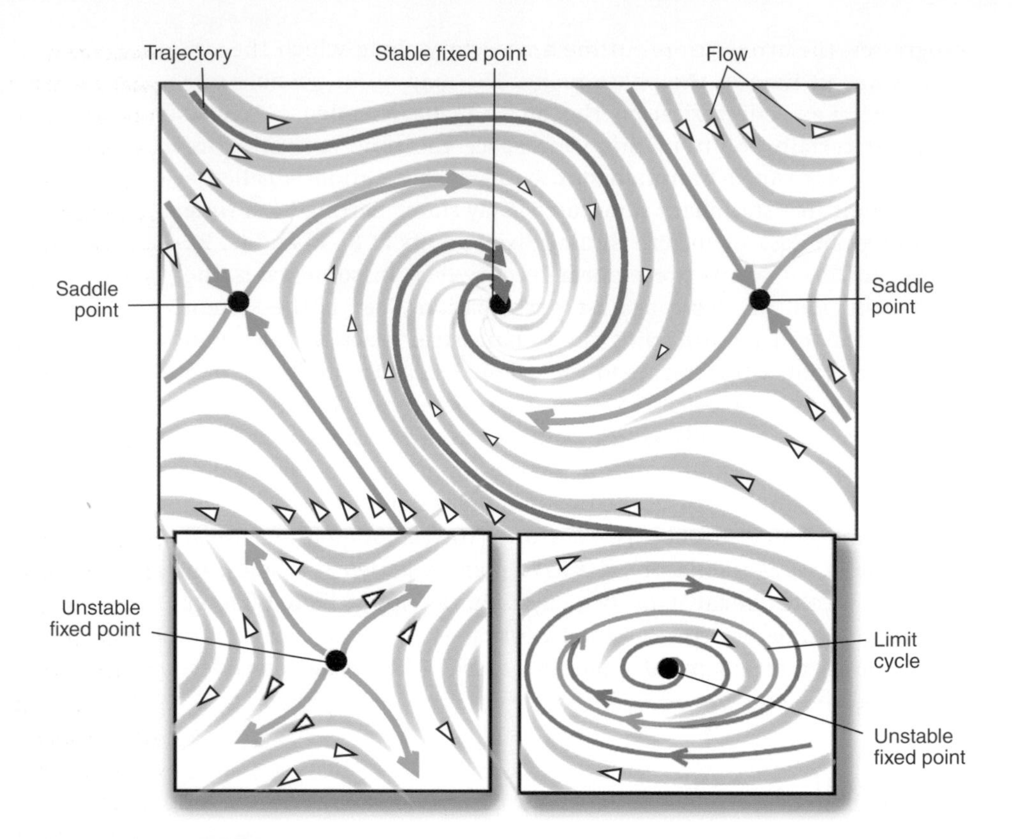

Fig. 3.1 Possible trajectories (flow lines) of the point representing the state of a dynamical system (the "system state"). The different scenarios are shown in a vector field superimposed on a two-dimensional phase plane. In addition to the overall geometry of the vector field (gray flow lines, open arrowheads), the figure shows representative trajectories (red, green, and blue) leading to special points: a stable fixed point or attractor, to which all local trajectories converge; an unstable fixed point or repeller, from which all local trajectories diverge; a saddle point, attracting some local trajectories and repelling others; and a stable limit cycle (a closed orbit surrounding an unstable point) upon which all local trajectories converge. (In an unstable limit cycle, not pictured, the closed orbit, from which all local trajectories diverge, surrounds a stable fixed point.) All these behaviors have analogues in dynamical state spaces of higher dimensionality. Regions of state space in which the local dynamical behaviors differ are known as "basins of attraction" (see the upper panel) and are divided from each other by separatrices, lines in the phase plane or, in an n-dimensional state space, surfaces of fewer than n dimensions.

The examples in Eqs. 3.1 and 3.2, which were chosen for their heuristic value rather than for their plausibility as models for biochemical interactions, show three possible dynamical behaviors for a two-component system (i.e., a single stable or unstable fixed point and a limit cycle). More generally, a system can have more than one stable node or orbit. Systems of more than two variables can even wander around in a not strictly recurrent fashion in a limited region of the state space (a "strange attractor"). In cases in which there is more than one stable behavior, each will have its own "basin of attraction," that set of system states for which the trajectories passing through them will wind up at that particular node or orbit (Fig. 3.1). A boundary curve between adjacent basins of attraction is called a *separatrix*. A consequence of this situation is that there will be some distinct, but close, system states that will be in the basins of attraction of different nodes, causing the dynamical system to behave like a "switch." A similar fundamental property of a nonlinear system, known as chaotic behavior, means that even when the system starts its evolution from nearly identical states it may terminate at unpredictable endpoints.

We have considered only systems with two interacting components. Living cells produce tens of thousands of different proteins and contain a myriad of sugars, amino acids, nucleotides, and other small molecules. The state space of a real cell has thousands of dimensions, and the vector field describing the system's behavior is spectacularly more complicated than the two-dimensional case. Fortunately, in many cases only a small number of interacting components are needed to determine the cell's fate at critical stages of development – other components are either uncoupled from the "main event" or change on such different time scales that they can be treated as constant, or considered only with regard to their steady-state values. In what follows we will describe cases in which the oscillatory or switching behaviors of cellular dynamics can drive developmental decisions.

Oscillatory processes in early development

In our description of the mechanics of cleavage in Chapter 2 we assumed that at a particular point in the lifetime of the zygote, cytokinesis is initiated by a signal that arises from the mitotic apparatus (the astral signal). We did not discuss what the initiating signal might be, however. It is clear that this signal must be periodic in nature, since once the initial cleavage division is complete each of the daughter cells undergoes the same process, and so on for the resulting daughters, until the egg cytoplasm is subdivided into the hundreds or thousands of cells of the blastula, depending on the species. Experiments with extracts of fertilized eggs have indicated that the basis of this cyclical control process is a biochemical oscillator, since it is capable of occurring spontaneously in a soluble mixture of egg cytoplasmic components, with no organized cytoskeleton or nuclei present (Murray and Hunt, 1993).

Before we can consider the mechanism by which the cell cycle is controlled during blastula formation and later development, we must deal with one peculiarity of cell division prior to fertilization. Recall from the previous chapter that each cell in a diploid organism contains two *versions* (maternal and paternal) of each chromosome. Moreover, once DNA synthesis has occurred in preparation for cell division the cell contains two *copies* of each of these two versions.

Because fertilization must lead to a zygote that is itself a diploid cell, the generation of the egg and sperm ("oogenesis" and "spermatogenesis") must involve a cell division step that reduces the number of chromosomes to one version of each in each daughter cell. This is called the first meiotic division, or meiosis I (see Fig. 9.1). After this, one daughter cell (the other, called the "polar body," receives very little of the cytoplasm and has no further developmental role) contains two physically attached copies (sister chromatids) of a single version of each chromosome. This cell then undergoes a division in which the sister chromatids of the replicated chromosomes (Fig. 2.1) are partitioned into another polar body and a daughter cell which therefore now contains only *one* copy of *one* version of each chromosome. This division step, called meiosis II, is mechanically identical to mitosis, in which sister chromatids generated during the S phase are similarly separated. Fertilization, which brings together the chromosomes of the egg and the sperm, yields a cell, the zygote, which thus contains two versions of each chromosome.

The frog, *Xenopus laevis*, has provided an experimentally useful system for studying the control of the cell cycle in early development. During oogenesis (Chapter 9) in the frog, once the immature egg, or *oocyte*, reaches about 1 mm in diameter, the DNA undergoes one round of replication and then arrests before the first meiotic division. After that the oocytes are stimulated by the hormone progesterone, produced in the mother's body, to undergo the two meiotic divisions, arresting once again, this time at the metaphase of meiosis II (meiosis in mammals, such as the human, is under similar hormonal control). After fertilization, the egg nucleus completes meiosis II and fuses with the sperm nucleus. The zygote then undergoes 12 rapid, synchronous, mitotic cycles to form a hollow blastula of 4096 cells. At this stage, called the midblastula transition, cell division in the embryo slows down and gastrulation movements begin.

The 14 (two meiotic and 12 mitotic) cell divisions that produce the frog blastula are triggered by an M-phase-promoting factor (MPF), a protein kinase (an enzyme that phosphorylates other proteins) consisting of two subunits, Cdc2 (the catalytic subunit) and cyclin B (the regulatory subunit). MPF phosphorylates an array of proteins involved in nuclear-envelope breakdown, chromosome condensation, spindle formation, and other events of meiosis and mitosis. Whereas Cdc2 is present at a constant level throughout the cell cycle, the concentration of cyclin B and thus the MPF activity varies in a periodic fashion, rising to a peak value just before M-phase (see Chapter 2) and dropping to a basal value as a cell exits M-phase (Fig. 3.2).

Although MPF controls nuclear properties, the oscillation in its levels can occur in the absence of nuclei. Indeed, in frog embryos no transcription of any genes (including those specifying the components of MPF) in zygotic or blastomere nuclei occurs until the midblastula transition. In nucleus-free cytoplasmic extracts of immature frog eggs there are spontaneous oscillations of MPF with a period of about 60 min. If sperm nuclei are added to the extract these nuclei

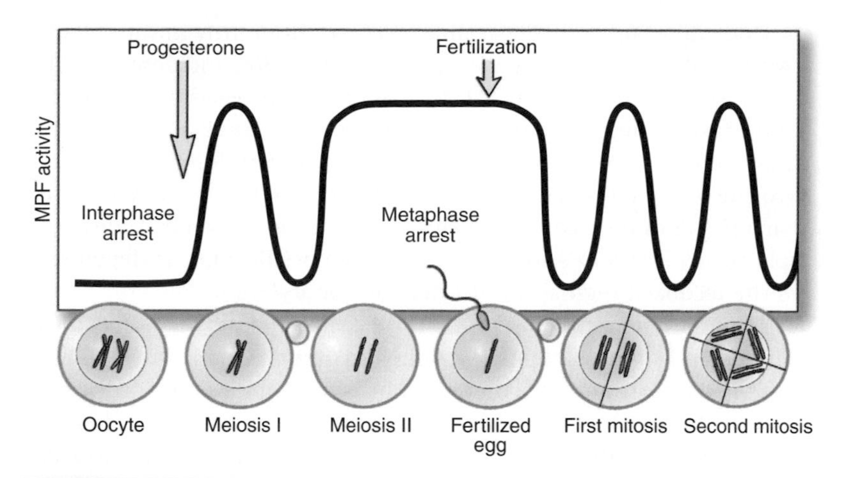

Fig. 3.2 Variation in MPF activity in the meiotic and mitotic cycles of frog eggs. Between the oocyte stage and meiosis I, a reduction division leaves the egg in a haploid state with one member of each homologous pair of chromosomes and most of the cytoplasm (one homologous pair is shown). The other members of each pair are allocated to the first polar body (pictured as a small empty circle). During meiosis II (which is completed after fertilization) the egg divides once more, this time separating identical replicated copies of each remaining chromosome into the zygote and the second polar body. Fertilization restores the diploid state, thus now two versions (i.e., a homologous pair) of each chromosome are present (see also Fig. 9.1). For illustrative purposes, the chromosomes are represented in a condensed state even though the cells are mainly in interphase during these events. Once the egg is fertilized and starts cleaving (the figure shows the first two cleavages) periodic MPF fluctuations continue through the 12 mitotic divisions that produce the frog blastula. (After Borisuk and Tyson, 1998.) See the main text for a discussion of the relationship of progesterone and fertilization to interphase and metaphase arrest.

undergo periodic mitoses whenever the MPF concentration is high. Biochemical analysis (Murray and Kirschner, 1989) showed that the cyclin protein is periodically degraded in the extract and resynthesized in a manner that depends upon the presence of its mRNA.

Mitotic control in frog eggs: the Tyson–Novak model

It is reasonably straightforward to devise a simple set of coupled nonlinear differential equations along the lines of Eqs. 3.1 and 3.2 for the breakdown and synthesis of cyclin in a cell-free extract and to propose the resulting limit-cycle oscillation as a model for cell-cycle control in the cleaving frog egg (Norel and Agur, 1991; Tyson, 1991; Goldbeter, 1996). But the cell-free system differs in several important features from the intact egg, and it is the properties of the latter that any realistic model must explain. In the first place, the free-running MPF cycle in the extract is about twice as long as the MPF-cell-cycle period in intact eggs (Murray and Kirschner, 1989), suggesting that additional important processes are at work in the living system.

One important control variable that acts differently in the cell-free extract and in the intact egg is the phosphorylation and dephosphorylation of Cdc2. In the extract, Cdc2 goes through cycles of phosphorylation and dephosphorylation on one of its tyrosine residues, whereas after the first cleavage in the intact egg Cdc2 is no longer found in a tyrosine-phosphorylated form (Ferrell *et al.*, 1991). It is likely that the dynamics of mitotic control in the early embryo, which includes multiple steady states in addition to sustained oscillations, is dependent on the feedback regulation of Cdc2 phosphorylation.

Taking into account what is known experimentally from both the cell-free and intact systems, Tyson, Novak, and their coworkers have presented a more complex model than a simple limit cycle for the generation of active MPF in the early frog embryo. This model comprises nine coupled nonlinear differential equations (for the concentration of nine regulatory proteins), governing, among other factors, the regulation of cyclin B levels and the state of Cdc2 phosphorylation (Novak and Tyson, 1993; Borisuk and Tyson, 1998) (see Fig. 3.3). Using computational methods they found solutions to these equations corresponding to several physiological or experimental states.

(i) A steady state with low MPF activity (as in the immature oocyte prior to progesterone activation).
(ii) A steady state with high MPF activity (as in the mature egg prior to fertilization).
(iii) A limit-cycle oscillation of MPF concentration accompanied by cyclic changes in the phosphorylation state of Cdc2 having maxima between peaks of MPF. (This feature of the cell-free extract must characterize the system dynamics under some parameter choices. However, prediction by the model of realistic Cdc2 dynamics in the egg requires parameter choices that suppress this behavior once cleavage begins).
(iv) A limit-cycle oscillation of MPF concentration with little tyrosine phosphorylation of Cdc2 (as in the cleaving egg).

The numerous concentrations, rate constants, and other parameters that enter into such a complex system are often poorly characterized and continually undergoing experimental evaluation and reevaluation. Therefore it is important to know whether behaviors seen with different parameter values are tied to those particular values or are "robust" (that is, capable of maintaining their integrity) in the face of changes. By exploring the structure of the system's vector field using numerical methods, it is possible to test features of the model despite empirical uncertainties.

Borisuk and Tyson (1998) performed a "bifurcation analysis" of the Tyson–Novak model using the AUTO computer program, which was specifically designed to handle such problems (Doedel and Wang, 1995). Bifurcation refers to an abrupt change in the type or number of attractors due to variation in one or more parameters, which leads to qualitative alterations in the system's behavior. Figure 3.4 gives an example of one of the bifurcation diagrams obtained. In this case,

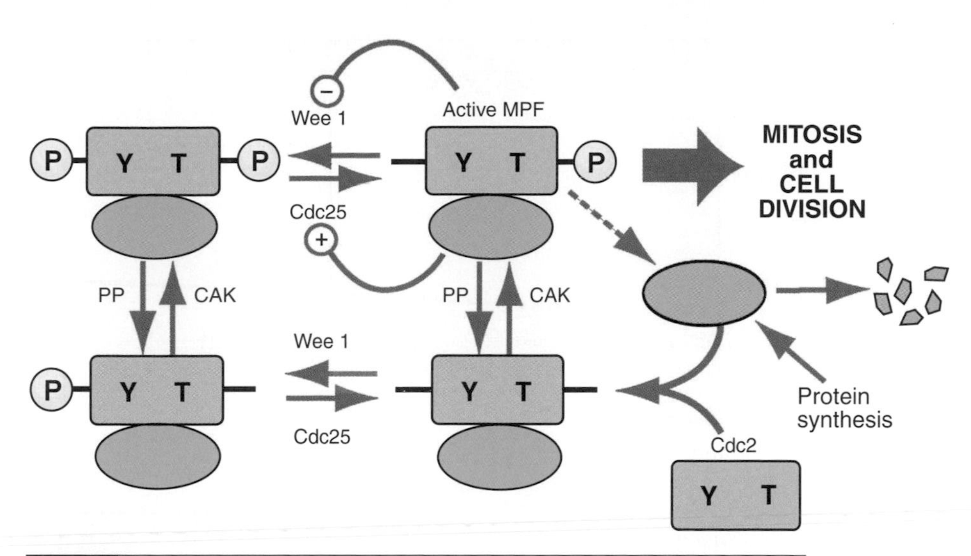

Fig. 3.3 Relationships among regulatory proteins in the cell-cycle model of Borisuk and Tyson (1998). Active MPF consists of a molecule of Cdc2 (blue rectangle) that is phosphorylated on threonine-161 (T), but not on tyrosine-15 (Y), and dimerized, i.e., associated, with a molecule of cyclin (green oval). The yellow circles represent phosphate groups that are attached by the kinases Wee 1 (acting on tyrosine) and CAK (acting on threonine) and are removed by protein phosphatases operating on tyrosine (Cdc25) and threonine (PP, most likely the protein phosphatase 2C; Cheng *et al.*, 1999; De Smedt *et al.*, 2002). Active MPF promotes the phosphorylation of Wee 1 and Cdc25, which respectively inactivates (−) and activates (+) these enzymes and indirectly promotes the degradation of cyclin (broken-line arrow). (After Borisuk and Tyson, 1998.)

MPF activity is plotted against the rate constant k_1 for cyclin synthesis, which is one of the "adjustable" parameters of the system. There is a range of values for this rate constant that dictate distinct stable steady states (depending on k_1) for MPF activity and a range of values that dictate limit-cycle behavior. These limit cycles also vary in their properties, both period and amplitude (maximum MPF activity minus minimum MPF activity) being functions of k_1. Thus while details such as the absolute value of MPF concentration cannot be predicted with certainty, the qualitative behavior of the system is robust to fairly wide changes in the parameters. Borisuk and Tyson (1998) also carried out bifurcation analysis with two simultaneously varying parameters and came to similar conclusions with respect to robustness.

The system of equations in the Tyson–Novak model also exhibits unstable steady states and limit cycles: at particular points in the bifurcation diagram there is a qualitative change in the character of the solution with, for example, a small change in k_1 causing the system to jump from a stable steady state to a limit cycle (Fig. 3.4). This is known as a Hopf bifurcation (Strogatz, 1994). Because changes in a parameter can be triggered from outside (e.g., by exogenous chemicals), such bifurcations provide a model for alterations in the behavior of

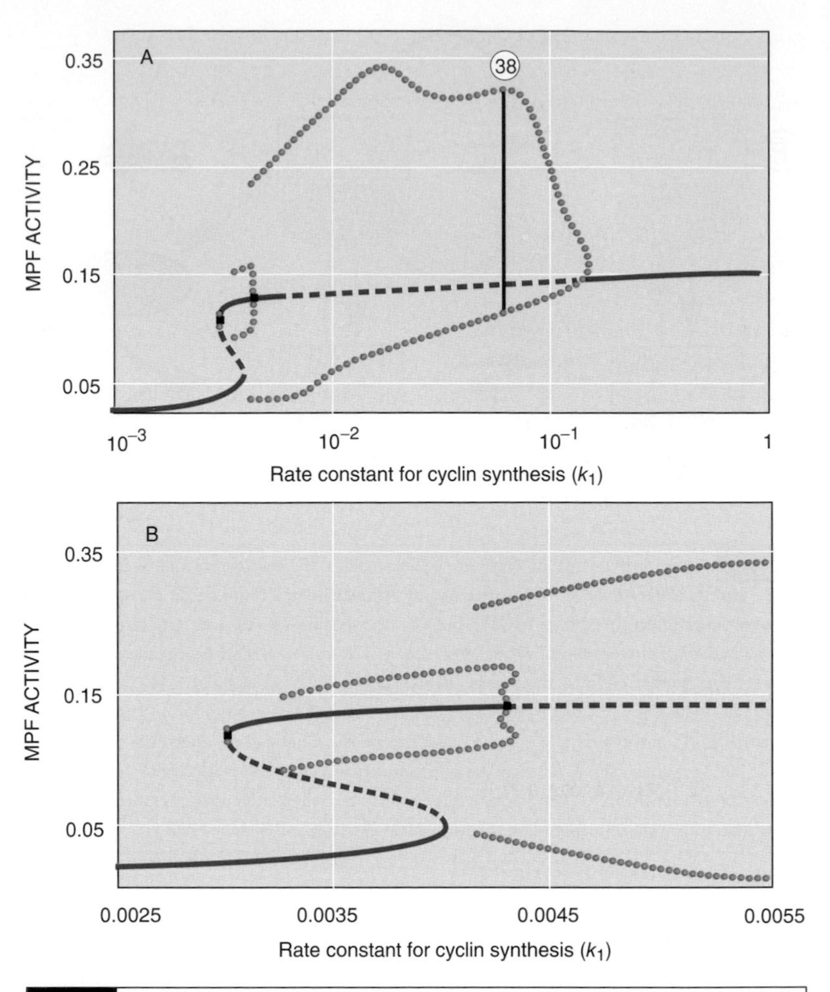

Fig. 3.4 One-parameter bifurcation diagram for the Tyson–Novak model. The character of the biological state changes as a function of the rate constant k_1 for cyclin synthesis. Solid red line, stable steady state; broken red line, unstable steady state; mauve dots, stable limit cycle; blue dots, unstable limit cycle; black square, Hopf bifurcation. At the Hopf bifurcations the system abruptly moves from a stable steady state to an unstable steady state or limit cycle. Consider the value of k_1 at the black vertical line. For this particular value of k_1, the MPF concentration varies periodically between the two points at the ends of the line, the upper and lower being its maximum and minimum values, respectively. The encircled number indicates the period of oscillation. All the pairs of dots above and below the red line should be interpreted analogously. Panel B is a horizontal enlargement of part of panel A. (After Borisuk and Tyson, 1998.)

the egg when the immature oocyte is activated by progesterone or the mature egg is fertilized.

Does a physical model of the frog-egg mitotic oscillator need to be so complicated that it consists of nine differential equations and 26 adjustable parameters? Borisuk and Tyson pointed out that while simpler models can exhibit robust oscillations, the full range of dynamical behaviors seen in the frog egg, which also includes multiple steady states, seem to depend on many of the positive and negative feedback loops known to exist in the MPF regulatory system. And the story becomes even more complex when we consider the cell cycle in cells of the later embryo (e.g., the post-midblastula-transition frog embryo) or of more mature tissues.

The cell divisions of the cleaving egg occur with essentially no time gaps between mitosis and the round of DNA synthesis that follows it, or between the ensuing DNA synthesis and mitosis. In contrast, the division of mature cells occurs with several checkpoints at which the cell cycle is halted. These checkpoints are overcome by the irreversible signals "Start," in which cyclin synthesis is induced and its degradation inhibited, and "Finish," in which the anaphase-promoting complex (APC) is activated (Novak *et al.*, 1999). APC degrades the proteins that tether the sister chromatids together and targets cyclins for degradation (Zachariae and Nasmyth, 1999). Other complexities involve the existence of more than one family of cyclins and numerous enzymes of the cyclin-dependent kinase (Cdk) class, of which Cdc2 is only one (Kohn, 1999). As shown by Tyson, Novak, and their coworkers (Sveiczer *et al.*, 2000; Tyson and Novak, 2001), the overall effect of the interactions among these components is to change the cell cycle from a free-running process, based on a chemical oscillation such as that seen in the egg, to a recurrent process based on a sequence of stationary states in which the cell is conveyed from one point to another by a series of stage-specific triggers. These two ways of generating a cell cycle, each of which occurs at appropriate stages of development, have been referred to as the "clock" and the "domino" mechanisms (Murray and Kirschner, 1989).

Multistability in cell-type diversification

The hypothesis that the cell cycle is controlled by the sort of dynamical system studied by physicists implies that when a cell divides it inherits not just a set of genes and a particular mixture of molecular components but also a dynamical system in a particular *dynamical state*. A limit-cycle oscillator like the one thought to underlie the cell cycle has a continuum of dynamical states that define a stable orbit surrounding an unstable node. A particular dynamical state on such an orbit corresponds to a particular *phase* of the limit cycle, for instance the phase at which MPF activity is at its minimum. If a given dynamical state of the cell-cycle regulatory network can be inherited, this should also be possible in other biochemical networks. Elowitz and Leibler (2000) used genetic engineering techniques to provide the bacterium *Escherichia coli* with a set of feedback circuits involving transcriptional repressor proteins in such a way that a biochemical oscillator not previously found in this organism was produced. Individual cells displayed a chemical oscillation with a period longer than the cell cycle. This implied that the dynamical state of the artificial oscillator was inherited across cell generations (if it had not, no periodicity would have been observed). Because (unlike in the MPF example described above) the biochemical oscillation was not tied to the cell cycle, newly divided cells in successive generations found themselves at different phases of the oscillation.

The ability of cells to pass on dynamical states (and not just "informational" macromolecules such as DNA) to their progeny has important implications for developmental regulation, since the continuity and stability of a cell's biochemical identity is a key factor in the performance of its role in a fully developed organism. Inheritance that does not depend directly on genes is called "epigenetic inheritance" and the biological or biochemical states inherited in this fashion are called "epigenetic states." The ability of cells to undergo transitions among a limited number of discrete, stable, epigenetic states and to propagate such "decisions" from one cell generation to the next is essential to the capacity of the embryo to generate diverse cell types.

Let us consider a dynamical system that, instead of exhibiting limit-cycle behavior, displays alternative stable steady states. Like the examples above, a dividing cell might transmit its particular system state to each of its daughter cells, but it is also possible that some internal or external event accompanying cell division could push one or both daughter cells out of the basin of attraction in which the precursor cell resided and into an alternative state. Thus, just as a limit cycle triggers particular biological activities (mitosis, entering interphase) by its distinct dynamical states (i.e., the phases of the oscillation), multistable dynamical systems can similarly provide the physical basis for developmental changes in cell state – what were referred to as "determination" and "differentiation" earlier in this chapter.

Not every molecular species needs to be considered simultaneously in modeling a cell's transitions between alternative biochemical states. Changes in the concentrations of the small molecules involved in the cell's "housekeeping" functions, such as energy metabolism and amino acid, nucleotide, and lipid synthesis, occur much more rapidly than changes in the pools of macromolecules such as RNAs and proteins. The latter represent the cell's "gene expression" profile, the changes in which can therefore be considered against an average "metabolic background." Moreover, a cell's active genes can be partitioned into different categories, and this leads to a simplification of the gene regulatory dynamics. Many of the cell's housekeeping genes (see above) are kept in the "on" state during the cell's lifetime. The pools of these "constitutively active" (i.e., always "on") gene products can be considered constant, so that their concentrations enter into the dynamic description of the developing embryo as fixed parameters rather than variables. In a similar fashion, those housekeeping genes whose expression varies in a cyclical fashion keyed to the cell cycle are typically not dynamically connected to the genetic circuitry that determines cell type. It is primarily the developmentally regulated genes that are important to consider in analyzing determination and differentiation. And of these genes, the most important are those whose products control the activity of other genes.

Epigenetic multistability: the Keller autoregulatory transcription-factor-network model

Transcription factors are proteins responsible for turning genes on and off. They do this by binding to specific sequences of DNA called "enhancers." Enhancers are usually (but not always) located next to the "promoter," itself a regulatory DNA sequence directly adjacent to a gene's transcription start site. Frequently the term "promoter" is used as short-hand for the true promoter plus the enhancer sequences. We will follow this practice in the present text. Important developmental transitions are controlled by the relative levels of different transcription factors (Wilkins, 1992; Davidson, 2001). Because the control of most developmentally regulated genes is a consequence of the synthesis of the factors that regulate their transcription, transitions between cell types during development can be driven by changes in the relative levels of a fairly small number of transcription factors (Bateman, 1998). We can thus gain insight into the dynamical basis of cell-type switching (i.e., differentiation) by focusing on molecular circuits, or networks, consisting solely of transcription factors and the genes that specify them. Such circuits, in which the components mutually regulate one another's expression, are termed autoregulatory. Cells make use of a variety of autoregulatory transcription-factor circuits during development. It is obvious that such circuits, or networks, will have different properties depending on their particular "wiring diagrams," that is, the patterns of interaction among their components.

It is also important to recognize that transcription factors that control cell differentiation not only initiate RNA synthesis on promoters of template-competent genes (see above) but also are capable of remodeling chromatin so as to transform their target genes from the template-incompetent into the template-competent configuration (Jimenez *et al.*, 1992; Rupp *et al.*, 2002).

Transcription factors can be classified as either *activators*, which bind to a site on a gene's promoter and enhance the rate of that gene's transcription over its basal rate, or *repressors*, which decrease the rate of a gene's transcription when bound to a site on its promoter. The basal rate of transcription depends on constitutively active transcription factors, which are distinct from those in the cell-type-determining autoregulatory circuits that we will consider below. Repression can be competitive or noncompetitive. In the first case, the repressor will interfere with activator binding and can only depress the gene's transcription rate to the basal level. In the second case, the repressor acts independently of any activator and can therefore potentially depress the transcription rate below basal levels.

Keller (1995) used computational methods to investigate the behavior of several autoregulatory transcription-factor networks with a range of wiring diagrams (Fig. 3.5). Each network was represented by a set of n coupled differential equations – one for the concentration of each factor in the network – and the steady-state behaviors of the

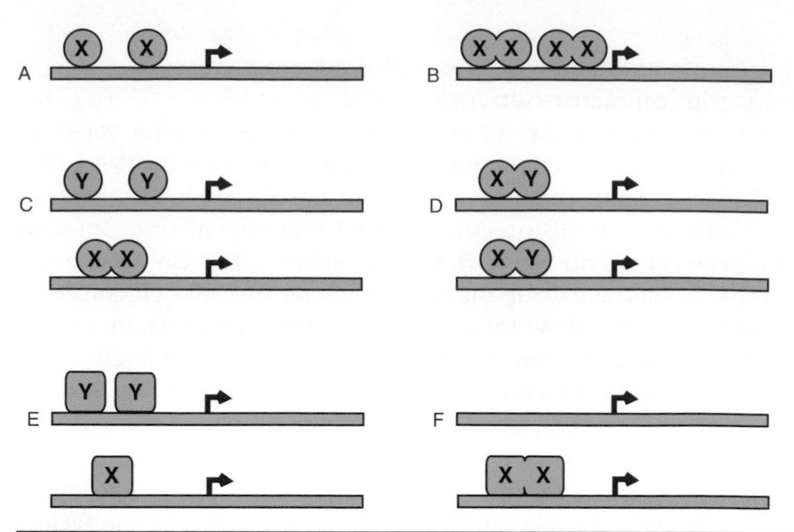

Fig. 3.5 Six model genetic "circuits" encoding autoregulatory transcription factors. (The diagrams represent circuits in the sense that there is feedback between their components.) The green and blue horizontal bars represent the genes specifying the transcription factors X (green symbols) and Y (blue symbols), respectively. The arrow denotes the transcription start site. The portion of the gene to the left of the arrow is the promoter. (A) Autoactivation by monomer X. (B) Autoactivation by dimer X_2. (C) Mutual activation by monomer Y and dimer X_2. (D) Autoactivation by heterodimer XY. (E) Mutual repression by monomers X and Y. (F) Mutual repression by dimer X_2 and heterodimer XY (with no binding site for the heterodimer on either promoter; see the main text for details). Activating and repressing transcription factors are denoted by circular and square symbols respectively.

systems were explored. The questions asked were the following. How many stationary states exist? Are they stable or unstable?

To illustrate Keller's approach, we discuss in more detail one such network, designated as the "mutual repression by dimer and heterodimer" (MRDH) network, shown in Fig. 3.5F. It comprises the gene encoding the transcriptional repressor X and the gene encoding the protein Y, and thus represents a two-component network. Below we summarize the salient points of the MRDH model.

1. The rate of synthesis of a transcription factor (i.e., $d[X]/dt$ and $d[Y]/dt$, [X] and [Y] being the respective concentrations of X and Y) is proportional to the rate of transcription of the gene encoding that factor.
2. The transcription of a gene depends, in turn, on the specific molecules that are bound to sites in the gene's promoter. Molecules can bind either as monomers (single protein molecules, denoted by X, Y, etc.) or dimers (complexes of two protein molecules). Both homodimeric complexes (X_2, Y_2) and heterodimeric complexes (XY) may be active. Thus a promoter can be in various configurations, with respective relative frequencies, depending on its "occupancy," that is, on how the various potential factors bind to it. Specifically, in the case of the MRDH network, as characterized by Keller (Fig. 3.5F), the

promoter of gene X has no binding site for any molecule, either activator or repressor: its only configuration is the empty state, whose relative frequency therefore can be taken as 1. (Note that the relative frequency of a promoter configuration is the probability of this configuration, thus the sum of all possible relative frequencies must be 1.) The (activator-independent or basal) rate of synthesis of gene X is denoted by S_{X_B}. The promoter of gene Y contains a single binding site for the dimeric form, X_2, of the non-competitive repressor X. The monomeric forms of proteins X and Y cannot bind DNA. Furthermore, proteins X and Y can form a heterodimer XY that is also incapable of binding DNA. Thus, while protein Y does not act directly as a transcription factor, it affects transcription since it antagonizes the repressor function of X by interfering with the formation of X_2 (by forming XY). The promoter of gene Y can therefore be in two configurations: its binding site for X_2 is either occupied or not with respective relative frequencies $K_X K_{X_2}[X]^2/(1 + K_X K_{X_2}[X]^2)$ and $1/(1 + K_X K_{X_2}[X]^2)$. (Note that the sum of these two relative frequencies is 1.) Here K_X and K_{X_2} are the binding affinity of X_2 to the promoter of gene Y and the dimerization rate constant for the formation of X_2, respectively, and we have used the relationship $K_X[X_2] = K_X K_{X_2}[X]^2$. The synthesis of Y in both configurations is activator-independent and is denoted by S_{Y_B}. To incorporate the fact that X_2 reduces the rate of transcription of Y in a noncompetitive manner, in the occupied Y-promoter configuration S_{Y_B} is replaced by ρS_{Y_B}, with $\rho \leq 1$.

3. The overall transcription rate of a gene is calculated as the sum of products. Each term in the sum corresponds to a particular promoter-occupancy configuration and is represented as a product of two factors, namely, the frequency of that configuration and the rate of synthesis resulting from that configuration. In the MRDH network this rate for gene X is thus $1 \times S_{X_B}$ (with S_{X_B} as introduced above), because it has only a single promoter configuration. The promoter of gene Y can be in two configurations (with rates of synthesis S_{Y_B} and ρS_{Y_B}, see above); therefore its overall transcription rate is $(1 + \rho K_X K_{X_2}[X]^2)S_{Y_B}/(1 + K_X K_{X_2}[X]^2)$.
4. The rate of decay of a transcription factor is a sum of terms, each proportional to the concentration of a particular complex in which the transcription factor participates. This is equivalent to assuming exponential decay. For the transcriptional repressor X in the MRDH network these complexes include the monomer X, the homodimer X_2, and the heterodimer XY. If the corresponding decay constants are denoted respectively by d_X, d_{X_2}, and d_{XY} then the overall decay rate of X is given by $d_X[X] + 2d_{X_2}K_{X_2}[X]^2 + d_{XY}K_{XY}[X][Y]$, K_{XY} being the rate constant for the formation of the heterodimer XY (i.e., $[XY] = K_{XY}[X][Y]$). The analogous quantity for protein Y is $d_Y[Y] + d_{XY}K_{XY}[X][Y]$.
5. By definition, the steady-state concentration of each transcription factor is determined by the solution of the equation that results from setting its rate of synthesis equal to its rate of

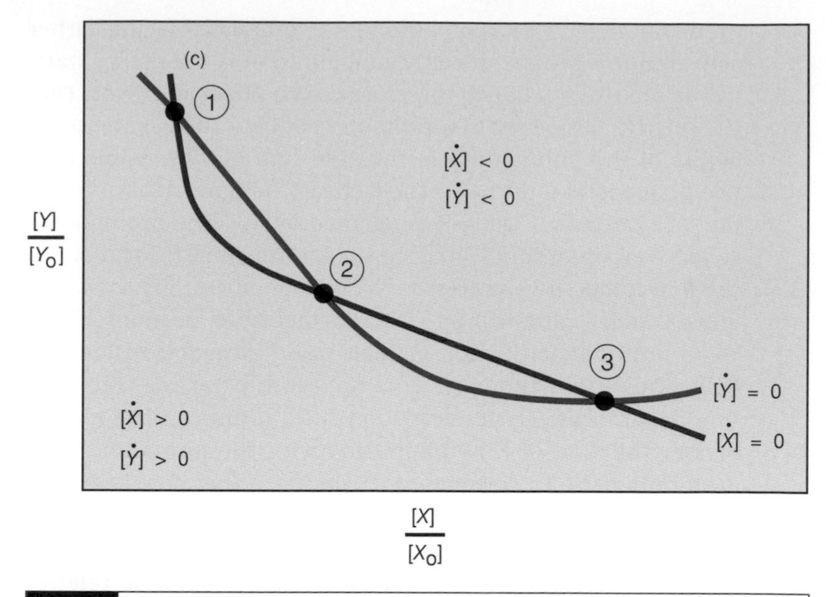

Fig. 3.6 The solutions of the steady-state Eqs. 3.3 and 3.4, given in terms of the steady-state solutions $[X_0]$ and $[Y_0]$. Here $[X_0]$ is defined as the steady-state cellular level of monomer X produced in the presence of a steady-state cellular level $[Y_0]$ of monomer Y, assuming a rate of transcription S_{X_0}. Thus, by definition (see Eq. 3.3), $S_{X_0} = d_X[X_0] + 2d_{X_2}K_{X_2}[X_0]^2 + d_{XY}K_{XY}[X_0][Y_0]$. Since along the red and blue lines, respectively, $d[X]/dt \equiv [\dot{X}] = 0$ and $d[Y]/dt \equiv [\dot{Y}] = 0$, it is the intersections of these curves that correspond to the steady-state solutions of the system of equations 3.3 and 3.4. Steady states 1 and 3 are stable, whereas steady state 2 is unstable.

decay. With the above ingredients, for the MRDH network one arrives at

$$\frac{d[X]}{dt} = S_{X_B} - \{d_X[X] + 2d_{X_2}K_{X_2}[X]^2 + d_{XY}K_{XY}[X][Y]\} = 0, \tag{3.3}$$

$$\frac{d[Y]}{dt} = \frac{1 + \rho K_X K_{X_2}[X]^2}{1 + K_X K_{X_2}[X]^2} S_{Y_B} - \left\{d_Y[Y] + d_{XY}K_{XY}[X][Y]\right\} = 0. \tag{3.4}$$

Keller found that if in the absence of the repressor X the rate of synthesis of protein Y is high then in its presence the system described by Eqs. 3.3 and 3.4 exhibits three steady states, as shown in Fig. 3.6. Steady states 1 and 3 are stable and thus could be considered as defining two distinct cell types, while steady state 2 is unstable. In an example using realistic kinetic constants, the steady-state values of [X] and [Y] at the two stable steady states differ substantially from one another, showing that the dynamical properties of these autoregulatory networks of transcription factors can provide the basis for generating stable alternative cell fates during early development.

The validity of Keller's approach depends on the assumption that the steady-state levels of transcription factors determine cell-fate choice. In most cases this appears to be a reasonable assumption (Bateman, 1998; Becker *et al.*, 2002). Sometimes, however (e.g., at certain stages of early sea urchin development), transitions between cell types appear to depend on the initial rates of synthesis of transcription factors and not on their steady-state values (Bolouri and Davidson, 2003). Under conditions in which Keller's assumption holds, however, one can ask whether switching between these alternative states is predicted to occur under realistic biological conditions. This can be answered affirmatively: the external microenvironment of a cell that contains an autoregulatory network of transcription factors could readily induce changes in the rate of synthesis of one or more of the factors via signal-transduction pathways that originate outside the cell ("outside–in signaling," Giancotti and Ruoslahti, 1999). The microenvironment can also affect the activity of transcription factors in an autoregulatory network by indirectly interfering with their nuclear localization (Morisco *et al.*, 2001). In addition, cell division may perturb the cellular levels of autoregulatory transcription factors, particularly if they or their mRNAs are unequally partitioned between the daughter cells. Any jump in concentration of one or more factors in the autoregulatory system can bring it into a new basin of attraction and thereby lead to a new stable cell state.

In several well-studied cases, transcription factors regulating differentiation in a mutually antagonistic fashion are initially co-expressed in progenitors before one is upregulated and others downregulated. Using a framework similar to Keller's, Cinquin and Demongeot (2005) have explored how known interactions among such factors may give rise to multistable systems for cell-type determination.

Dependence of differentiation on cell–cell interaction: the Kaneko–Yomo "isologous diversification" model

We have seen that the existence of multiple steady states in a cleaving egg makes it possible, in principle, for more than one cell type to arise among the descendents of such cells. However, this capability does not, by itself, provide the conditions under which such a potentially divergent cell population would actually be produced and persist as long as it is needed. In later chapters we will see how the early embryo and various organ primordia use chemical gradients and other short- and long-range signaling processes to ensure that the appropriate cell types arise at the correct positions to make a functional structure. In certain cases (possibly including the early mammalian embryo) the first step in the developmental process may be to generate a mixture of different cell types in a relatively haphazard fashion, just to get things going. As we will see in the next chapter, sometimes it does not matter what the initial arrangement of cells is – differential affinity can cause cells to "sort out" into distinct layers and lead to organized structures from a randomly mixed population.

Consider the following observations: (i) during muscle differentiation in the early *Xenopus* embryo, the muscle precursor cells must be in contact with one another throughout gastrulation (the set of rearrangements that establish the body's main tissue layers, see Chapter 5) in order to develop into terminally differentiated muscle (Gurdon, 1988; Standley *et al.*, 2001); (ii) if the cells in the neuron-containing layer of the retina (the neural retina or NR cells) of the developing chicken eye are placed in culture they "dedifferentiate" away from the neuronal type and can "transdifferentiate" into a different cell type, that of the pigmented layer of the retina (the pigment retina or PR cells). PR cells do not arise individually but as small random groups within sheets of NR cells. If the PR cells that form in this fashion are isolated and exposed to the appropriate molecular signals they will transdifferentiate back into NR cells (Opas *et al.*, 2001).

The need for cells to act in groups in order to acquire new identities during development has been termed the "community effect" (Gurdon, 1988). This phenomenon is a developmental manifestation of the general property of cells and other dynamical systems of assuming one or another of their possible internal states in a fashion that is dependent on inputs from their external environment. But in the two cases noted above the external environment consists of other cells of the same type, and in discussions of the community effect there is the strong implication that dynamical interactions among similar cells, and not just exposure to secreted factors, is responsible for the emergence of new cell types (Carnac and Gurdon, 1997).

Kaneko, Yomo, and coworkers (1994; 1997; 2003, and references therein) described a previously unknown physical process, termed "isologous diversification," by which replicate copies of the same dynamical system (e.g., cells of the same type) can undergo stable differentiation simply by virtue of exchanging chemical substances with one another. This differs from the Keller model that we have just considered in that the final state achieved in the Kaneko–Yomo model exists only in the state space of the collective "multicellular" system. Whereas the distinct local states of each cell within the collectivity are mutually reinforcing, these local states are not necessarily attractors of the dynamical system representing the individual cell, as they are in Keller's model. The Kaneko–Yomo system thus provides a model for the community effect.

The general scheme of the model, as described in Kaneko and Yomo (1999), is shown in Fig. 3.7A. Here we present the model in its simplest form (Kaneko and Yomo, 1994) to illustrate how dynamical system analysis at the multicellular level can describe differentiation. Improvements and generalizations of the simple model do not change its qualitative features.

Kaneko and Yomo considered a system of originally identical model cells (called "cells" in what follows) with intracell and intercell dynamics that incorporate cell growth, cell division, and cell death. The dynamical variables (spanning the state space of the system) are the concentrations of molecular species ("chemicals") inside and outside the cells. The criterion by which differentiated cells are distinguished is the average of the intracellular concentrations of these chemicals (over the cell cycle). As a vast simplification, only three chemicals, P, Q, R with respective time-dependent intracellular concentrations in the ith cell ($x_i^P(t)$, $x_i^Q(t)$, x_i^R (t) and intercellular

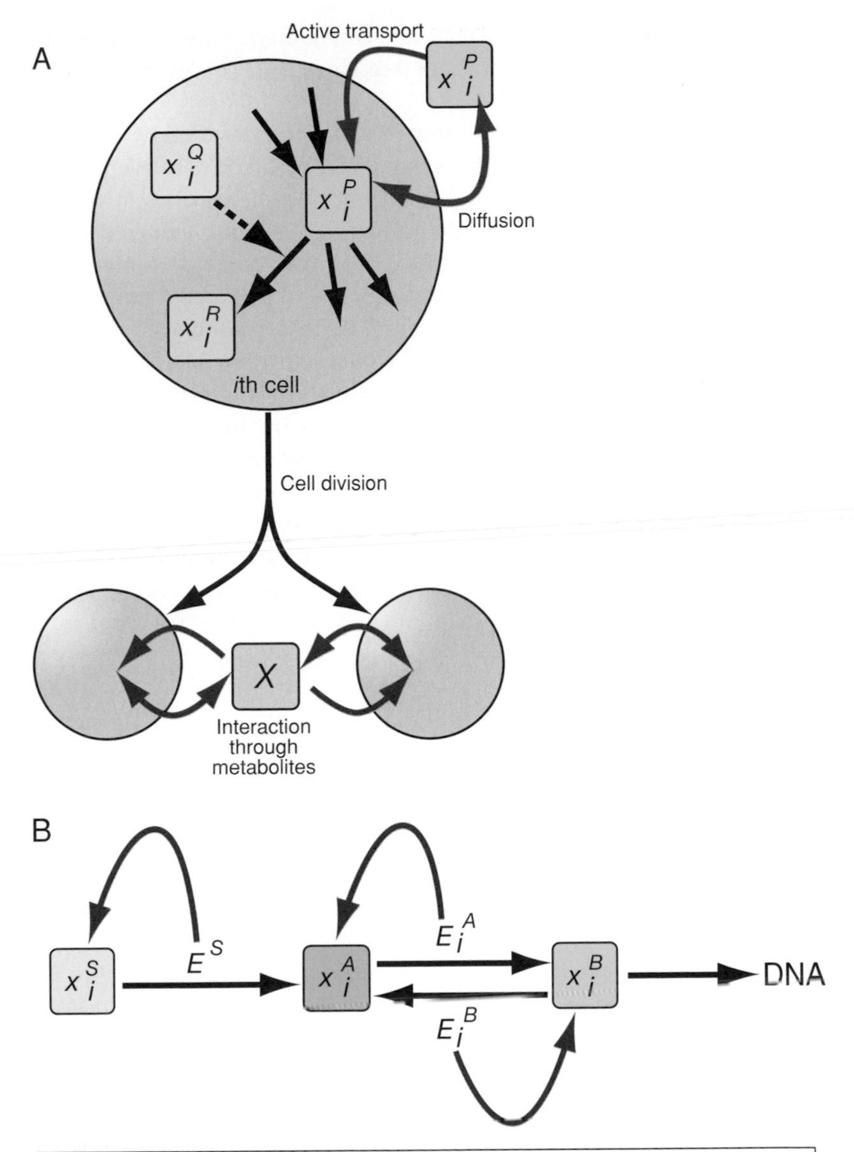

Fig. 3.7 (A) General scheme of the model of Kaneko and Yomo. The biochemical state of a cell is characterized by chemicals P, Q, R, etc., with corresponding intracellular concentrations x_i^P, x_i^Q, x_i^R, etc., which vary from cell to cell (cell i is shown in the upper center). Chemicals may exit cells (chemical P in the example), their extracellular concentrations being X_i^P, X_i^Q, X_i^R, etc., and enter other cells either via passive diffusion or active transport; this establishes interactions between cells (as shown at the bottom for two daughter cells). The concentration of any chemical inside a given cell depends on chemical reactions in which other chemicals are precursors or products (solid arrows) or cofactors (e.g., enzymes; broken-line arrow). A cell divides when the synthesis of DNA (one of the "chemicals") exceeds a threshold. (B) Schematic representation of the intracellular dynamics of a simple version of the Kaneko–Yomo model. Red arrows symbolize catalysis. The variables $x_i^A(t)$, $x_i^B(t)$, $x_i^S(t)$ and E_i^A, E_i^B, E^S denote respectively the concentrations of chemicals A, B, and S and of their enzymes in the ith cell, as described in the text. Note that the letters P, Q, R are used for the general scheme of the Kaneko–Yomo model whereas the letters A, B, S are used for the specific mathematical representation given in the text.

concentrations ($X^P(t), X^Q(t), X^R(t)$) are considered. One of those (P) serves as the source for the others (see Fig. 3.7A). The model has the following features.

1. *Intracellular dynamics.* The source chemical, denoted by S in the simplified scheme of Fig. 3.7B, is catalyzed by a constitutive (always active) enzyme (concentration E^S) to produce chemical A, which in turn is catalyzed by a regulated ("inductive") enzyme (concentration E_i^A) to produce chemical B. Chemical B on one hand is catalyzed by its own inductive enzyme (concentration E_i^B) to produce A and on the other hand controls the synthesis of DNA. This sequence of events is shown schematically in Fig. 3.7B. The concentration of the constitutive enzyme is assumed to have the same constant value E^S in each cell, whereas those of the inductive enzymes in the ith cell, E_i^A and E_i^B are both taken to be proportional to the concentration x_i^B of the chemical B in that cell (and therefore to be dependent on time), so that $E_i^A = e_A x_i^B$ and $E_i^B = e_B x_i^B$ (e_A and e_B are constants). Thus, in terms of chemicals A and B the intracellular dynamics is described by

$$\begin{aligned} \frac{dx_i^A}{dt} &= \left(e_B x_i^B\right) x_i^B - \left(e_A x_i^B\right) x_i^A + E^S x_i^S, \\ \frac{dx_i^B}{dt} &= \left(e_A x_i^B\right) x_i^A - \left(e_B x_i^B\right) x_i^B - k x_i^B. \end{aligned} \tag{3.5}$$

Here the factor k accounts for the decrease in B due to its role in DNA synthesis (see Fig. 3.7B). Note the nonlinear character of these equations. (Parentheses are used to indicate the inductive enzymes.)

2. *Intercellular dynamics.* Cells are assumed to interact with each other through the changes in the intercellular concentrations of the chemicals A and B. Chemicals are transported in and out of the cells. The rate of transport of a chemical into the cell is proportional to its concentration outside. However, it also depends on the internal state of the cell, which we have characterized in terms of the intracellular concentrations of the chemicals A and B. This dependence is typically complicated. Kaneko and Yomo assumed that the rate of import of chemical M (i.e., A, B, or S) into the ith cell, denoted by $Transp_i^M$, has the form

$$Transp_i^M(t) = p\left(x_i^A + x_i^B\right)^3 X^M. \tag{3.6}$$

Here p is a constant. As long as the dependence of $Transp$ on the intracellular concentrations is nonlinear, any choice of exponent (taken to be 3 above) leads to the same qualitative result.

Besides the mechanism of active transport described by Eq. 3.6, chemicals also enter the cells by diffusion through the membrane. The corresponding rate is taken as

$$Diff_i^M(t) = D\left[X^M(t) - x_i^M(t)\right], \tag{3.7}$$

where D is a (diffusion) constant.

Combining intracellular (Eq. 3.5) and intercellular (Eqs. 3.6 and 3.7) dynamics, the rate equations for the intracellular chemicals become

$$\frac{dx_i^S}{dt} = -E\,x_i^S + Transp_i^S + Diff_i^S,$$
$$\frac{dx_i^A}{dt} = \left(e_B x_i^B\right) x_i^B - \left(e_A x_i^B\right) x_i^A + E\,x_i^S + Transp_i^A + Diff_i^A,$$
$$\frac{dx_i^B}{dt} = \left(e_A x_i^B\right) x_i^A - \left(e_B x_i^B\right) x_i^B - kx_i^B + Transp_i^B + Diff_i^B. \tag{3.8}$$

It is further assumed that only the source chemical is supplied by a flow from an external tank to the chamber containing the cells. Since it must be transported across the cell membrane to produce chemical A (see Eqs. 3.5), the intercellular dynamics of the source chemical is described by

$$\frac{dX^S}{dt} = (\overline{X^S} - X^S)f - \sum_{i=1}^{N} \left(Transp_i^S + Diff_i^S\right). \tag{3.9}$$

Here $\overline{X^S}$ is the concentration of the source chemical in the external tank, f is its flow rate into the chamber, and N is the total number of cells in the system.

3. *Cell division.* Kaneko and Yomo considered cell division to be the result of the accumulation of a threshold quantity of DNA. DNA is synthesized from chemical B and therefore the ith cell, born at t_i^0, will divide at $t_i^0 + T$ (T defines the cell-cycle time), when the amount of B in its interior (proportional to x_i^B) reaches a threshold value. (Mathematically this condition is expressed as $\int_{t_i^0}^{t_i^0+T} x_i^B(t)dt \geq R$ in the model, R being the threshold value.)

4. *Cell death.* To avoid infinite growth in cell number, a condition for cell death also has to be imposed. It is assumed that a cell will die if the amount of chemicals A and B in its interior is below the "starvation" threshold S, which is expressed as $\left[x_i^A(t) + x_i^B(t)\right] < S$.

Simulations based on the above model and its generalizations using a larger number of chemicals (Kaneko and Yomo, 1997, 1999; Furusawa and Kaneko, 2001), lead to the following general features, which are likely to pertain also to real, interacting, cells:

- **(i)** The state of a cell, defined as a point in its chemical state space (see Fig. 3.1), tends to recur over time in a periodic or quasi-periodic fashion, analogous to the cell cycle.
- **(ii)** As cells replicate (by division) and interact with one another, eventually multiple biochemical states corresponding to distinct cell types appear.
- **(iii)** The different types are related to each other by a hierarchical structure in which one cell stands at the apex, cells derived from it stand at subnodes, and so on (Fig. 3.8). Such pathways of generation of cell types, which are seen in real embryonic systems, are referred to as developmental lineages.

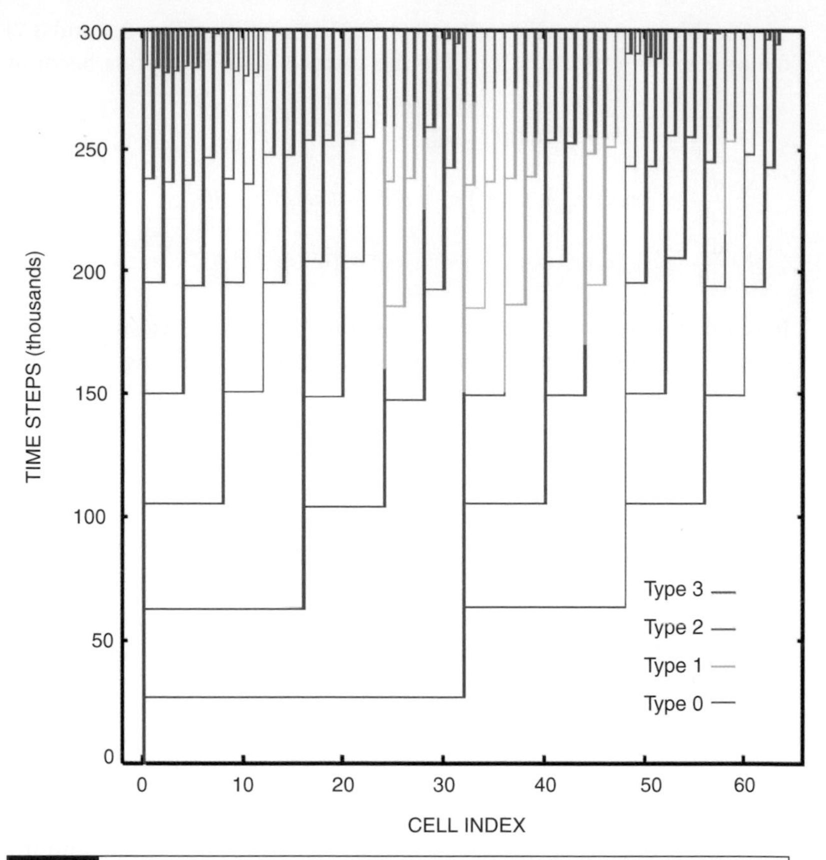

Fig. 3.8 Cell-lineage diagram generated from a simulation of cell dynamics in the model of Kaneko and Yomo. The system simulated contains more chemical components than the one described in the text, but otherwise has similar features. Each cell in the final population is assigned a subscript number (its index *i*; see Fig. 3.7). The differentiation of cells is plotted with time along the vertical axis. In this diagram, each branch point corresponds to a cell division, while the color indicates the cell type. (Reprinted from Furusawa and Kaneko, 1998, with permission from Elsevier publishers.)

(iv) The hierarchical structure appears gradually. Up to a certain number of cells (which depends on the model parameters), all cells have the same biochemical state (i.e., $x_i^A(t)$, $x_i^B(t)$, and $x_i^S(t)$ are independent of i) (Fig. 3.9A). When the total number of cells rises above a certain threshold value, the state with identical cells is no longer stable. For example, the synchrony of biochemical oscillations in different cells of the cluster may break down (by the phases of $x_i^A(t)$, $x_i^B(t)$, $x_i^S(t)$ becoming dependent on i). Ultimately, the population splits into a few groups ("dynamical clusters"), the phase of the oscillator in each group being offset from that in other groups, like groups of identical clocks in different time zones (Fig 3.9B).

(v) When the ratio of the number of cells in the distinct clusters falls within some range (depending on model parameters),

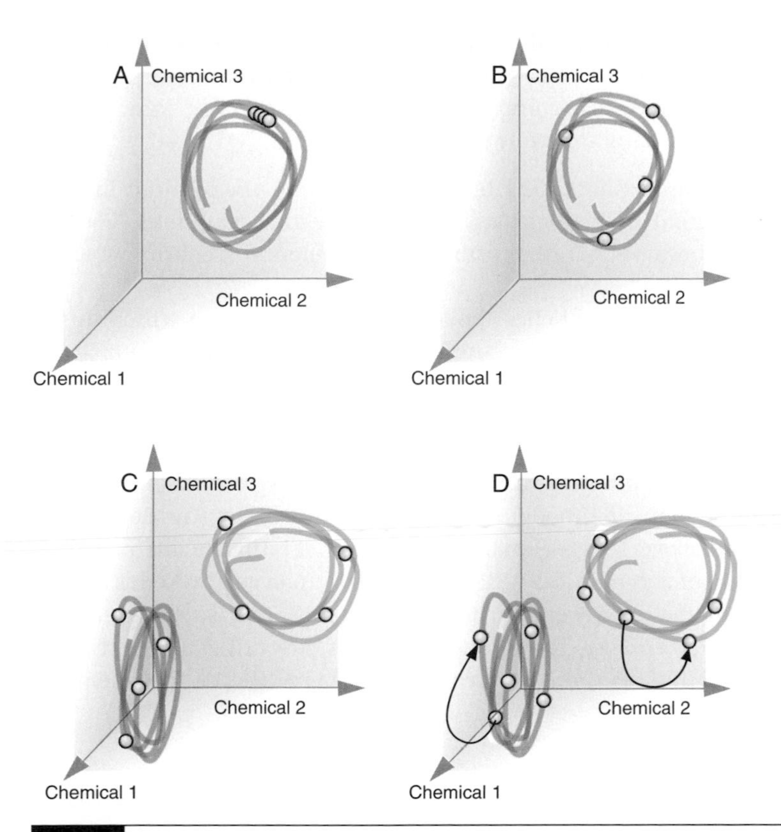

Fig. 3.9 Schematic representation of the differentiation scenario in the isologous diversification model of Kaneko and Yomo. When there are N cells and C chemicals, the state space of the multicellular system is NC-dimensional. The three axes in the figure represent the state of an individual cell in the multicellular system for $C = 3$. A point in this three-dimensional space corresponds to a set of instantaneous values of the three chemicals. As long as the biochemical states of the replicating cells are identical, a point along the orbit could characterize the synchronous states of each cell. This is illustrated in panel A, where the four circles, representing cells with the same phase and magnitude of their chemicals, overlap. With further replication, cells with differing biochemical states appear. First, chemicals in different cells differ only in their phases; thus the circles in panel B still fall on the same orbit, albeit well separated in space. With further increase in cell number, differentiation takes place: not only the phases but also the magnitudes (i.e., the averages over the cell cycle) of the chemicals in different cells will differ. The two orbits in panel C represent two distinct cell types, each different from the original cell type shown in panels A and B. Panel D illustrates the "breeding true" of the differentiated cells. After the formation of distinct cell types, the chemical compositions of each group are inherited by their daughter cells. That is, the chemical compositions of cells are recursive over subsequent divisions, as the result of stabilizing interactions. Cell division is represented here by an arrow from a progenitor cell to its progeny. (Adapted, with changes, from Kaneko, 2003.)

the differences in intracellular biochemical dynamics are mutually stabilized by cell–cell interactions.

(vi) With further increase in cell number, the average concentrations of the chemicals over the cell cycle become different. That is to say, groups of cells come to differ not only in the phases of the same biochemical oscillations but also in their average chemical composition integrated over the entire lifetimes of the cells (Fig. 3.9C). After the formation of cell types, the chemical compositions of each group are inherited by their daughter cells (Fig. 3.9D).

The addition of fluctuating terms (representing the random fluctuations inherent in biological systems) to the rate equations, Eqs. 3.8 and 3.9, does not change the above developmental scenario, which is therefore robust.

In contrast to the Keller model described earlier, in which different cell types represent a choice among basins of attraction for a multi-attractor system, external influences having the potential to bias such preset alternatives, in the Kaneko–Yomo model interactions between cells can give rise to stable intracellular states that would not exist without such interactions. Isologous diversification thus provides a plausible model for the community effect (Gurdon, 1988), described above. It is reasonable to expect that both the intrinsic multistability of a dynamical system of the sort analyzed by Keller as well as the interaction-dependent multistability of other such systems, as described by Kaneko, Yomo, and coworkers, are utilized in initiating developmental decisions in various contexts in different organisms.

Perspective

Living cells are immensely complex, multicomponent, entities, which are subject, in principle, to random fluctuations and chaotic behavior. Despite such complications, important cellular subsystems act as well-behaved dynamical systems. The physical laws governing such systems ensure that the concentrations of key regulatory molecules that trigger important events in the early embryonic cell cycle undergo regular, self-organizing temporal variations. It is plausible that alternative, stable, biochemical states, intrinsic to the individual cell or dependent on interactions between cells, are utilized in adding layers of control (in terms of regulated checkpoints) to the mature cell cycle and initiating pathways of cell diversification – determination and differentiation – during subsequent development. Many processes of embryogenesis discussed in this book from this point onward are based on the existence and action of cellular mechanisms of oscillatory and multistable behaviors. Although such mechanisms are inevitably complex, and interconnect with other subcellular mechanisms, their underlying physical bases (and evolutionary origins) are likely to be the kind of generic dynamical systems discussed in this chapter.

Chapter 4

Cell adhesion, compartmentalization, and lumen formation

In Chapter 2 we followed development up to the first nontrivial manifestation of multicellularity, the appearance of the blastula. To describe the mechanism of blastula formation we needed a model for cleavage, as well as for the collective behavior of a large number of cells in contact with one another. We based our model on physical parameters such as surface tension, cellular elasticity, viscosity, etc. and when possible related these quantities to experimentally known information such as the expression of particular genes. However, we have so far not dealt with the fundamental question concerning multicellularity: what holds the cells of a multicellular organism together?

As cells differentiate (see Chapter 3) they become biochemically and structurally specialized and capable of forming multicellular structures, with characteristic shapes, such as spheroidal blastulae, multilayered gastrulae, planar epithelia, hollow ducts or crypts. The appearance and function of these specialized structures reflect, among other things, differences in the ability of cells to adhere to each other and the distinct mechanisms by which they do so. During development certain cell populations need to bind, to varying extents, to some of their neighbors but not to others. In mature tissues the nature of the cell–cell adhesion contributes to their functionality: the manner in which two cells bind tightly to one another in the epithelial sheet lining the gut, for example, must be different from the looser attachment between the endothelial cells lining blood vessels. Differentiated and differentiating cells must therefore possess distinct adhesion molecules and apparatuses that reflect biological specificity. But cell adhesion also has common features across cell types and classes of adhesion molecules, and these can often be studied by physical methods and theories.

The major classes of cell adhesion molecules include the calcium dependent *cadherins* (Takeichi, 1991, 1995; Gumbiner, 1996; Wheelock and Johnson, 2003), the *immunoglobulin superfamily* (which also includes antibodies) an example of which is N-CAM (Williams and Barclay, 1988; Hunkapiller and Hood, 1989), the *selectins* (Bevilacqua *et al.*, 1991), and the *integrins* (Hynes, 1987; Hynes and Lander, 1992). The molecules in the first three classes are primarily responsible for direct cell–cell adhesion and are called cell adhesion molecules or CAMs. Cadherins, for example, enter into homophilic interactions in which a molecule on one cell binds to an identical molecule on a second cell. Integrins mainly mediate cell-extracellular-matrix (ECM) interactions and are called substratum adhesion molecules or SAMs. All CAMs and SAMs have distinct structures and specialized morphogenetic roles (Edelman, 1992).

The formation of complex multicellular structures is determined not only by the chemical nature of the cells' adhesion molecules but also by the distribution of these molecules on the cell surface. Epithelial cells forming strictly two-dimensional sheets must have predominantly CAMs along their lateral surfaces, whereas SAMs must populate their basal surfaces, along which interaction with the supporting specialized extracellular matrix, the basal lamina, occurs (Fig. 4.1). Such a distribution of CAMs and SAMs renders epithelial cells *polarized*.

Adhesion and differential adhesion in development

Adhesion enters into the phenomena of embryonic development in several distinct ways. The first, and most straightforward, way is in simply holding tissues together. This, of course, is the default role of cell–cell adhesion in all tissues, embryonic or adult. Mature tissues contain the definitive, relatively long-lived forms of the CAM-containing junctions represented in Fig. 4.1. These comprise desmosomes and adherens junctions, hemidesmosomes and focal adhesion complexes, as well as tight junctions, which provide a transepithelial seal impermeable to ions and polar molecules, and gap junctions, which allow ions and other small molecules to pass directly from one cell to another (Alberts *et al.*, 2002). During early development the CAM-containing junctions are present in apparently immature forms (DeMarais and Moon, 1992; Kofron *et al.*, 1997, 2002; Eshkind *et al.*, 2002) consistent with the provisional arrangement of cells and their capacity to rearrange during this period.

The other roles for adhesion during development are based on its modulation – the phenomenon of *differential adhesion*. As we will see in this and following chapters, the regulated spatiotemporal modulation of adhesion is an important driving force for major morphogenetic transitions during embryogenesis. The simplest form of this is the detachment of cell populations from existing tissues. This is usually a prologue to their relocation, as in gastrulation and the formation of the neural crest (see Chapters 5 and 6). But the modulation of adhesive strength without complete detachment also has morphogenetic consequences, whether it occurs locally, on the scale of the individual cell surface, or more globally, on the scale of groups of cells within a common tissue.

The polar expression of CAMs can lead directly to morphogenetic change, as illustrated in Fig. 4.2. In the process of differentiation some

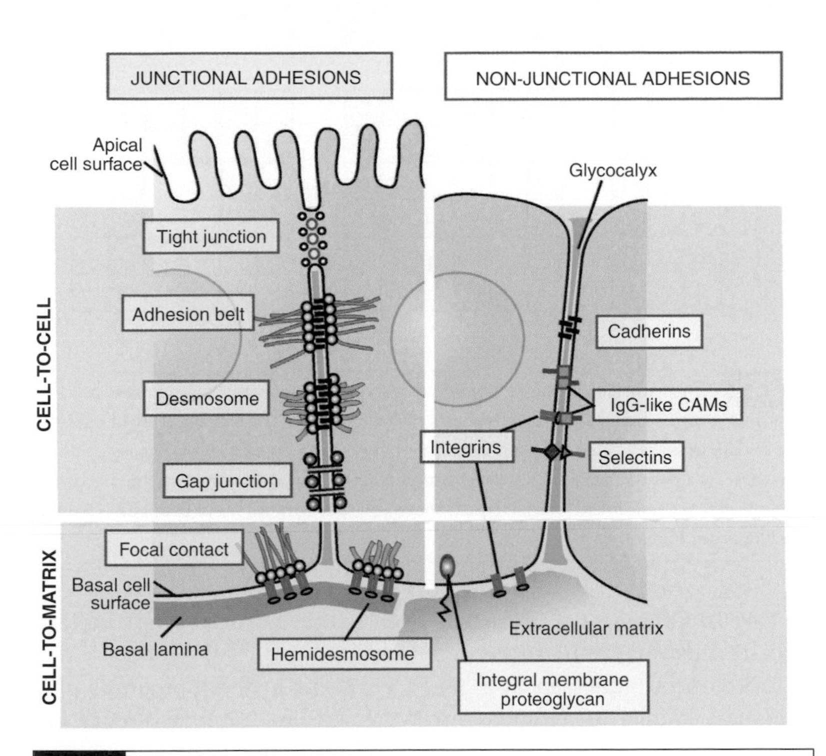

Fig. 4.1 Schematic representation of various cell–cell and cell–substratum adhesion complexes. On the left, a *tight junction* seals neighboring cells together in an epithelial sheet to prevent the leakage of molecules between them. An *adhesion belt*, also known as an *adherens junction*, joins an actin bundle (green strands) in one cell to a similar bundle in a neighboring cell via integral membrane proteins called cadherins and accessory proteins on the cytoplasmic faces. A *desmosome* anchors intermediate filaments (blue strands) in one cell to those in its neighbor via cadherins and accessory proteins. A *gap junction* provides tubular channels made up of connexin proteins, which allow the passage between cells of small water-soluble ions and molecules. A *hemidesmosome* anchors intermediate filaments at the basal surface of the cell to the basal lamina, via integral membrane proteins called integrins and cytoplasmic accessory proteins. A *focal adhesion complex* or *focal contact* anchors actin bundles to the extracellular matrix via integrins and cytoplasmic accessory proteins.

On the right, epithelioid cells in early embryos, which typically do not contain mature junctional complexes, and certain cells in the adult organism such as lymphocytes and platelets, utilize cadherins, immunoglobulin(IgG)-like CAMs (e.g., PECAM), or selectins, adhesive interactions that are homophilic (pairs of the same molecule) or heterophilic (pairs of different molecules). These are termed non-junctional adhesions. Integrins and integral membrane proteoglycans mediate attachment to the ECM. (After Alberts *et al.*, 2002.)

of the originally nonpolarized cells may lose their CAMs along part of their surfaces, thus preventing them from adhering to one another at those sites (Tsarfaty *et al.*, 1992, 1994). If the cells move around randomly, maximizing their contacts with one another in accordance with this constraint (see below), a hollow region or *lumen* will naturally arise. Cells polarized in their expression of CAMs arise during the formation of the blastula; they are necessary to this process, as

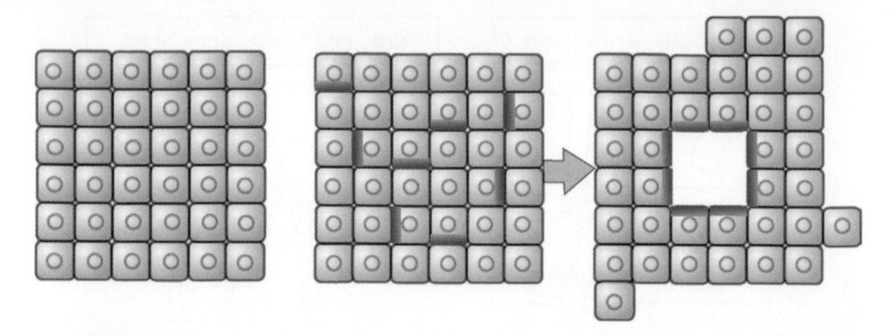

Fig. 4.2 Schematic illustration of lumen formation when polarized cells express cell adhesion molecules only on restricted parts of their surface. The shading in eight cells in the middle panel represents the lack of adhesion molecules on corresponding regions of the cells. As a consequence, if the minimization of configurational energy is the driving force in cell rearrangement, then a lumen, shown on the right, is bound to appear. (Based on Newman, 1998a.)

we saw in Chapter 2. This is true even if the blastula is not hollow, since it must maintain a free outer surface.

Differential adhesion also plays a role in a developmental phenomenon known as "compartment formation." Compartments are regions of embryonic tissue, which, though similar in other respects, do not mix (exchange cells) across a common boundary (Crick and Lawrence, 1975; Garcia-Bellido *et al.*, 1976). In certain cases (particularly as tissues mature) such interfaces of immiscibility are not dependent on adhesive differentials alone but are reinforced by specialized structures (Heyman *et al.*, 1995). Nonetheless, the initial establishment of such boundaries during embryogenesis typically depends on differential adhesion (Guthrie and Lumsden, 1991; Heyman *et al.*, 1995; Godt and Tepass, 1998; Gonzalez-Reyes and St Johnston, 1998; Hayashi and Carthew, 2004).

In order to understand the physical basis of the role of differential adhesion in developmental morphogenesis it is necessary to have some knowledge of the structure and composition of the cell surface, which we present in the following section.

The cell surface

The majority of CAMs and SAMs are transmembrane proteins, which attach to the actin cytoskeleton at their intracellular ends. Cadherins, for example, accomplish this through a protein complex involving alpha and beta catenin (Gumbiner, 1996). Moreover, if the cytoskeletal attachment is damaged or missing, cadherins' ability to establish an adhesive junction (called the *adherens junction*) is severely compromised (Nagafuchi and Takeichi, 1988). Similarly, if integrins are unable to attach to the cytoskeleton, the mechanical integrity of the focal adhesion complex (by which epithelial cells attach to the basal lamina, or moving cells assemble on the substrate along which they translocate) is strongly reduced (Wang *et al.*, 1993).

The extracellular portion of most membrane proteins (and of lipids in the outer layer of the plasma membrane) is "decorated" by either short chains of sugars (oligosaccharides) or longer polysaccharides containing amino sugars (glycosaminoglycans). Proteins linked to oligosaccharides are referred to as *glycoproteins*, whereas those with glycosaminoglycan attachments are called *proteoglycans*. These sugar-containing molecules collectively form the *glycocalyx* (Fig. 4.1), a coating that extends several nanometers from the plasma membrane on its noncytoplasmic face. The characteristic glycocalyx of each specialized cell represents the basis by which it is recognized by cells like it and by other cell types. For the basal portions of epithelial cells, and of mesenchymal and connective tissues (which will be discussed in Chapter 6), the microenvironment is even more elaborate (see Fig. 4.1), constituting an ECM that extends beyond the glycocalyx and contains additional proteins, glycoproteins, proteoglycans, and in some cases minerals. In this chapter we will confine our discussion to epithelial and epithelioid tissues, in which cells contact one another more or less directly via their glycocalyces.

As a result of the complex nature of the glycocalyx, cell recognition and subsequent adhesion involve a variety of chemical bonds and forces: covalent bonds, e.g., those between protein ligands and oligosaccharide side chains; ionic bonds or electrostatic interactions, between charged regions of membranes; van der Waals forces between induced and permanent dipole moments due to polar molecules; and hydrogen bonds between membrane proteins. Each of these bonds and the associated forces have a characteristic distance and strength. In addition, undulations in the membrane cause steric hindrance to adhesion, an effect that may reduce the magnitude of the overall adhesive force.

From these considerations it is clear that membrane interactions and cell adhesion have many facets. While we will not attempt to present an exhaustive discussion of either the biology or the physics of membrane interactions, an enormous field of research that extends well beyond developmental biology, we will provide enough details in both areas to demonstrate how adhesion and its modulation control the changes in tissue form that occur during key developmental processes. For more on the physical basis of membrane interactions, see Israelachvili (1991) and Boal (2002).

Cell adhesion: specific and nonspecific aspects

The preceding discussion indicates that cell–cell and cell–substratum adhesion have both nonspecific and specific aspects. As with many of the other phenomena considered in this book, the nonspecific, or generic, interactions represent the physical basis that is likely to have been built upon and embellished over the course of evolution to generate biological specificity (see Chapter 10).

Before we attempt to construct a physical model of cell adhesion we will briefly review the composition of a typical membrane and its

but were not designed to take into account important aspects such as the cytoskeletal attachment of the proteins, which is essential to the generation of stable cell–cell bonds (Adams *et al.*, 1996, 1998; Gumbiner 1996). Other studies that utilize the ability of cells with differing adhesive properties to sort themselves out within a cell mixture (Beysens *et al.*, 2000) estimate the strength of binding indirectly, but in a more physiologically authentic context.

In what follows, we first describe the classical Bell model of cell adhesion (Bell, 1978), which serves as a basis for more sophisticated models. We then apply the relevant aspects of this model, along with the physical notions introduced earlier, to cell sorting and to certain related morphogenetic processes underlying organ formation.

The kinetics of cell adhesion

There are many situations in development in which a cell establishes new contacts with its surroundings. At one extreme this can happen by default: after cytokinesis two daughter cells may remain in contact (at least for some time) and the cells do not need to move in order to find new attachments. This typically happens in the tightly adhering cells of epithelial tissues. At the other extreme cells may detach from their progenitors (this happens with the neural crest, see Chapter 6), translocate to new sites, and form connections with other cells in an environment that is itself changing with time. For generality we will develop a model in which cells are rather mobile. The starting point is a mixture of cells (as in a sorting assay) in which the cells move around until a configuration corresponding to a minimum (free) energy is established. In the course of this process, cells must break and reform adhesive bonds with one another. We thus consider two cells as they approach each other and eventually attach.

To describe this process in a relatively simple physical way, we will make use of the fluid mosaic model of the cell membrane (Singer and Nicholson, 1972; Jacobson *et al.*, 1995; see Fig. 4.3). This model assumes that the membrane's structural components such as phospholipids and proteins (in particular CAMs) can undergo two-dimensional diffusion in the plane of the lipid bilayer. Such diffusion has been studied extensively and diffusion coefficients have been measured (Saxton and Jacobson, 1997; Jacobson *et al.*, 1997; Fujiwara *et al.*, 2002). Their magnitudes depend on the permeability of the lipid bilayer, on the mass of the diffusing molecules, and, in the case of proteins, on possible cytoskeletal attachments. Typical values (for receptors on the surface of T lymphocytes, for example) are from 10^{-10} to 10^{-12} cm^2/s. (Compare this with the much larger diffusion coefficient of a G actin molecule in an aqueous medium (see Chapter 1), which is around 10^{-6} cm^2/s).

If there are no CAMs on the surfaces of the two cells, nonspecific binding may take place. Given the types of force involved this can only be transient. More typically, the two cells would not adhere at all since the surfaces of most cells are negatively charged and this would lead to repulsion. In the presence of CAMs, however, more long-lived bonds can be formed and these will be energetically more favorable than the nonspecific bonds. There are several reasons for this. First, owing to mobile ions in the vicinity of the membrane the net negative charge is strongly screened. As a result, beyond about a nanometer from the cell surface the electrostatic repulsion is ineffective. Second, once the two cells are sufficiently close that specific bonds between them can form, diffusion-dependent clustering of the CAMs is induced and highly cooperative multimolecular binding interactions can take place.

Formation of an adhesive bond

In order to describe the process of bond formation in a rigorous fashion a mathematical formalism is required that simultaneously describes chemical reactions and diffusion. We postpone a full presentation of reaction–diffusion systems until Chapter 7 and here, following Bell, consider a simplified two-step model for bond formation, schematically shown by the following relation

$$A + B \underset{d_-}{\overset{d_+}{\rightleftarrows}} AB \underset{r_-}{\overset{r_+}{\rightleftarrows}} C\,. \tag{4.1}$$

Here A and B are the two CAMs, AB is an "encounter complex" (in which, by definition, the two molecules are close enough to allow the chemical reaction between them to take place), and C is the bound state of A and B. The encounter complex arises purely by diffusion. Thus the rate constants d_+ and d_- are related to $D(A)$ and $D(B)$, the diffusion coefficients of A and B within the cell membrane (Dembo *et al.*, 1979), by

$$d_+ = 2\pi[D(A) + D(B)] \tag{4.2a}$$

$$d_- = 2[D(A) + D(B)]R_{AB}^{-2}. \tag{4.2b}$$

It is assumed here that the reactants form an encounter complex whenever they are separated by R_{AB}, the encounter distance. Note that the units of d_+ and d_- correspond to the concentrations of A, B, and C measured in numbers of molecules per unit area. Thus the equilibrium constant K_e for the encounter step is $K_e = d_+/d_- = \pi R_{AB}^2$, which is the area of a disc of radius R_{AB}.

Equation 4.1 represents two simple processes, the second of which corresponds to the purely reactive part of bond formation:

$$\frac{dN_{AB}}{dt} = d_+ N_A N_B - d_- N_{AB} - r_+ N_{AB} + r_- N_C\,, \tag{4.3a}$$

$$\frac{dN_C}{dt} = r_+ N_{AB} - r_- N_C\,, \tag{4.3b}$$

N_A, N_B, N_{AB}, and N_C being the concentrations of A, B, AB, and C, respectively. Following Bell's simplified picture, we assume that the reaction complex is in equilibrium with the reactants and the product (i.e., $dN_{AB}/dt = 0$). Then the rate constants in the overall reaction

$$A + B \underset{k_-}{\overset{k_+}{\rightleftarrows}} C \tag{4.4}$$

can be expressed as

$$k_+ = \frac{d_+ r_+}{d_- + r_+} \tag{4.5a}$$

$$k_- = \frac{d_- r_-}{d_- + r_+}. \tag{4.5b}$$

The above expressions show that if the encounter complex is much more likely to react than to dissociate (i.e. $r_+ \gg d_-$) then $k_+ \approx d_+$, that is, the forward reaction is diffusion limited (i.e., its progression in time is controlled by the rate of diffusion). In this case $k_- \approx d_- r_-/r_+$; from Eq. 4.2 we can see that it also depends on the diffusion coefficients. Thus if the diffusion coefficients are small, as for receptors in the viscous membrane, both forward and reverse rate constants are small. (Note, however, that the equilibrium constant $K_e = k_+/k_- = R_{AB}^2 r_+/r_-$, which is independent of the diffusion coefficients.)

We now apply the above model to describe the formation of homophilic bonds between identical CAMs (such as cadherins) on juxtaposed cells. Let the total number of cadherins per unit area of the membrane be N_1 and N_2 and the corresponding numbers for unbound (free) molecules be N_{1f} and N_{2f}. The number of bound molecules N_b is then determined by

$$N_i = N_{if} + N_b \qquad (i = 1, 2). \tag{4.6}$$

The kinetic equation that describes the bond formation reaction (4.4) where $A \to N_{1f}$, $B \to N_{2f}$, $C \to N_b$ has the form

$$\frac{dN_b}{dt} = k_+ N_{1f} N_{2f} - k_- N_b, \tag{4.7}$$

k_+ and k_- being given by Eqs. 4.5a, b. Using Eq. 4.6 this can be rewritten as

$$\frac{dN_b}{dt} = k_+(N_1 - N_b)(N_2 - N_b) - k_- N_b \tag{4.8}$$

and solved for the time evolution of $N_b(t)$. Equation 4.8 is a nonlinear first-order differential equation whose explicit solution is complicated (Bell, 1978) and is not essential for understanding the relevant features of bond formation. It is clear from Eq. 4.8 that the rate of bond formation is maximal when the cells are first brought together ($t = 0$), so that $N_b(t = 0) = 0$ and

$$\left(\frac{dN_b}{dt}\right)_{max} = k_+ N_1 N_2. \tag{4.9}$$

At long times equilibrium will be approached and the right-hand side of Eq. 4.8 will approach zero, resulting in

$$N_b^{eq} = \frac{1}{2}\left(N_1 + N_2 + \frac{1}{K}\right) - \frac{1}{2}\left[\left(N_1 + N_2 + \frac{1}{K}\right)^2 - 4N_1N_2\right]^{1/2}, \quad (4.10)$$

K being the equilibrium constant.

While the mathematical manipulations leading to Eq. 4.10 may seem involved, the following simple application demonstrates its usefulness and shows the value of such analysis. If K is known, Eq. 4.10 can be used to determine the number of homophilic bonds per unit area of the cell surface connecting it to adjacent cells. For a rough estimate we consider cells of the same type and therefore take $N_1 = N_2 = N$. Since the cases $1/K \gg N$ or $1/K \ll N$ lead to the unrealistic results of either no adhesive bonds ($N_b^{eq} \approx 0$) or all CAMs being in bound pairs ($N_b^{eq} = N$), we assume that $1/K \approx N$. Using Eq. 4.10, we then obtain $N_b^{eq} \approx 0.38N$, indicating that at any given time about 40% of the CAMs form adhesive bonds.

Strength of adhesive bonds

The stability of adhesive bonds is determined by the free-energy changes associated with the electrostatic, van der Waals, or hydrogen-bond interactions between CAMs. Such bonds are reversible and each will be broken and reformed given sufficient time. However, since, as we have seen above, up to 40% of CAMs might be bound at any given instant of time, the probability of all the receptors being simultaneously unbound is very small. Thus the separation of two cells from one another requires a force that is capable of fairly rapidly (see below) rupturing every bond.

The formation of an adhesive bond is accompanied by the lowering of the system's free energy. Moreover, at equilibrium a bond has a well-defined length. Thus in order to disrupt a bond, work must be done to increase the separation between the molecules. The situation is represented schematically in Fig. 4.4. The minimum in the free energy at r_b corresponds to the equilibrium bound state of two neighboring cells. Performing work on the system by "pulling" on the bond allows the free-energy barrier to be overcome, after which the two cells separate and eventually settle into new energy minima. Clearly a force applied over a range r_0 from the minimum will rapidly rupture the bond. This force can thus be written as $f_0 = E_0/r_0$ (remember, the energy change equals the work done and is given by force times distance). Taking the binding energy between cadherins lacking cytoskeletal attachments (which such CAMs have *in vivo*) as about 1 kcal/mole (as measured by Sivasankar *et al.*, 1999) and r_0 as about 1 nm (the typical linear dimension of the binding cleft on an antibody, Pecht and Lancet, 1976), as a rough estimate we obtain $f_0 \approx 50$ pN, which is comparable with the results of studies on single molecules by Baumgartner *et al.* (2000) using atomic force microscopy. The latter authors found that about 40 pN was required to separate a single adhesion dimer of vascular

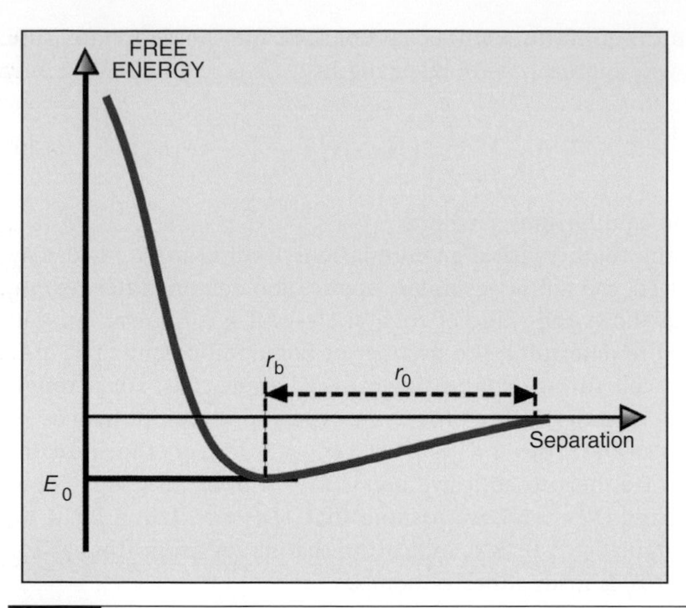

Fig. 4.4 Typical variation of the adhesive free energy of two neighboring cells with the separation between opposing cell surfaces. For simplicity we assume the connection is established via a single homophilic pair of CAMs. For small distances the cells repel each other owing to electrostatic or mechanical forces. Repulsion corresponds to large positive values of the free energy. The minimum at r_b can be considered as the equilibrium binding energy between the two CAMs. Bond rupture between the two CAMs and thus the separation between the cells takes place when the adhesive free energy changes sign, turning from negative to positive. (Compare with the curve shown in Fig. 2.12).

endothelial (VE) cadherins. Since the cadherins in this experimental system also lacked cytoskeletal attachments, the measured force of separation is probably an underestimate of the biological value.

In the absence of external forces two CAMs will dissociate in a time $t > \tau_0$, where τ_0 is the average lifetime of the adhesive bond. To describe the situation in which a constant force f stresses the bond, Bell, applying results from the theory of solids, suggested the following equation for the modified lifetime τ, the *Bell equation*:

$$\tau = \tau_0 \exp(-\gamma f/k_B T). \tag{4.11}$$

Here γ is a parameter (with units of length) whose value has to be determined empirically; in the case of a solid γ accounts for its structure and imperfections.

We now are in a position to make the phrase "fairly rapid rupturing" of the bond more quantitative. Consider an adhesive cadherin dimer between two cells, one of which is attached to a substratum and the other to a spring. By stretching the spring to a given length quickly (in the ideal case, instantaneously), one can measure the time it takes for the cadherin–cadherin bond to dissociate under a constant force. Since bond lifetime is a statistical quantity, performing such a measurement a number of times yields a probability distribution for $\tau(f)$. Fitting the average of this distribution to Eq. 4.11 provides values for τ_0 and γ (Marshall *et al.*, 2003).

Another application of Eq. 4.11 is based on pulling the spring with varying velocity. Baumgartner *et al.* (2000) created cadherin adhesion dimers in the following way. Cadherins derived from the vascular endothelial cells lining the inner surface of blood vessels (VE cadherins) were adsorbed onto mica and also covalently attached to the tip of an atomic force cantilever, which acted as the spring. The cantilever was brought close to the mica and adhesion between the cadherins was established. Subsequently the cantilever was retracted with velocity v, and the rupture force $f_c(v)$ was measured. It was found that f_c depended logarithmically on v. This is consistent with the Bell equation if it is assumed that $\tau(v) = z/v$ (z is the distance over which the spring has to be pulled before the bond ruptures), since in this case Eq. 4.11 yields $f_c = kT/\gamma \ln(v\tau_0/z)$. From the fit to the experimental data the authors deduced $\tau_0 = 0.55$ s and $\gamma = 0.59$ nm. This result implies that for VE cadherins (lacking cytoskeletal attachments) the reverse rate constant in Eq. 4.8 is $k_- = 1/\tau_0 = 1.8\ \mathrm{s}^{-1}$.

The preceding discussion pertains to the force needed to rupture a single adhesive bond on the cell surface. An extension of this analysis to estimate the force required to separate two cells attached to each other by N_b complementary CAMs is presented in Box 4.1.

The Bell model has a number of limitations. For example, it ignores the possible interaction between receptors on the same cell surface (cis interactions) or the role of the cytoskeleton in adhesion, factors that have been shown important for cadherin function (Wheelock and Johnson, 2003). It treats bond formation as a simple chemical reaction (between the members of the encounter complex), whereas in reality it is an intricate multistep process (Adams *et al.*, 1996, 1998; Gumbiner, 2000). Despite these deficiencies the Bell model remains the basis of our physical understanding of cell–cell adhesion.

Box 4.1 | Separating two adhering cells

We consider two cells adhering to each other through N_b complementary CAMs. If the force needed to separate the two cells is F then, assuming that each bond is equally stretched, the force per bond is F/N_b. Applying Bell's equation, Eq. 4.11, to this situation, the reverse rate constant in Eq. 4.8 (with $N_1 = N_2 = N$) should be replaced by $k_- \exp(\gamma F/k_B T N_b)$ to yield

$$\frac{dN_b}{dt} = k_+(N - N_b)^2 - k_- N_b \exp(\gamma F/k_B T\, N_b). \qquad \text{(B4.1a)}$$

Since both terms in Eq. B4.1a are positive, as long as the second term is smaller than the first, N_b will increase in time. If the force is large and the second term exceeds the first, N_b will rapidly go to zero; there will be no finite solution to $dN_b/dt = 0$, which now yields an equation similar to Eq. 4.10 in which K has been multiplied by $\exp(-\gamma F/k_B T N_b)$. Thus, at a critical value of the force F_c the equilibrium number of adhering CAMs, N_b^{eq}, will vanish. This happens when the curves (as functions of N_b) determined by the first term and the second term (without the minus sign) in Eq. B4.1a become tangential.

Mathematically, this condition is expressed by equating the derivatives of the two terms with respect to N_b, yielding

$$2K(N - N_b) = \left(1 - \frac{\gamma F_c}{k_B T N_b}\right) \exp\left(\frac{\gamma F_c}{k_B T N_b}\right). \tag{B4.1b}$$

The solution of this equation for N_b is then inserted into the right-hand side of Eq. B4.1a, which when equated to zero determines the critical detachment force F_c. Owing to the complicated form of these equations this program can be carried out only numerically. Using for example $\gamma = 0.59$ nm (Baumgartner *et al.*, 2000), for a large range of $K\,N$ values one obtains for the critical force per bond $f_c = F_c/N \approx 40$ pN.

For forces larger than F_c, bonds break rapidly and Bell's model provides the time at which N_b becomes zero. This time is obtained from Eq. B4.1a, in which the first term can now be neglected. Within a good approximation (for $F > F_c$)

$$t_{detach}(F) \approx \frac{\exp(-\gamma F/k_B T N)}{k_-(1 + \gamma F/k_B T N)}. \tag{B4.1c}$$

The above analysis can be used to estimate the critical forces and times of detachment of cells for any experimental situation.

Differential adhesion of embryonic tissues

One of the most dramatic morphogenetic processes exhibited by embryonic cells is sorting, the ability of cells of distinct types to segregate into distinct tissues that do not intermix at their common boundary. Cells of the prospective central nervous system, for example, must differentiate, segregate from prospective skin cells, and, in a concerted fashion, fold inward along the dorsal surface (i.e., the back) of the embryo. If this process fails to occur properly the spinal cord remains open to the body surface, resulting in a condition known as "spina bifida." A half-century ago Holtfreter and colleagues provided insight on how this is accomplished (Townes and Holtfreter, 1955). They dissociated tissues such as nervous system and skin primordia into single cells and then mixed them together randomly. Although all the cells were able to adhere to one another, they did so to different degrees. After random wandering within the cell mixture, skin and neural progenitor cells found their respective counterparts and formed first islands and then two distinct, uniform, regions, in which the skin cells surrounded a central sphere of neural cells. In a similar fashion, an initially random mixture of endodermal and ectodermal cells, derived from the two tissue layers of the invertebrate *Hydra*, sorted themselves out in such a way that the endodermal cells wound up at the center of the aggregate and the ectodermal cells came to surround them, precisely their relationship in the intact organism (Technau and Holstein, 1992; Rieu *et al.*, 2000).

Steinberg (1963) postulated that cells of different origin adhere to each other with different strengths. Furthermore, in analogy with immiscible liquids such as oil and water, mixtures of such cells undergo a process of phase separation in which the final configuration corresponds to the minimum of interfacial and surface free energies. (Here the cells play a role analogous to the molecules of a liquid). This "differential adhesion hypothesis" (DAH) was expressed in quantitative terms, by Phillips and coworkers (Phillips, 1969; Heintzelman *et al.*, 1978), based on a geometric analysis of the surface and interfacial tension of immiscible droplets of liquid (Israelachvili, 1991; see Chapter 2 for a more complete version of this analysis). According to the DAH, the final "phase-separated" state of two adjacent tissues is an equilibrium configuration not dependent on the pathway by which it was reached. That is, it will be the same whether arrived at by the fusion of two intact fragments of tissue or by the sorting out of their respective cells from a binary mixture (Fig. 4.5A). Another implication of the DAH is that tissue engulfment relationships should form a hierarchy (Fig. 4.6): if tissue *A* engulfs tissue *B* and *B* engulfs *C* in separate experiments, it follows that *A* will engulf *C* if that experiment is performed. Finally, the DAH predicts that the values of tissue surface tensions (see below) should fall into a quantitative order that corresponds to the engulfment hierarchy. Each of these predictions of the DAH has been amply confirmed experimentally (Steinberg, 1963; 1978; 1998; Armstrong, 1989; Mombach *et al.*, 1995; Foty *et al.*, 1994, 1996; Duguay *et al.*, 2003).

According to the DAH, any pair of tissues, not only those that contact each other in normal development, will undergo phase separation and sorting, provided that their cells are capable of rearranging and that one of the tissues is more cohesive than the other (see Figs. 4.6 and 4.7). This was confirmed in a decisive fashion in an experiment in which mouse L cells, which do not normally express CAMs, were genetically engineered to express P-cadherin (i.e., placental cadherin; Steinberg and Takeichi, 1994). Two populations of cells were prepared – high expressers and low expressers – and when intermixed they sorted themselves out, the high expressers ending up in the interior of the aggregate as predicted on thermodynamic grounds. (The result of a similar experiment is shown in Fig. 4.7.)

Embryonic tissues as liquids

In Chapter 2 we noted that while the shape of an individual cell could be described in terms of a surface-tension-like quantity termed the "cortical tension," it is inaccurate to consider individual cells as exhibiting a true surface tension. The reason for this is that the cell surface – its plasma membrane – does not consist of the same material as the interior of the cell and will not therefore automatically

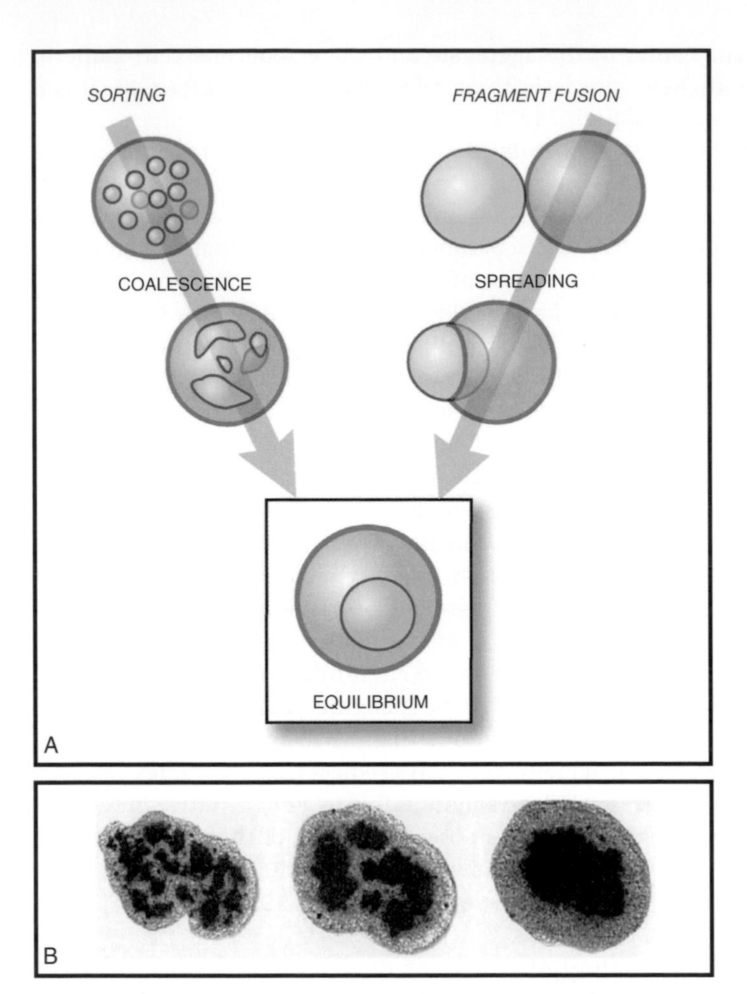

Fig. 4.5 (A) Different paths by which two immiscible liquids or cell aggregates (composed of cells with differing adhesive properties) may arrive at the same equilibrium state. The path on the left shows sorting, or nucleation, which proceeds through the gradual coalescence of groups of cells. The path on the right corresponds to engulfment, which occurs through spreading. (After Steinberg, 1978.) (B) The time evolution of sorting in a mixture of chicken embryonic pigmented epithelial (dark) and neural retinal (light) cells. The images show the equatorial section of a three-dimensional spheroidal aggregate. The left-hand, middle, and right-hand panels correspond respectively to 17, 42, and 73 hours after the initiation of sorting. The diameter of the sorted configuration is around 200 microns. (After Beysens *et al.*, 2000.)

increase or decrease its area as the cell experiences external forces. With tissues the story is different: in many cases a true surface tension can be defined. An aggregate of cells coheres by virtue of adhesive forces between its cells rather than by a distinct bounding layer. New surface can potentially be created from the interior by the movement of cells to the surface and surface can be lost by cells moving inward.

A tissue will have the ability to increase or decrease its surface in this fashion if its cells are individually mobile and can easily slip

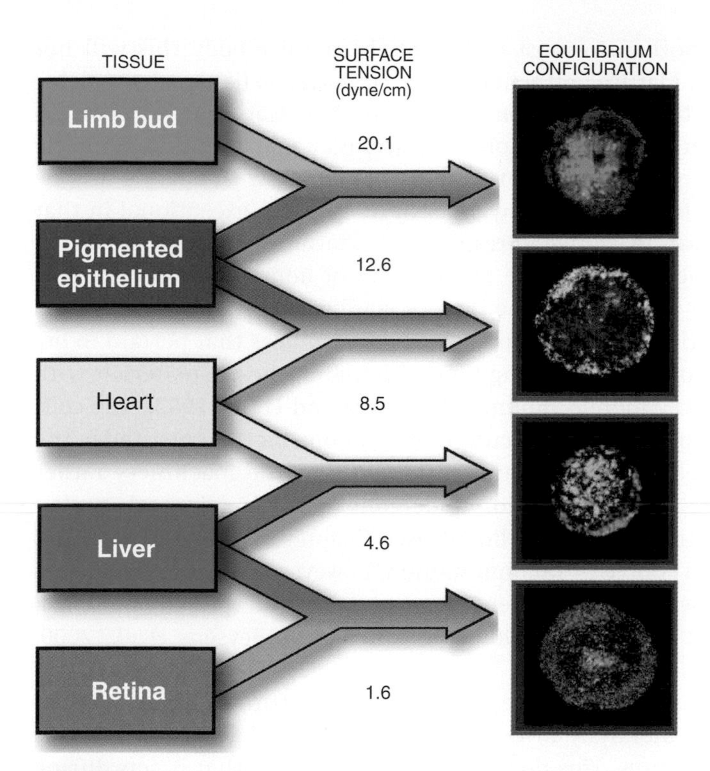

Fig. 4.6 The correspondence between equilibrium cellular patterns and values of surface tension for five embryonic chicken tissues. On the right are shown the configurations generated by cell sorting or aggregate fusion that occur when adjacent tissues in the surface tension hierarchy are combined and allowed to rearrange *in vitro* (see Fig. 4.5). Cells from the two tissue sources were stained with contrasting fluorescent markers. In each case, the more cohesive cell population, as quantified by its surface tension, was engulfed by the less cohesive cell population (Foty *et al.*, 1996).

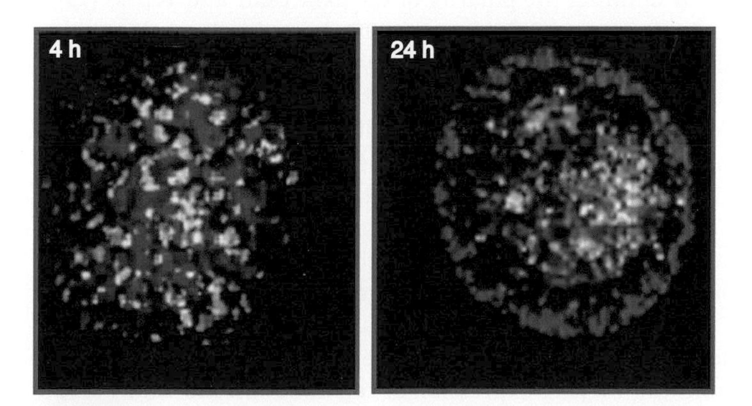

Fig. 4.7 Sorting of two genetically transformed Chinese hamster ovary cell populations with ~50% difference in N-cadherin expression. On the left, the configuration four hours after mixing. On the right, the fully sorted configuration after 24 hours (Forgacs and Foty, 2004).

past one another. In most tissues of the mature body this will not be the case. Mature epithelial tissues consist of cells that have elaborate, specialized, connections to one another. Mature connective tissues are embedded in complex extracellular matrices that similarly impede their mobility. Embryonic tissues (along with healing and regenerating tissues, and many types of tumors) are exceptional in that their cells can rearrange in response to external stresses on the tissue mass, although it may take of the order of hours for them to do so and for the tissue to assume its new equilibrium shape. With such a capability, tissues can be modeled by physical laws that pertain to liquids since, like any nonliving liquid, they are cohesive materials with independently mobile subunits (Steinberg and Poole, 1982). The cells within "liquid" tissues execute their random movements under the power of cell metabolism and an intrinsic motile machinery rather than by the thermal agitation undergone by the smaller-scale molecular subunits of nonliving liquids (see Chapter 1). These distinctions in scale and source of random motion, however, turn out to make no difference when sorting phenomena are being interpreted (Mombach *et al.*, 1995; Foty *et al.*, 1996; Rieu *et al.*, 1998; Beysens *et al.*, 2000; Duguay *et al.*, 2003).

Differential adhesion clearly dictates the sorting behavior and the engulfment hierarchy in experimentally manipulated tissue fusions and cell mixtures. This does not necessarily mean that it acts during development to guide tissue assembly. That it does in some cases has been demonstrated in a series of *in vivo* experiments (Godt and Tepass, 1998; Gonzalez-Reyes and St Johnston, 1998). The anterior–posterior axis of the *Drosophila* embryo originates from two symmetry-breaking steps during early oogenesis. Each oocyte arises within a cyst of 16 interconnected cells that are formed by four incomplete cell divisions. The one cell of these 16 that becomes the oocyte then comes to lie posterior to the other 15 cells of the cyst in an enclosure called the follicle, thereby defining the polarity of the axis. Godt and Tepass (1998) and Gonzalez-Reyes and St Johnston (1998) showed independently that during this cell rearrangement the oocyte adheres to the cells of the follicle that express the highest amounts of DE-cadherin. The positioning of the oocyte, moreover, requires cadherin-dependent adhesion between these two cell types, since the oocyte is frequently misplaced when DE-cadherin is removed from either the "germ-line" cells that give rise to the oocyte or the posterior follicle cells. Analogous studies of the development of the *Drosophila* retina similarly demonstrate the role of differential adhesion in cell patterning (Hayashi and Carthew, 2004).

As these and other experiments have shown, differential adhesion results from the varying expression of cell adhesion molecules in different cell types (Friedlander *et al.*, 1989). There are numerous cases in development in which boundaries are established in response to patterning signals and in which the cells on one side of such a boundary do not mix with cells on the other side (Blair, 2003). Examples include the development of compartmental subdivisions within the wings and other surface structures of *Drosophila* (Garcia-Bellido, 1975; Crick and Lawrence, 1975), the formation of segmental boundaries during development of the vertebrate body axis (Meier, 1984; Palmeirim *et al.*, 1997) and hindbrain (Guthrie and Lumsden, 1991; see Chapter 7), and the formation of mesenchymal condensations during vertebrate limb development (Newman, 1977; Newman and Tomasek, 1996; Hall and Miyake, 2000) (see Chapter 6). In each case, local sorting appears to be involved in keeping the boundaries distinct when they are first established. The challenge for a model of sorting is to relate the physical quantities, such as surface tension, to biological properties characterizing the adhesion complex in terms of specific CAMs.

The physics of cell sorting

The change over time of the configuration of an originally random mixture of embryonic chicken neural retinal and pigmented epithelial cells is shown in Fig. 4.5B. The evolution of the cellular pattern and the final equilibrium state qualitatively resemble the phase separation of immiscible fluids and have been shown to follow that process quantitatively as well (Beysens *et al.*, 2000). Similar experiments have been performed with numerous pairs of different embryonic tissues (Armstrong, 1989; Foty *et al.*, 1996). In each case the sorted pattern corresponded to a configuration in which the more cohesive tissue was surrounded by the less cohesive one. The cohesivity of many of these tissues has been quantified by measuring their surface tension (i.e., the interfacial tension with the surrounding tissue-culture medium) with a specifically designed parallel plate tensiometer (Foty *et al.*, 1996). (See Chapter 2 for the relation between surface tension and cohesivity.)

The sorted patterns exhibit a hierarchical relationship consistent with the values of the tissue surface tensions (Fig. 4.6) as predicted by the DAH. The sorting of two cell populations that are identical except for having been genetically manipulated to express differing numbers of N-cadherin molecules on their surfaces (as their only CAMs) (Fig. 4.7) illustrates how a 50% difference in the quantitative expression of cell-surface adhesive molecules leads to a dramatic difference in tissue behavior – i.e., the establishment of immiscible tissue layers. Cell sorting as a liquid-like phenomenon is thus readily studied *in vitro*. Can we use this physical phenomenon to extract useful quantitative insights into cell behaviors that are relevant to early developmental processes such as the *Drosophila* oocyte–follicle-cell interaction described above?

In the state of lowest energy in a sorting experiment, cells with the higher number of CAMs must be surrounded by adhesively similar cells, because such a configuration allows for the maximum number of CAM bonds to be formed (Fig. 4.8). To reach this state, cells in

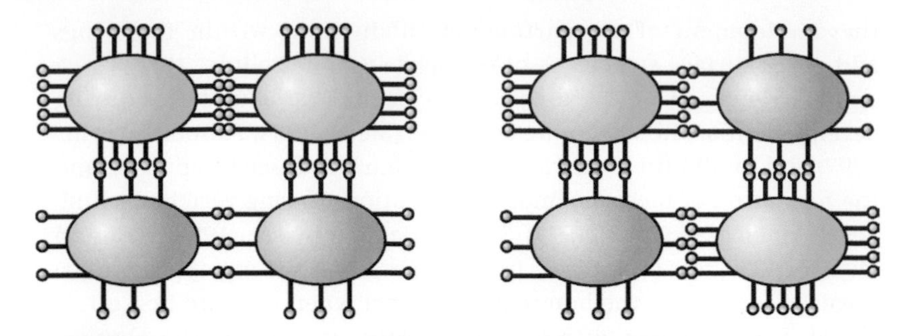

Fig. 4.8 Schematic illustration of adhesion in a mixture of two cell populations with differential expression of homophilic CAMs. The small bars ending in small circles symbolize CAMs. To minimize the configurational energy, high expressers tend to group together to benefit from all available bonds. The arrangement on the left has lower energy than that on the right. Compare with Fig. 4.7.

the initial random mixture of the two cell types must move around until they find their respective partners. This motion is driven by cytoskeletal rearrangement powered by metabolic energy, in contrast with the motion of liquid particles (which is driven by thermal fluctuations, see Chapter 2). The surprising outcome is that this biological mechanism, despite the complexity of the underlying machinery, drives the cellular system to its lowest energy state, as characterized by purely physical parameters such as surface and interfacial tensions. For cells to move, they have to break adhesive bonds with their immediate neighbors and reform bonds as they find new partners. This amounts to a frictional force experienced by the cells. The corresponding friction coefficient, μ, is a characteristic property of the cellular environment, which relates the velocity v of the cell's motion to the force F under which the motion takes place (see Chapter 1), $\mu = F/v$. In the sorting process this force is generated by the energy difference between the sorted and unsorted configurations. Using dimensional analysis (as introduced in Chapter 1), and the Bell model, we can now relate μ, a physical parameter whose value can be experimentally determined, to important variables that characterize the biological state of cells in an aggregate and which are all but impossible to measure directly (Forgacs *et al.*, 1998).

From its definition, the unit of μ is N s/m. The more CAMs a cell has on its surface the stronger is its tendency to adhere to its neighbors and the more difficult it will be for it to arrive at a lower energy state by changing its neighbors. Thus $\mu \propto N$, N being the number of CAMs per unit area of the membrane. The stronger the adhesive bonds the cell forms the more difficult it will be to break them, so we have also $\mu \propto E$, E being the energy of a single bond. The longer these bonds last the more they hinder the motion of a cell; hence furthermore $\mu \propto \tau$, where τ is the lifetime of a bond (see the Bell equation, Eq. 4.11). Combining the above observations, we arrive at

$\mu \propto NE\tau$. The units of the three quantities on the right-hand side of this relation are, respectively, m^{-2}, $J = N\,m$, and seconds; thus dimensional analysis suggests that in the combination $NE\tau$ we have taken into account all the relevant parameters that may influence the friction coefficient, and so $\mu = aNE\tau$ where a is a dimensionless constant.

The Bell model, through Eq. 4.10, allows us to fix the value of a, since a must be related to the fraction of receptors that are bound, the only ones we need to consider in the above analysis (see also Howard, 2001). A similar dimensional analysis for tissue surface tension leads to $\sigma = bNE$, where b is a constant related to the difference in the fraction of bound CAMs between cells at the surface and in the bulk of the tissue (Forgacs *et al.*, 1998). In relation to the *Drosophila* oocyte–follicle-cell interaction described earlier in this chapter, we now have predictive criteria, in terms of measurable parameters such as the relative number of DE-cadherin molecules on the two cell types, that could rigorously establish whether the observed terminal cell arrangement is fully accounted for by differential adhesion.

In the case of sorting involving the two genetically engineered cell populations shown in Fig. 4.7, the surface concentrations of receptors are not equal, $N_1 \neq N_2$, and therefore μ is not uniform across the aggregate. The time evolution of the sorting pattern in this case will be governed by $\mu = a\max(N_1, N_2)\,E\,\tau$, that is, by the more cohesive cells.

Surface tensions and friction coefficients have been determined experimentally for a number of natural tissue types (Foty *et al.*, 1994; 1996) and tumors (Foty and Steinberg, 1997; Steinberg and Foty, 1997; Forgacs *et al.*, 1998), as well as for aggregates of genetically transformed cells with varying levels of cadherin or integrin expression (Ryan *et al.*, 2001; Robinson *et al.*, 2003; Foty and Steinberg, 2005). Sorting experiments (Figs. 4.5–4.7) have been performed to verify the consistency of the obtained surface tension values (Foty *et al.*, 1996; Beysens *et al.*, 2000). Using the results of dimensional analysis in conjunction with physical models, the measured tensions and friction constants provide quantitative information on such molecular parameters as the lifetime of the bonds between CAMs and the energy of these bonds. Since this type of measurement involves large numbers of cells the results obtained reflect conditions comparable to those in tissues. Therefore when compared with single molecular studies, in which typically CAMs are considered under non-physiological conditions, the significance of factors such as the cytoskeleton or the interaction between CAMs on the same cell can be assessed.

Perspective

Cell adhesion is the defining characteristic of multicellular organisms and the nature and strength of cell bonding is a major determinant of tissue properties. Cells in embryos have the unique feature of being bound to each other by forces that are neither so strong as to resist relative movement (as in mature organisms) or so weak as to disperse, and they therefore constitute tissues that behave like viscoelastic materials. Physical models can account for many aspects of the reversible cell–cell interactions seen in developing systems, as well as for certain self-organizing consequences of differential adhesion such as boundary formation via sorting, engulfment behavior, and the internal spaces (i.e., lumens) within tissues containing polarized cells.

Chapter 5

Epithelial morphogenesis: gastrulation and neurulation

In the previous chapters we have followed the process of rapid cell divisions in the early embryo until the formation of the blastula, initially a solid mass of cells poised to develop into the structurally and functionally differentiated organism. Adhesive differentials along the surfaces of individual cells of the early blastula, the blastomeres, drive the formation of spaces or lumens (see Fig. 4.2) within the embryos of most species. As a result, the typical blastula acquires a geometrically simple closed spheroidal structure that consists of a single cell layer enclosing the hollow blastocoel.

By the time the blastula has developed, the embryo already contains, or begins to generate, a number of differentiated cell types (see Chapter 3). Insofar as these cell types have or acquire distinct physical (adhesive, contractile) properties, compartmentalization or other forms of regional segregation start taking place. This regionalization, accompanied by the collective movement of the resulting cell masses, gives rise in most cases to embryos consisting of two major cell layers, referred to as "germ layers," along with some subsidiary populations of cells.

The various modes of cell rearrangement by which a solid or single-layered blastula becomes multilayered are known collectively as *gastrulation*. In "diploblastic" animals, such as sponges and coelenterates (hydra, jellyfish), gastrulation is complete when the two germ layers, the outer *ectoderm*, and inner *endoderm*, are established. Further cell specialization occurs within these two main layers and any subsidiary cell populations. For other, "triploblastic," animals such as insects, echinoderms (e.g., sea urchins, starfish), and vertebrates (e.g., frogs, humans), the initial binary segregation results in a "pre-gastrula," which sets the stage for the next phase of the developmental process, establishment of a third germ layer, the *mesoderm*, and, in those triploblasts that contain one, the primordium of the axial skeleton (the vertebral column). This later set of processes is often referred to, narrowly, as gastrulation, but we will use the term to encompass all the cell and tissue movements leading to both two and three layers. Gastrulation is followed, in species with an axial nervous system (e.g., vertebrates, the subphylum to which humans and other mammals belong), by *neurulation*, the formation of the tubular rudiment of the nerve cord.

Both gastrulation and neurulation involve the folding and reshaping of epithelial sheets. During gastrulation the blastula is deformed and reorganized in a sequence of steps, often involving the narrowing and elongation of internal and external embryonic tissues. (This latter set of tissue-reshaping movements is termed "convergence and extension.") The resulting forms have a central "primary" axis. In neurulation, the central zone of the gastrula's surface ectoderm forms an elongated "neural tube" that parallels the primary axis. Although the tissue movements and rearrangements underlying gastrulation and neurulation lead to dramatically different outcomes, from a physical standpoint both sets of processes employ similar mechanisms and are subject to similar constraints.

Physical properties of epithelia

In Chapter 4 we described the physical bases of adhesion and the consequences of differential adhesion in *epithelioid* tissues, that is, tissues made up of cells in direct contact with one another via their membrane-adhesion proteins or associated surface coats (glycocalyces). This chapter is concerned with a subset of these tissues, referred to as "epithelia." In these tissues cells still make direct contact with one another, but only along their lateral surfaces. The cells are therefore "polarized." The apical and basal surfaces of such cells do not adhere to adjacent cells of the same type but rather to extracellular matrices known as basal laminae (see Fig. 4.1) or to other cell types, tissue fluids, or acellular matrices. For example, the outer surfaces of many blastulae are in contact with specialized matrices called egg envelopes or "jelly coats" (Dumont and Brummett, 1985), and their inner surfaces are in contact with blastocoelic fluid. By expressing adhesive proteins in a polar fashion (that is, on some portions of the cells but not others), epithelioid tissues can generate internal spaces (lumens) or, under appropriate conditions, form themselves into "two-dimensional" epithelial cell sheets. The apical and basal surfaces of epithelia are sometimes referred to as "free surfaces." (In physics this term is usually reserved for surfaces in contact with the vacuum, something never encountered in biology).

As we discussed earlier, the fact that attachments between embryonic epithelioid cells are weak and short-lived causes the tissues they comprise to behave like liquids. This is the case for many tissues in the embryo, including planar epithelia. The basal laminae underlying many epithelia, however, are stiff and have decided elastic properties. The epithelia of early embryos and developing organs are therefore unique in that they behave like liquids in the plane but like elastic sheets when deformed out of the plane. This unusual combination of properties largely accounts for the ability of these tissue sheets to undergo a wide range of morphological changes, including bending, eversion, invagination, and placode, cyst, and tubule formation (Gierer, 1977; Mittenthal and Mazo, 1983; Newman 1998a; see below).

Epithelial folding is perhaps the most typical morphogenetic phenomenon in early development, giving rise to the complex shapes and forms of the early embryo and eventually the organs (Gierer, 1977). The parameter that quantifies the extent of folding of a sheet is its local average curvature C, introduced in Chapter 2 (see the text following Eq. 2.2). Bending a sheet from its equilibrium configuration (determined by its spontaneous curvature C_0) requires energy, the bending or curvature energy (see Eq. 2.5). Thus the initiation of gastrulation and neurulation, which involve epithelial folding at precise locations along the blastula and gastrula, imply alterations in specific physical properties of cells at those locations relative to their neighbors.

It is clear that the physical characteristics of individual cells (e.g., spontaneous elongation or flattening) are determined by the molecular content and organization of their cytoplasm and microenvironment. The molecular composition of cells is the consequence of differential gene expression, which itself is influenced by many factors (e.g., intercellular communication and cell adhesion). But molecules by themselves do not determine cell and tissue shape; physical processes do. Morphogenetic processes such as gastrulation and neurulation can only arise from the coordinated interplay of physical and genetic mechanisms. For such processes to occur, the biochemically excitable cells that constitute the epithelia involved must develop spatial variations in their physical properties, i.e., "self-organize." They do this by interacting with one another directly and/or with diffusible or otherwise propagating signals, thereby establishing patterned arrangements of differentiated cell types (see Chapter 7).

Our objective in this chapter is to characterize the physical principles that come into play in the behavior of embryonic epithelia. By considering the viscoelastic and adhesive properties of cells and tissues, described in previous chapters in the geometric context of the early embryo, we will show (within the limits of necessarily simplified models) how certain morphological changes emerge during the early stages of development. We will rely heavily on the well-justified assumption that an epithelial structure, the blastula for example, corresponds to the state, of a collection of cells, with particular physical characteristics and subject to specified physical constraints, that has the lowest energy, at least temporarily. As cells differentiate in a regional fashion and change their physical properties the original structure will no longer correspond to the lowest energy state of the embryo or primordium. Time-dependent physical processes will be triggered (i.e., shape changes) that will steer the system into a new equilibrium (or steady) state, which corresponds to a new shape, again temporary. The altered shape of the embryo, or portion thereof, may in turn influence the course of subsequent development.

We have already carried out such a program in Chapter 2 when discussing cleavage and blastula formation. We used Monte Carlo simulations to demonstrate how a model system representing the early embryo changed from one equilibrium state to another. In the case of a single cleavage event it was the astral signal (causing an imbalance in pressure) that triggered the departure from the initial spherical equilibrium state of the embryo and led eventually to the final equilibrium state with two daughter cells. In the case of blastula formation it was the sequence of cleavages that resulted in new equilibrium states and finally in a hollow sphere.

As powerful as the methods of energy minimization are, Monte Carlo simulation being one such, they have a severe drawback: the final equilibrium state of the system, which corresponds to the lowest (free) energy under specified conditions is arrived at through postulated displacements of its components (or in simple cases by explicit minimization of the mathematical expression for the energy). In the case of the Drasdo–Forgacs model, for example, the energy is calculated after computer-selected shape deformations of randomly chosen cells. Such an approach provides no information about the true dynamics of the system, however, and makes no attempt to explain how it reaches one equilibrium state from another.

The explicit time evolution of a process (e.g., epithelial folding) characterized by a set of parameters $x_1, \ldots, x_n$ is described by the set of equations

$$\frac{dx_i}{dt} = f_i(x_1, \ldots, x_n) \qquad (i = 1, \ldots, n). \tag{5.1}$$

In most physicochemical applications the parameters x_i represent amounts or concentrations of substances. To apply this formalism to shape changes, the parameters should also include those characterizing the geometric properties of the embryo (e.g., the linear dimensions, radii of curvature, etc.). With appropriate choice of the functions f_i the process may drive the system to one or several steady states, which are stable against perturbations and capable of being reached from any initial state nearby (see the discussion on dynamical systems in Chapter 3). The stable states correspond to local minima in the energy landscape of the system.

In what follows we use both energy minimization and the dynamical approach to capture different aspects of epithelial folding and rearrangement. Most of what we will present relates to gastrulation and neurulation, in keeping with our focus on tracking the processes of early development in approximate temporal order. We will begin, however, by discussing a model for epithelial folding and shape change that, while motivated by the problem of leg morphogenesis in insects, is also applicable to the phenomena that concern us here. To do this we first have to introduce the notion of the "work of adhesion."

Work of adhesion

Consider a unit interfacial area between two materials A and B immersed in a medium, denoted by M (which in particular could be the

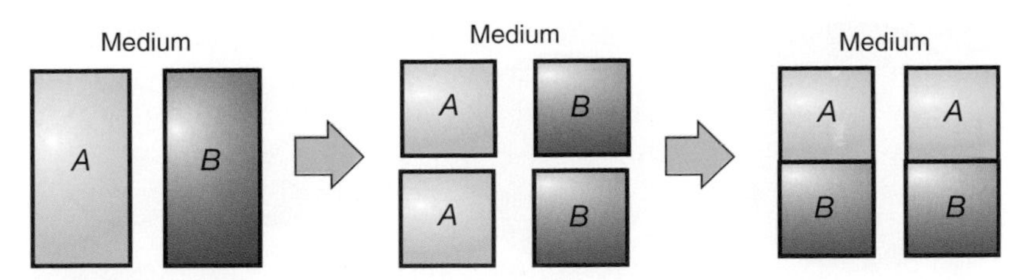

Fig. 5.1 Schematic illustration of the creation of an interface between two materials *A* and *B* immersed in a common medium. In the two-step process shown, the first free interfaces of materials *A* and *B* are produced (middle panel), which requires the separation of their corresponding subunits (molecules, cells) from one another and thus involves the works of cohesion w_{AA} and w_{BB}. In the second step the free interfaces are combined to form *AB* interfaces (by rearranging the *A* and *B* blocks). The separation of *A* and *B* requires the work of adhesion w_{AB}. If the cross-sectional area of columns *A* and *B* is of unit magnitude then the operation shown results in two units of interfacial area between *A* and *B*.

vacuum or, in the case of tissues, the extracellular matrix or tissue culture medium). We define the work of adhesion w_{AB} as the energy input required to separate *A* and *B* across a unit area in medium *M*. We can imagine such a unit area to be formed in the following way. First, we separate a rectangular column of *A* and *B* to produce two free unit surfaces of each substance (Fig. 5.1). This requires amounts of work w_{AA} and w_{BB}, respectively. These quantities are called the works of cohesion. (Note that the magnitudes of the works of adhesion and cohesion depend on the medium. To avoid the use of clumsy notation, the explicit dependence on *M* is not indicated here.) We then combine these pieces to end up with two unit interfacial areas between *A* and *B*, as shown in Fig. 5.1. Thus the total work, Δ_{AB}, needed to produce a unit interfacial area between *A* and *B* is given by

$$\Delta_{AB} = \tfrac{1}{2}(w_{AA} + w_{BB}) - w_{AB}. \tag{5.2}$$

The quantity Δ_{AB} is called the interfacial energy and Eq. 5.2 is known as the Dupré equation (Israelachvili, 1991). If the interface is formed by two immiscible liquids then Eq. 5.2 can readily be expressed in terms of liquid surface and interfacial tensions. By definition the surface tension is the energy required to increase the surface of the liquid by one unit of area (see Chapter 2). Since an amount of work $w_{AA} + w_{BB}$ creates two units of area, we obtain

$$\sigma_{AB} = \sigma_{AM} + \sigma_{BM} - w_{AB}, \tag{5.3}$$

where σ_{AM}, σ_{BM}, and σ_{AB} are respectively the surface tensions of liquids *A* and *B* and their mutual interfacial tension (Israelachvili, 1991). (Whereas for solids Δ_{AB} depends on the amount of interfacial area between *A* and *B*, for liquids, σ_{AB} does not; see Chapter 2.) Note that the immiscibility of *A* and *B* implies that $\sigma_{AB} > 0$. If, on the contrary, $\sigma_{AB} \leq 0$ then it is energetically more beneficial for the molecules of liquid *A* to be surrounded by molecules of liquid *B*, and vice versa; that is, *A* and *B* are miscible.

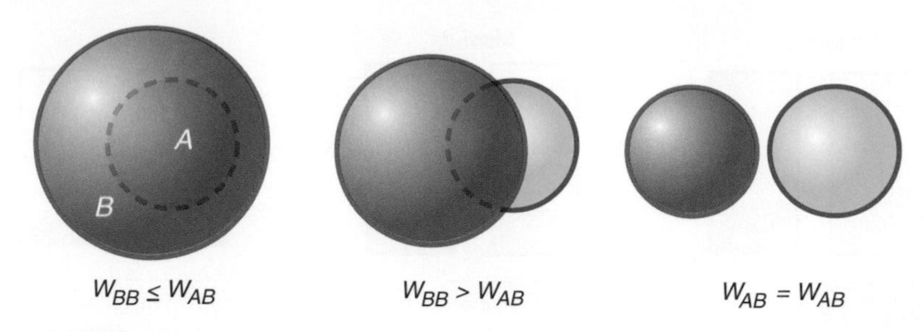

Fig. 5.2 Geometric configurations of immiscible liquids *A* and *B*, and the corresponding relations between the works of cohesion and the work of adhesion. It is assumed that *A* is more cohesive than *B*, so that $w_{AA} > w_{BB}$ and $\sigma_{AM} > \sigma_{BM}$ (*M* denotes the surrounding medium).

If we now invoke the liquid-like behavior of tissues, we can apply Eqs. 5.2 or 5.3 to obtain the conditions for sorting in terms of the *w*s or the σs. We imagine the cells of tissues *A* and *B* to be initially randomly intermixed and surrounded by tissue culture medium and we then allow them to sort, as discussed in Chapter 4. Let us assume that tissue *A* is the more cohesive. This implies that w_{AA} is the largest of the three quantities on the right-hand side of Eq. 5.2. In the energetically most favorable configuration at the end of the sorting process, cells of tissue *A* form a sphere, the configuration in which they have minimal contact with their environment and maximal contact with each other. Then, depending on the relative magnitudes of the *w*s, the sphere of tissue *B* may completely or partially envelope the sphere of tissue *A*, or the two spheres may separate (Fig. 5.2) (Steinberg, 1963, 1978; Torza and Mason, 1969).

When liquid *B* just fully envelopes liquid *A*, $\sigma_{AB} = \sigma_{AM} - \sigma_{BM}$ (see Chapter 2, Eq. B2.1c). Combining this result with Eq. 5.3 and the relationships $\sigma_{AM} = w_{AA}/2$, $\sigma_{BM} = w_{BB}/2$ yields $w_{AB} = w_{BB}$. Thus, complete envelopment takes place when *B* adheres more strongly to its partner than to itself (Fig. 5.2, panel on the left). Partial envelopment occurs when the less cohesive liquid, *B*, adheres more strongly to itself than to its partner, and therefore tries to minimize its contact with *A* (Fig. 5.2, middle panel). To sum up, partial or complete envelopment corresponds respectively to $w_{BB} > w_{AB}$ and $w_{BB} \leq w_{AB}$. When $w_{AB} = 0$, there is no energy gain from the contact of the two liquids and they separate (Fig. 5.2, panel on the right).

Applying Eqs. 5.2 and 5.3 to actual tissues results in quantitative expressions for the extent of differential adhesion and its effect on cell sorting.

The Mittenthal–Mazo model of epithelial shape change

A particularly simple and elegant quantitative model of epithelial morphogenesis was constructed by Mittenthal and Mazo (1983). The model accounts for the generic shape changes that occur when an anchored epithelial sac (an "imaginal disc" on the surface of a fruit fly larva) is transformed into a series of tubular structures, such as leg segments in the adult insect. Since these authors base their analysis on tissue liquidity and elasticity, notions with which we are already familiar, we will start our discussion of the physics of epithelial sheets with this model. The Mittenthal–Mazo analysis can be summarized by the following points.

1. In the course of transformation of the two-dimensional leg imaginal disc into a series of elongated three-dimensional leg segments (which takes place with negligible cell division), the leg segments telescope out from the disc, leading to a progressive narrowing and elongation of the structure. (Note that the disc's thickness is the height of an epithelial cell.) As Mittenthal and Mazo noted, there are two limiting mechanisms that could contribute to the reshaping of an epithelial sheet such as the imaginal disc. In one mechanism, cells would deform without rearrangement as an elongated segment forms through the bulging out (i.e., evagination or eversion) of the sheet. In the second mechanism, the reshaping of the tissue takes place by the rearrangement of its cells; specifically, the cells exchange neighbors without changing their own shape. In the first case the epithelial sheet behaves as an elastic medium (Fig. 5.3, panel on the left), in the second case as a liquid (Fig. 5.3, panel on the right). The first basic assumption of the Mittenthal–Mazo model is that the epithelium of the imaginal disc, the hypodermis, is entirely liquid in the tissue plane, exactly as we have assumed in the Steinberg differential adhesion model in Chapter 4.

2. If the two-dimensional imaginal disc were an ideal liquid it would resemble a soap film. Certain experimental findings, however (Fristrom and Chihara, 1978), led Mittenthal and Mazo to make a second basic assumption: unlike ordinary two-dimensional liquids such as soap films, in which the two faces are mechanically equivalent, embryonic epithelia have a globally distributed elastic component that resists bending outside the plane of the tissue. As noted earlier in this chapter, the cells of epithelia are polarized and their basal surfaces adhere to basal laminae consisting largely of dense sheets of collagenous protein (Yurchenco and O'Rear, 1994). These planar structures provide a resistance to the bending of the tissue sheet. Mittenthal and Mazo thus treated the epithelium as a *fluid elastic shell*, fluid-like in its capacity for in-plane rearrangement of cells but resembling an elastic sheet when bending.

3. It was further assumed that the imaginal disc contains several epithelial cell types, cells of each type having different adhesive properties. When Steinberg's differential adhesion hypothesis for cell sorting in a binary mixture is applied to multiple cell types on a two-dimensional sheet, the predicted pattern is reminiscent of a planar bulls-eye target consisting of several concentric bands, as shown in Fig. 5.4A. In a two-dimensional system, this pattern is a straightforward consequence of the conditions discussed in connection with the complete envelopment of tissue *A* by tissue *B* (Fig. 5.2, left-hand panel).

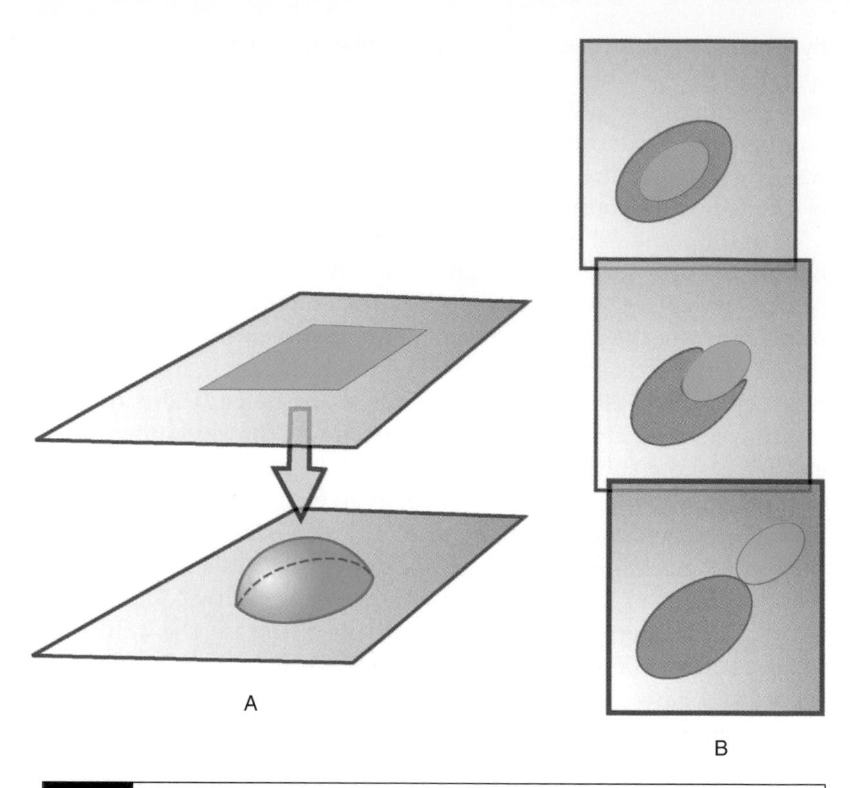

Fig. 5.3 Deformation of an epithelial sheet in the Mittenthal–Mazo model. (A) If a square patch of more cohesive tissue (e.g., imaginal disc tissue, shown as green) is grafted onto a less cohesive host substratum (e.g., the surface of a fruit fly larva, shown as pink), it will tend to minimize its contact region (i.e., perimeter) with the host. This can be accomplished by a rounding, or, if the patch is more elastic than liquid, by a bulging out from the host. In the course of bulging the individual cells in the patch must change their shape. (B) If two patches, which are more liquid than elastic, from different donor sites and thus of differing cohesive properties are grafted adjacent to each other at a third host site, they assume one of the configurations expected from the differential adhesion hypothesis for sorting: (top to bottom) complete envelopment, partial envelopment, or separation (compare with Fig. 5.2). In these configurations the individual cells may retain their original shape. (After Mittenthal and Mazo, 1983.)

Thus if the bands are numbered sequentially with numbers ($j = 1, 2, \ldots$) increasing toward the center, the bulls-eye pattern is stable if

$$\Delta_{j,j+1} > 0, \tag{5.5a}$$

$$w_{j+1,j} > w_{j,j}, \tag{5.5b}$$

$$w_{j,k} > w_{j+1,k} \quad \text{for} \quad k < j. \tag{5.5c}$$

Conditions 5.5a–c ensure, respectively, the stable segregation of cell types, the complete envelopment of each band by its outer neighbor, and the maintenance of a particular sequence of bands. (In Eq. 5.5a we have used the more general quantity $\Delta_{j,j+1}$ instead of the interfacial tension $\sigma_{j,j+1}$ in order to emphasize the mixed elastic–liquid nature of the epithelium.)

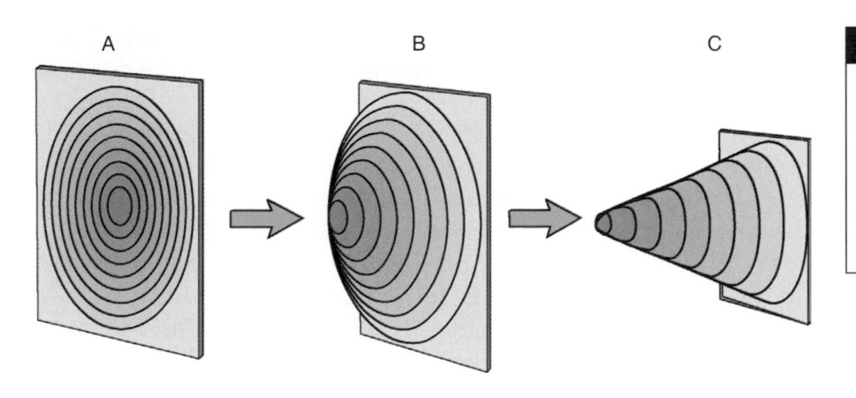

Fig. 5.4 Deformation of a planar bull's-eye target into a hollow cone. (A) Each annulus, proceeding toward the center, contains successively more cohesive tissue. (B, C) Successive stages of deformation. (After Mittenthal and Mazo, 1983.)

4. The bulls-eye target of bands can reduce its energy by deforming into a hollow tube, as shown in Fig. 5.4B, C. This shape change would occur because it reduces the area of contact between adjacent bands. But, because of its elasticity, the epithelium resists out-of-plane bending. Its final equilibrium shape, therefore, corresponds to the minimum of the total energy, which incorporates both the adhesive and elastic contributions.

5. The result of the interplay of forces described above is a "leg" consisting of a tubular epithelium arranged in a sequence of bands. The shape of this tube represents a compromise between an elongate shape that minimizes contact between bands (that is, maximizes the total work of adhesion) and a squat shape that minimizes the strain in the hypodermis (Fig. 5.5). The adhesion energy is expressed in terms of the interfacial energy Δ defined in Eq. 5.2. Since no *a priori* information exists on the number of adhesion bands in the imaginal disc, the authors assumed Δ to be a function varying continuously along the developing leg segment. The elastic energy can be obtained by adding the contributions for each band using the curvature energy given in Eq. 2.5. Again, instead of a finite number of bands, a continuous approximation was used. Minimization of the total energy predicted that each arthropod leg segment has a cylindrical shape of length L and radius R, and led to an experimentally testable relationship between these quantities,

$$\frac{1}{R} = a\frac{R}{L} + b. \tag{5.6}$$

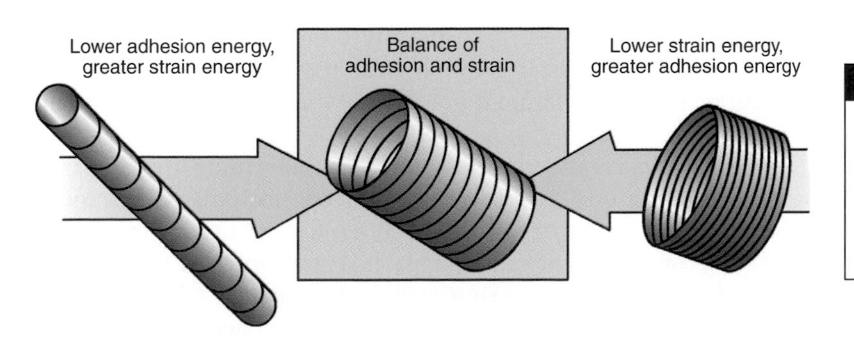

Fig. 5.5 Shaping of a leg segment in the Mittenthal–Mazo model. The middle panel represents a compromise between maximizing the work of adhesion and minimizing the strain. (After Mittenthal and Mazo, 1983.)

Here a and b are constants. This "scaling relationship" between L and R provides a relatively good fit to the measurements on the different leg segments in *Drosophila*.

As instructive as it is, the Mittenthal–Mazo model clearly neglects a number of considerations required for a satisfactory understanding of epithelial morphogenesis. In the model the imaginal disc is assumed to consist of bands of cells with differential adhesive properties, but the origin of this differential adhesion is not addressed. The model is based solely on the idea that the balance between strain and adhesion can determine the shape of the epithelium, which is treated as a continuous medium. Thus it ignores the possibility of active and passive shape changes of individual cells.

The following detailed description of certain features of gastrulation will indicate why more elaborate models of epithelial morphogenesis, which explicitly address some of the above issues, are required.

Gastrulation

The formation and rearrangement of distinct tissue layers during the establishment of the body plan in the early development of animals involves about half a dozen distinct kinds of cell behavior, only a subset of which occur in the embryo of any given species. Each set of movements that constitute gastrulation depends on a prior "pattern formation" event that designates a subpopulation of the blastula's cells as having a distinctive character relative to the others (see Chapter 7). Often, but not always, the distinctive character is an adhesive differential whereby the subpopulation is either more or less cohesive than the originating blastula cells. The pattern-formation process that brings differentially adhesive cells into a specific spatial arrangement in a pregastrula does not, in principle, require anything beyond straightforward physics. Even if these cells arise in a purely random fashion, cell sorting alone can cause them to localize in a single region of the embryo. This, indeed, is how the arrangement of cells that precedes morphogenesis is generated in the model of Mittenthal and Mazo (1983) discussed above. Gravity, another purely physical force, can also lead indirectly to a nonuniform pregastrulation arrangement of different cell subpopulations (Allaerts, 1991). In the *Xenopus* egg, for example, the sedimentation of dense yolk platelets creates a gradient that causes the cells that contain these platelets after cleavage (the "vegetal pole" cells) to have physical and biochemical properties that differ from the ("animal pole") cells that do not contain them (Gerhart *et al.*, 1981; Neff *et al.*, 1983, 1984) (see Chapter 7).

In most embryos, however, other pattern-forming mechanisms, some depending on the spatially asymmetric introduction of molecules into the forming egg by maternal tissues and others on the self-organizing properties of cell aggregates (based on their biochemical activities and ability to transmit signals; see Chapter 7), will guarantee that the cells of the late blastula are heterogeneous.

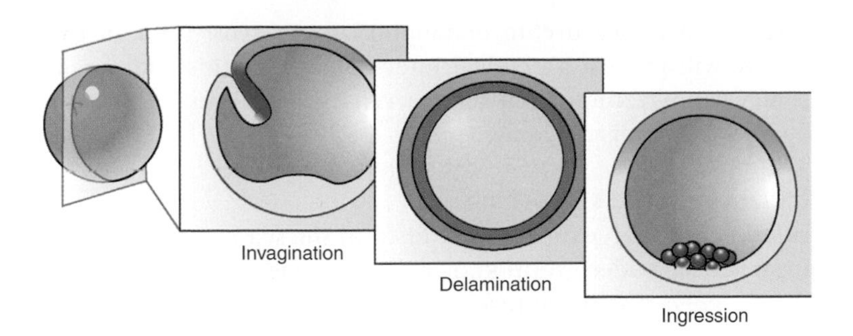

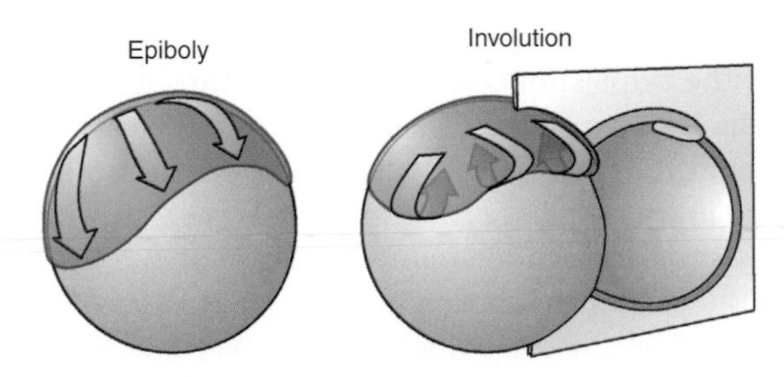

Fig. 5.6 Cell movements and tissue rearrangements involved in different forms of gastrulation. Most organisms employ a combination of one or more of these basic processes.

The basic categories of gastrulation-related cell rearrangements all follow from these initial regional differences in cell phenotype (Fig. 5.6). In *invagination*, a hollow single-layered embryo develops one or two additional layers by the folding inward of a portion of its surface under the guidance of a differentiated subpopulation of cells with distinct adhesive and/or motile properties. In *delamination*, a solid embryo that has thus far developed into a form with two distinct but attached cell layers acquires an internal space (shown in brown) by physical separation of the layers. In *ingression*, cells in a subpopulation having changed adhesive and/or motile properties move individually or in groups into the interior of a hollow embryo. In *epiboly*, one of the two embryo cell layers comes to partially envelop the other by actively spreading around it. In *involution*, cells move under the edge of an existing tissue layer, thereby forming a new layer.

Once gastrulation is under way, the layers of the now multilayered embryo are assigned specific names: the "ectoderm" refers to the outermost layer and the "endoderm" to the innermost. In diploblastic forms such as hydra and jelly fish, these two layers give rise to all subsequent differentiated tissues of the body. In vertebrates, for example, the ectoderm gives rise to the skin and nervous tissue and the endoderm gives rise to the lining of the intestine and its derivatives, such as the liver and pancreas. For triploblastic forms such as sea urchins and humans the movements of gastrulation result in a third layer, interpolated between the other two. This is called the "mesoderm," and in vertebrates is ultimately the source of the "middle" tissues of the body – the skeleton, blood and muscles.

Convergence and extension amount to a cell rearrangement, utilized in the gastrulation of some species, that encompasses several distinct cell activities. It involves the simultaneous narrowing and elongation of a tissue primordium, such as that initiated by the process of invagination in sea urchin or amphibian embryos. Later in this chapter we will present a physical interpretation of this widespread but somewhat puzzling (from both a cell-biological and a physical standpoint) tissue rearrangement.

Gastrulation in the sea urchin

Gastrulation has been extensively studied in the sea urchin, an invertebrate from the wider group known as echinoderms (Hardin, 1996; Ettensohn, 1999). The advantages of using this organism stem from the easy availability of experimental material (thousands of eggs are released from single females on stimulation and can be fertilized on demand) and the transparency of the embryo through many successive stages, which allows for easy visual access. In the early sea urchin blastula, at about the 60-cell stage, various pattern-forming events, involving multiple direct (rather than long-range) cell–cell interactions cued by preexisting molecular nonuniformities in the uncleaved zygote, lead to the generation (not irreversible at this point) of radially symmetric tiers of cells with distinct cell fates (Fig. 5.7). By the 128-cell stage the embryo has formed a single-layered blastula with an internal cavity, the blastocoel (see Chapter 2). Several cell cycles later (by the ninth or tenth cycle, depending on the species) the blastula hatches out of its confining fertilization envelope and, as a hollow ball consisting of about 1000 cells, is ready to begin gastrulation. The multiple cell types arising from these local interactions represent alternative states of gene regulation produced by autoregulatory networks of transcription factors like those discussed in Chapter 3 (Davidson, E. H., 2001; Davidson, E. H., *et al.*, 2002; Oliveri *et al.*, 2002).

Once hatched from its fertilization envelope, the spherical blastula undergoes a series of morphological changes comprising gastrulation. First the cells of its base, or "vegetal," end begin to thicken and flatten. Next, as a result of reduced adhesion to neighboring cells and increased affinity for the basal lamina lining the blastocoel (Wessel and McClay, 1987), a small population of cells at the center of this vegetal plate, the primary mesenchyme cells, begin to ingress into the blastocoel and extend long filopodia, which attach to the inner surface of the blastula. The branched structures formed by the primary mesenchyme cells, their filopodia, and their secreted extracellular matrix eventually become the skeleton of the sea urchin larva (Fig. 5.7).

The remaining cells of the vegetal plate then rearrange to fill in the gaps left by the ingressing primary mesenchyme (Fig. 5.7). A series of biosynthetic changes in this region cause both the inner and outer layers of the extracellular matrix to change their physical properties and, in consequence, the plate invaginates about one-fourth to one-half the way into the blastocoel, by a process that has been likened to the thermal bending of a bimetallic strip (Lane *et al.*, 1993).

In the following sections we describe simplified quantitative models for aspects of sea urchin gastrulation. Our objective is to illustrate how elementary physical considerations can lead to the observed complex cellular pattern. The combination of the various models in

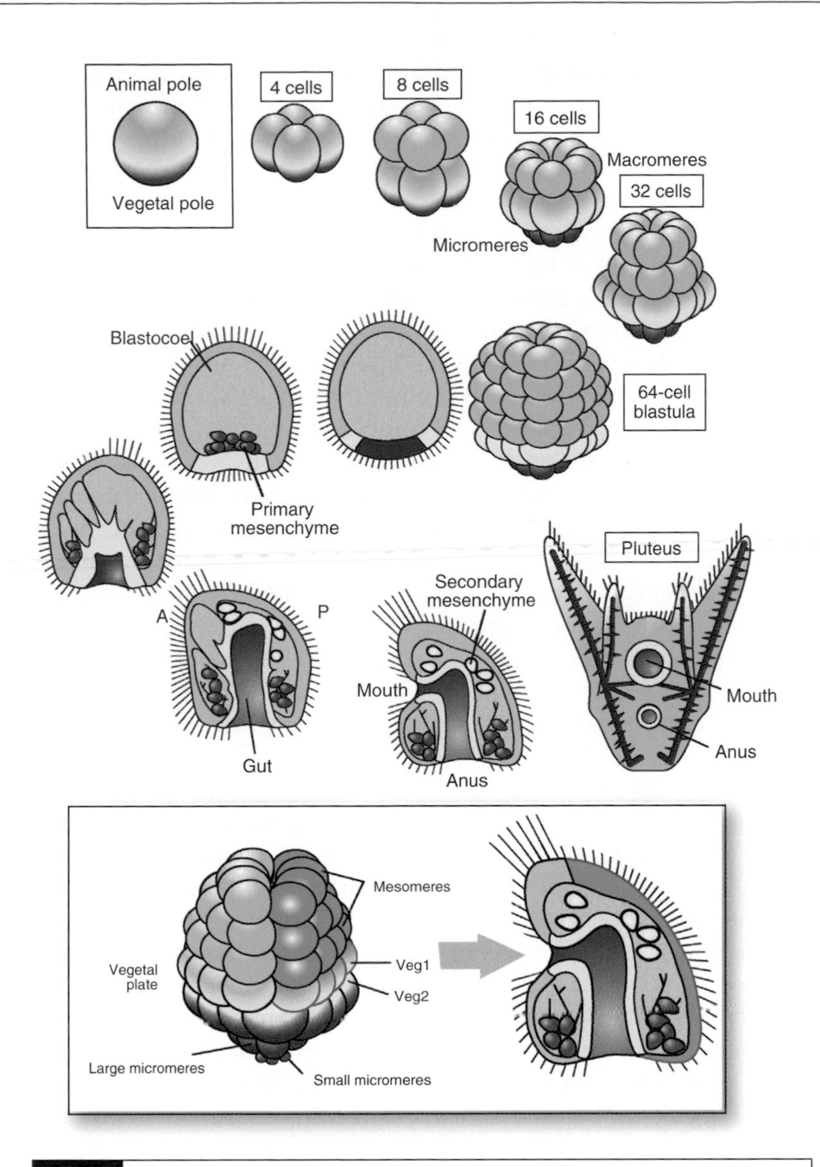

Fig. 5.7 Gastrulation in the sea urchin embryo, leading to the formation of the swimming larva, termed a "pluteus." The cell and tissue rearrangements that occur at the various stages are described in the text. Cells indicated by particular colors at early stages map into structures indicated by the same colors, at later stages. The lower box depicts a "fate map" in which the origin of tissues in the larva are traced back to specific cell populations at the 64-cell stage. (After Wolpert, 2002.)

conjunction with additional experimental details should eventually provide a biologically realistic description.

Modeling sea urchin gastrulation: an approach based on energy minimization

In Chapter 2 we presented a model of blastula formation in the sea urchin embryo (Drasdo and Forgacs, 2000) based on energy considerations. The model incorporated several key energy contributions characteristic of the developing embryo, and the resulting cellular pattern corresponded to the minimum of the total energy. In the description of development from the zygote to the hollow spheroidal blastula, the latter eventually became unstable and its folding unpredictable. The folding inwards corresponds to the appropriate biological outcome – invagination of the vegetal plate. The folding outward, in contrast, corresponds to "exogastrulation," which experiments have shown to be also in the physical repertoire of the living embryo (Hoshi, 1979; Kamei *et al.*, 2000). The origin of this instability is easy to understand. The growth and division of cells require that either individual cells or entire cell layers (epithelial sheets) be capable of migrating. Since in the simulation the irregular active motion of each cell within the layer was mimicked by random displacement, small stochastic differences in cell translocation must exist. These lead to undulations in the shape of the sheet. Such undulations are unfavorable because they increase the bending energy of the sheet and, as long as they remain small, are eliminated (since the Monte Carlo protocol is designed to drive the system toward lower energies). When cell proliferation is introduced, both the number and the extent of the undulations increase. Eventually these become too large to be controlled by the bending energy: the spheroidal cell arrangement ceases to correspond to the lowest energy configuration and folding takes place.

In the energy-minimization model, the direction of folding of the blastula, i.e., whether normal gastrulation or exogastrulation takes place, is subject to the influence of alterations in the local physical parameters of the embryo. Recent experiments have suggested that the suppression of the tendency to exogastrulate is under genetic control (Kamei *et al.*, 2000). One way in which this might be implemented in the Drasdo–Forgacs model is through the assumption that the increase in bending energy leads to local changes in gene activity along the blastula surface, which in turn result in the modification of the cells' physical properties. The authors thus postulated that a preexisting nonuniformity in the distribution of one or more gene products causes the sheet of cells near the vegetal pole to acquire nonzero spontaneous curvature (see the discussion following Eq. 2.5). This was sufficient to account for the observation, described above, that near the vegetal pole a distinct group of primary mesenchymal cells ingresses into the blastocoel shortly after the formation of the blastula. "Snapshots" of the developing system governed by the model, as implemented through Monte Carlo simulations, are shown in Fig. 5.8.

The Drasdo–Forgacs model describes morphogenesis in the sea urchin embryo from the first cleavage until the completion of gastrulation. It provides an explicit example of how an important set of morphological changes in early development can potentially be accounted for by an interplay between genetic and generic physical mechanisms. The most serious limitation of the model is that no information exists at this point on how to relate changes in the value of spontaneous curvature to the specific gene activity accompanying

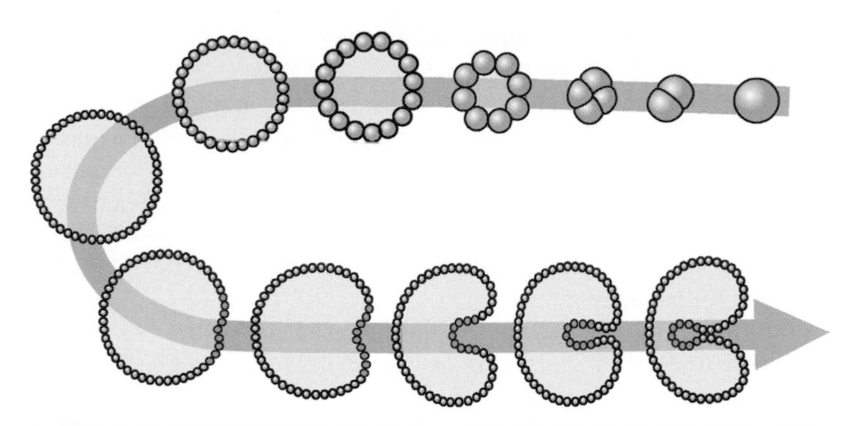

Fig. 5.8 A representation of early development in the Drasdo–Forgacs model. Events in the upper row illustrate the first cleavages leading to blastula formation and were discussed in Chapter 2. The lower row shows the simulation of gastrulation. The nine cells pictured in green develop a spontaneous curvature different from that of the remaining cells. It is this change that drives invagination in the model. Note that these simulations are performed with constant cell number, corresponding to the fact that cell division halts during gastrulation in the sea urchin. (After Drasdo and Forgacs, 2000.)

the onset of gastrulation (even though most of the other parameters have been measured, Davidson, L. A., *et al.*, 1999; see Chapter 2). The Drasdo–Forgacs model, similarly to the Mittenthal–Mazo model, describes morphogenetic transformations as processes that generate global equilibrium shapes corresponding to the minimum of some energy expression containing competing contributions. It is implicitly assumed that the system can arrive at these energy minima (instead of being locked into long-lived metastable states), a hypothesis that requires experimental verification. Other characteristic cell and tissue properties underlying gastrulation, such as excitability or shape transformations, have also been neglected so far and will be incorporated in the model presented below.

Modeling sea urchin gastrulation: an approach based on force balance

Cells of the developing embryo exert forces on each other. In doing so they undergo changes in both position and shape. Models of epithelial morphogenesis that are based on energy minimization, such as those of Mittenthal and Mazo (1983) and of Drasdo and Forgacs (2000), typically track pattern development by postulating changes in cell arrangement and comparing the energies of the original and modified patterns. The system reaches equilibrium when it arrives at the global energy minimum. Such models therefore do not consider the origin of cell-shape modifications explicitly.

According to Newton's second law, at equilibrium all the forces and all the torques acting in a system must balance at each point. This law also specifies how each volume element of the system must move to reach equilibrium. Thus combining such an approach with modern techniques that allow the following of individual cells during development (Czirok *et al.*, 2002; Kulesa and Fraser, 2000, 2002) may provide further insight into the mechanisms of early morphogenesis. In a series of papers, Oster and coworkers (Odell *et al.*, 1981; Davidson, L. A., *et al.*, 1995, 1999) used this approach to distinguish between several potential mechanisms of primary invagination in the sea urchin embryo. Here we focus on their model, based on apical constriction (Odell *et al.*, 1981), the major assumptions and properties of which are the following.

1. The cells making up the epithelial sheet of the sea urchin blastula undergo a shortening of their apical circumference by the contractile activity of actin filaments just beneath their plasma membranes (Fig. 5.9A). This constriction proceeds in a "purse-string" manner along the blastula similarly to the progression of the cleavage furrow in cytokinesis (see Fig. 2.9) and must overcome the intracellular viscous forces and the tractions exerted by neighboring cells. To model these viscoelastic forces, we represent each face and internal diagonal of a cell by a viscoelastic element (Fig. 5.9B). The linear characteristics, denoted by L, such as the circumference of a face or the length of a diagonal vary according to Newton's second law as follows:

$$m\frac{d^2L}{dt^2} = -k(L - L_0) - f\frac{dL}{dt} + F_{\text{load}}. \tag{5.7}$$

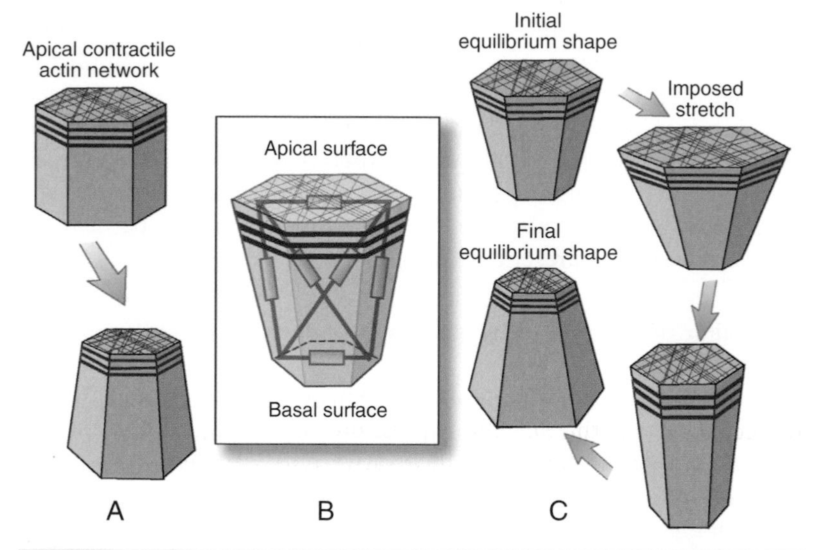

Fig. 5.9 Assumptions concerning the mechanism of cell-shape change in the apical constriction model of gastrulation. (A) Schematic representation of the contractile apical actin network. (B) The mechanical analogue of a trapezoidal cell in the blastula. External faces and internal diagonals are represented by viscoelastic elements, which are indicated by connected small rectangles. The apical element is excitable and thus differs from the others. (C) Illustration of the excitability of the apical surface: depending on the magnitude of its deformation it is capable of changing its equilibrium geometry. (After Odell *et al.*, 1981.)

Here the term on the left-hand side is the inertial force (mass times acceleration), whereas the first and second term on the right-hand side represent, respectively, the elastic and viscous restoring forces acting on a face or along the internal diagonal of a cell. L_0 is the equilibrium circumference of a face or the equilibrium length of an internal diagonal and k and f are material parameters characterizing the elastic and viscous properties (see Chapter 1, Eqs. 1.5 and 1.7). The third term on the right-hand side is the force exerted by the neighboring cells, and m is the net mass moved due to the change in L. The diagonal elements model the cells' internal viscoelastic properties and thus, in the corresponding equations, $F_{load} = 0$.

2. Equation 5.7 is valid for each face and internal diagonal. The apical face, however, is special in that the underlying contractile filaments constitute an active, excitable, system. If an apical fiber is stretched by a small amount (which happens when the vegetal plate starts flattening) it behaves as an elastic material: upon release of the stretch it returns to its original length (L_0). If, however, the stretch exceeds a critical value, it elicits an active response: the contractile system "fires" and does not return to its original length (Beloussov, 1998): it freezes in a new contracted state, with a changed equilibrium length resulting in an apical surface area smaller than before (Fig. 5.9C). Thus the apical viscoelastic element differs from the others (Fig. 5.9B) and the corresponding version of Eq. 5.7 has to be supplemented by another equation describing the variation of the equilibrium length (see below).

3. Inertial forces are insignificant. As discussed in Chapter 1, dimensional analysis indicates that for typical embryonic processes involving the motion and shape changes of cells such forces can be neglected. Thus Eq. 5.7 simplifies, and the evolution of the apical surface of an isolated cell can be described by the following equations:

$$\frac{dL}{dt} = -\frac{k}{f}(L - L_0), \tag{5.8a}$$

$$\frac{dL_0}{dt} = \frac{a}{\tau}\left(\frac{1}{16} - \frac{LL_0}{a^2}\right)\left[\frac{(4L/a - 3)^2}{2} + \frac{(4L_0/a - 3)^2}{2} - 1\right]. \tag{5.8b}$$

Here a and τ are positive constants (with units of length and time respectively) and the specific expression on the right-hand side of Eq. 5.8b has been chosen for illustrative purposes. As discussed by Odell *et al.* (1981), the system's behavior can be derived from certain of its qualitative features. Thus, if L_0 is a constant (i.e., $dL_0/dt = 0$) then a stretched apical face eventually returns to its original size. (The solution of Eq. 5.8a is $L = L_0 + (L_i - L_0)e^{-kt/\mu}$, where L_i is the initial circumference (at $t = 0$) of the stretched surface.) However, if L_0 varies with time then Eqs. 5.8a, b represent a simple dynamical system like those discussed in Chapter 3. It has two stable fixed points, $L = L_0 = a$ and $L = L_0 = a/4$, at which the left-hand sides of Eqs. 5.8a, b vanish; small deviations from these stationary points return the system to the same points. We can identify the untriggered equilibrium length with $L_0 = a$ and the rest length after firing with $L_0 = a/4$. A separatrix that divides the space in the LL_0-plane into the basins of attraction of the stable fixed points (Chapter 3) passes through the point $L = L_0 = a/2$, which is an unstable fixed point of the system defined in Eqs. 5.8a, b. This example illustrates how an excitable biological material can be modeled by a dynamical system. In the Appendix at the end of this chapter we use Eqs. 5.8a, b to demonstrate how the mathematical method of linear stability analysis can be employed to study the fixed-point structure of dynamical systems defined by differential equations.

4. Taking advantage of the near-spherical symmetry of the sea urchin blastula, Odell and coworkers (1981) performed their calculations in two dimensions. Thus, L in Eqs. 5.7 and 5.8a, b denotes the linear sizes (e.g., height or diagonals) of trapezoidal cells that are initially arranged along the perimeter of a circle, similarly to the Drasdo–Forgacs model (Fig. 5.8). Force and torque equilibrium is assumed along each face and corner of the initially tension-free cells.

Invagination is initiated in the model by the firing of a single cell at the middle of the vegetal pole. The apical contraction of this cell dilates the apical surface of its neighbors, an effect which, if sufficiently large, evokes their firing and subsequent apical contraction. This sequence of events leads to a spreading wave of contraction, which eventually generates an invagination in the cell layer, as illustrated in Fig. 5.10. The buckling of the vegetal plate, in this model, crucially depends on the unique excitable nature of these cells.

In contrast with the Drasdo–Forgacs model, on the one hand the model of Odell *et al.* postulates no biological basis for the unique behavior of vegetal plate cells. Its starting configuration is the spherical blastula, which it treats in isolation. It does not consider the preceding developmental processes. In particular, it does not address the issue of the folding instability disclosed by the energy-minimization-based Monte Carlo analysis. On the other hand, the dynamical nature of the model of Odell *et al.* permits it to capture certain characteristic shape changes of cells and tissue sheets associated with invagination. As can be seen, the various approaches to modeling sea urchin gastrulation by their very nature have restricted applicability. They can be viewed, however, as complementary and their eventual combination may provide a more comprehensive account of morphogenetic processes.

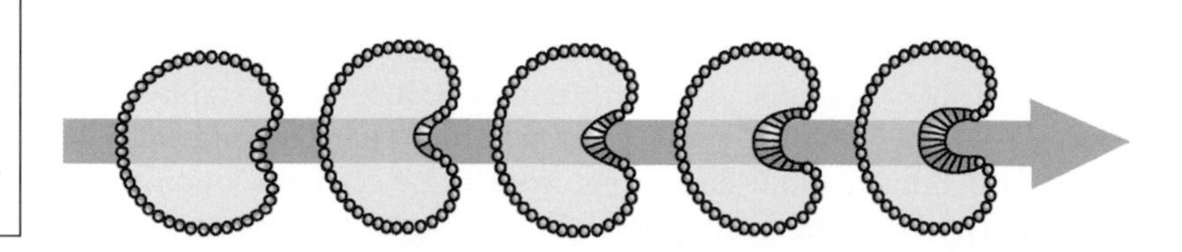

Fig. 5.10 Representation of gastrulation in the model of Odell and coworkers. Here invagination is triggered by changes in cell shape resulting from the excitable nature of the apical contractile actin network (compare with the Drasdo–Forgacs model in Fig. 5.8). (After Odell *et al.*, 1981.)

Convergence and extension

During the subsequent course of sea urchin development the primitive gut or "archenteron" formed by the invaginating blastula wall elongates and narrows. This is an example of the phenomenon of convergence and extension mentioned at the beginning of this chapter. This effect occurs across a wide range of animal phyla. In insects, for example, the surface ectodermal cells of the gastrula, along with the underlying mesoderm, elongate posteriorly in a process termed "germ-band extension." In chordates, the wider group of organisms that includes vertebrates, the forming mesoderm separates into two distinct subtissues, the broad somitic mesoderm that gives rise to the backbone and musculature and the notochord, a stiff central rod of mesoderm that first defines the axial skeleton and then induces the formation of the vertebral column and spinal cord. Both of these primordia also undergo convergence and extension (Keller *et al.*, 2000; Keller, 2002; see Fig. 5.11A).

On the basis of our earlier discussion of tissue behavior, these characteristic movements of late gastrulation might seem physically counterintuitive. In Chapter 4, for example, we saw how the liquid-like properties of tissues led to their "rounding up" to attain minimum surface-to-volume ratios. Because tissues undergoing convergence and extension violate this expectation, it was at one time thought that their movements might be dependent on external forces from adhesive substrata provided by adjoining tissues in the embryo. This was proved incorrect by experiments in which the dorsal sectors of two frog gastrula (that is, the back regions, containing the forming mesoderm) were cut out and their inner, deep-cell, surfaces sandwiched together. The composite tissue converged and extended purely on the basis of internally generated forces, with no mechanical or adhesive assistance from the rest of the embryo (Keller *et al.*, 1985; Keller and Danilchik, 1988).

What then, is the origin of these shaping forces? Some hints can be gained from the models described earlier in this chapter. In the Mittenthal–Mazo model, for example, epithelial tissues, using the passive mechanisms of differential adhesion and elastic response, bulged out and elongated into cones and tubes, reminiscent of convergent extension rather than rounding up. This effect arose because the cell sheet had a nonuniform distribution of cell adhesivity, a property presumed to depend on the active, nonequilibrium, processes of cell differentiation discussed in Chapter 3. However, because the frog mesoderm studied by Keller and coworkers consists of an essentially uniform population of cells this "adhesive prepattern" mechanism cannot by itself account for the classic cases of convergence and extension.

The model of Odell and coworkers (1981) accounted for non-passive tissue behaviors by invoking the dynamical properties of individual cells, such as apical constriction and mechanical excitability. That

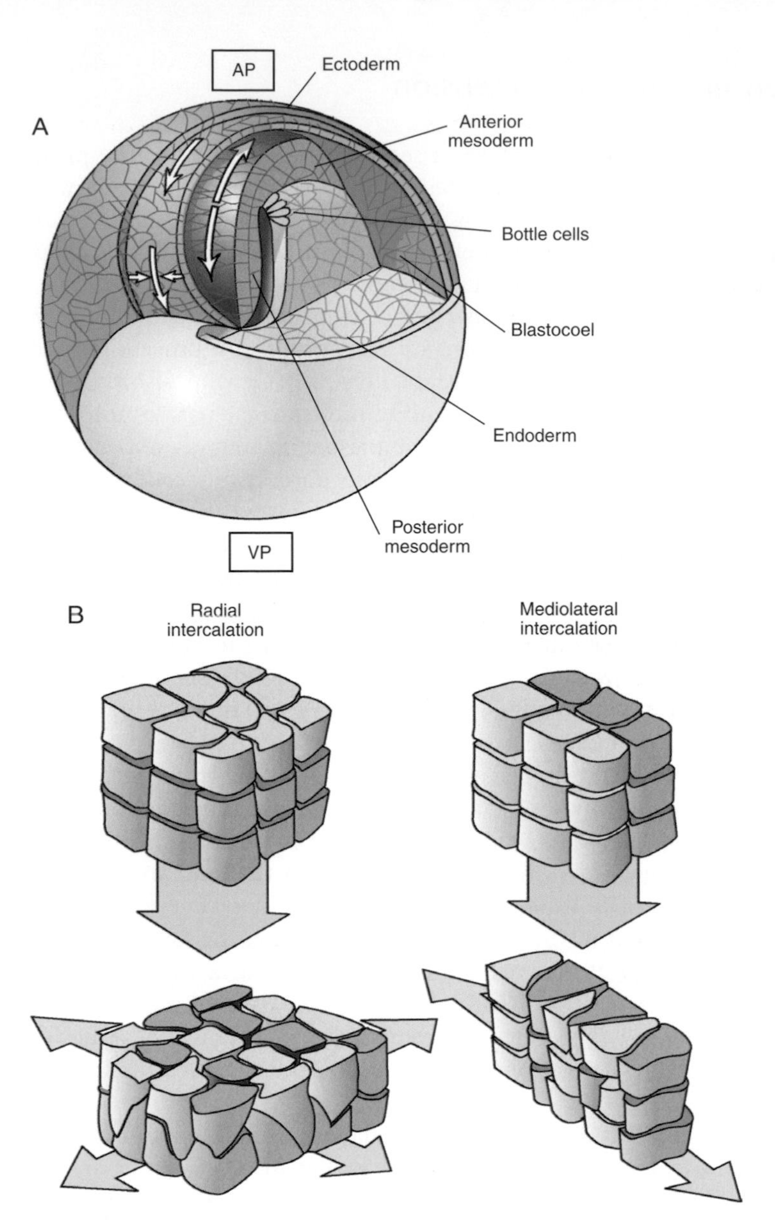

Fig. 5.11 Convergence and extension movements during gastrulation in *Xenopus*. (A) A view of the dorsal aspect of the gastrulating embryo. Both the mesodermal cells (red) and the ectodermal cells (blue) of the anteroposterior body axis first undergo *radial intercalation* (B, on the left, not shown in A), the rearrangement of several layers of deep cells along the radius of the embryo (normal to the surface) to form tissue masses having fewer cell layers; next, these same tissues undergo *mediolateral intercalation*. AP, animal pole; VP, vegetal pole. (B) On the right, the rearrangement of multiple rows of cells along the mediolateral axis (indicated by the small horizontal arrows in A) to form narrower tissue masses that are elongated along the anteroposterior axis (indicated by the long arrows in A and B). (A, after Gilbert, 2003; B, after Keller *et al.*, 2000.)

such dynamical properties might enter into the physics of convergence and extension is suggested by the fact that cells within the epithelial sheets undergoing these shape changes (Fig. 5.11A) manifest intercalation (Keller, 2002) in which they interdigitate among one another, first along the radius of the embryo, normal to its surface (Fig. 5.11B), and then along its mediolateral axis (e.g., from the center to the edges) (Fig. 5.11B) to produce a narrower, longer, and thicker array (Keller *et al.* 1989; Shih & Keller, 1992).

Recent evidence indicates that these intercalation movements depend on a dynamical property of cells termed "planar polarity" (Mlodzik, 2002). This phenomenon differs from the longer-known apical–basolateral polarity described earlier in this chapter. Planar polarization is a *dynamical* response of cells to extracellular effectors and homophilic adhesive interactions. In particular, it leads, via an intracellular signaling cascade, to cytoskeletal rearrangements and consequent elongation and flattening (Mlodzik, 2002), activities not specifically associated with apical–basal polarity. Once the planar polarization response "fires," the nonuniform localization of the intracellular components of several signaling pathways means that the cells acquire differing adhesive properties on their different surfaces. Zajac and coworkers (Zajac *et al.*, 2000, 2003) have developed a physical model for convergence and extension that makes use of precisely such triggered anisotropies in cell properties.

The Zajac–Jones–Glazier model of convergent extension

As described above, convergent extension is a morphogenetic process that transforms an epithelial sheet composed of cuboidal cells into a tissue that contains a more or less ordered array of elongated cells and has a characteristic shape, with the length along one direction considerably greater than that along the others (Fig. 5.12). Zajac and coworkers (2000, 2003) constructed a model for a cell population undergoing convergent extension based on differential adhesion. In this model, the cellular pattern arrived at through convergent extension corresponds to the energetically most favorable configuration of the system. The model is based on the following assumptions.

- **(i)** During convergent extension, cell division is negligible and the volume of individual cells is unchanged.
- **(ii)** Cells form a closely packed array with no internal empty spaces (Fig. 5.12).
- **(iii)** The original cuboidal cells (Fig. 5.12A) are "triggered" into elongated shapes by the acquisition of planar polarity mentioned earlier, with the result that differential adhesion occurs along the various faces (Fig. 5.12B).

We first show that under these assumptions the elongated cells prefer to arrange themselves in an ordered state in which they preferentially adhere along their similarly deformed surfaces (i.e., elongated–elongated and narrowed–narrowed, Fig. 5.12C). Since convergent extension can be manifested in the behavior of a cell sheet,

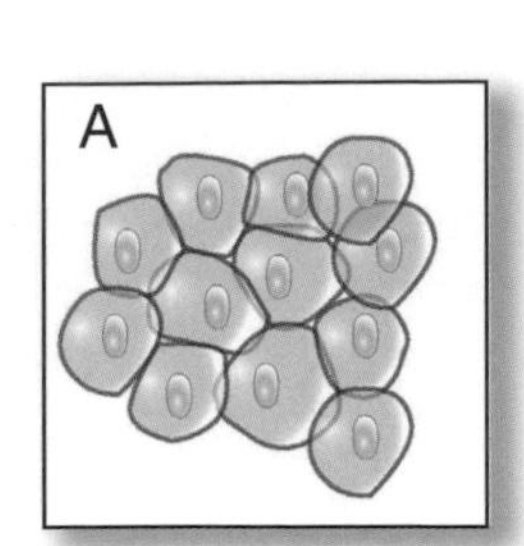

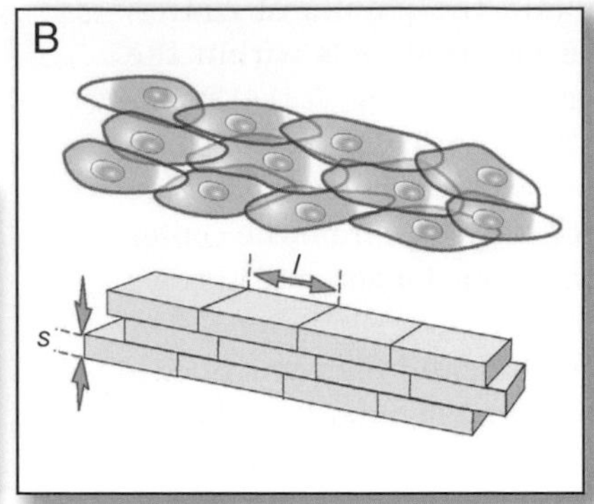

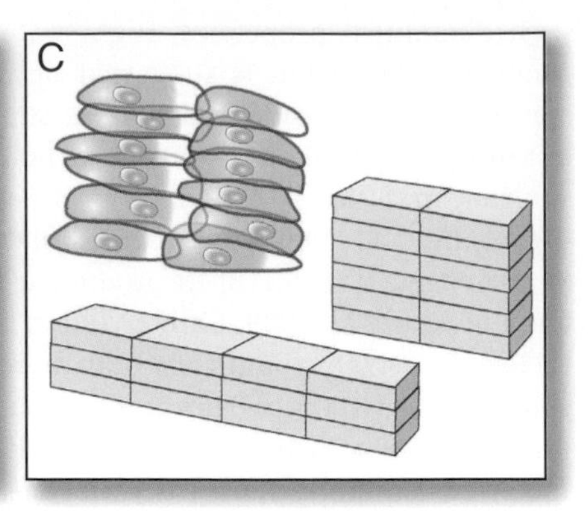

Fig. 5.12 Representation of convergent extension in the model of Zajac and coworkers. (A) Epithelial sheet, made of cuboidal cells, before the onset of convergent extension. (B) Top: Schematic representation of the epithelial sheet after the cells have elongated, become aligned, and begun to intercalate. Bottom: Representation of intercalation in the two-dimensional plane spanned by axes along the long sides *l* and the short sides *s* of rectangular model cells. (C) Top: Schematic representation of the cellular arrangement after the completion of convergent extension. Tissue elongation has occurred in the direction perpendicular to the long axes of the cells. Bottom and right: Two possible ordered cellular arrays predicted by the model. Under rather general conditions (see the main text) it is the taller pattern, on the right (resembling the cellular arrangement that results from convergent extension), that corresponds to the lowest energy configuration. (After Zajac *et al.*, 2000.)

we will construct the model in two dimensions. According to assumption 3, each elongated cell is represented by a rectangle with its longer and shorter sides denoted by l and s, respectively (Fig. 5.12B; the depth of the cells, perpendicular to the figure is added for illustrative purposes). Adjoining cells may in principle contact each other along their lengths (ll contact), along their width (ss contact), or form a mixed interface (ls contact). We denote the total contact length of each type as L_{ll}, L_{ss} and L_{ls}. The total contact length between the l and s sides and the surrounding medium at the boundaries is represented, respectively, by S_l and S_s. For N cells in the array the total contour length along the l and s sides (each cell having two l sides and two s sides) is $2Nl$ and $2Ns$, respectively, which can be expressed in terms of the above quantities as

$$2Nl = 2L_{ll} + L_{ls} + S_l, \tag{5.9a}$$

$$2Ns = 2L_{ss} + L_{ls} + S_s. \tag{5.9b}$$

If we denote the works of adhesion along the ll, ss, and ls contacts respectively by w_{ll}, w_{ss}, and w_{ls} (here they are energies per unit length, not per unit area as in Eq. 5.2), the total work necessary to disassemble an array with a given configuration (specified by the values of L_{ll}, L_{ss}, and L_{ls}) is $W = L_{ll}w_{ll} + L_{ss}w_{ss} + L_{ls}w_{ls}$. The most stable configuration clearly corresponds to that set of L_{ll}, L_{ss}, and L_{ls} which maximizes W. Using Eqs. 5.9a, b we eliminate L_{ll} and L_{ss} from W and arrive at

$$W = -L_{ls}\Delta_{ls} + N(lw_{ll} + sw_{ss}) - \tfrac{1}{2}(S_l w_{ll} + S_s w_{ss}). \tag{5.10}$$

where $\Delta_{ls} = \frac{1}{2}(w_{ll} + w_{ss}) - w_{ls}$ is the work needed to create a contact of unit length between the long and short sides of two cells, a quantity analogous to the interfacial energy introduced in Eq. 5.2. For close-packed cellular arrays (such as those shown schematically in Fig. 5.12A, B, C) the boundary lengths are proportional to $\sqrt{N}$; thus for large N the third term in the above equation is negligible. Equation 5.10 then reduces to $W = -L_{ls}\Delta_{ls} + C$, where C is a constant (it denotes the second term on the right-hand side of Eq. 5.10, which contains only fixed model parameters). Finally, assuming that $\Delta_{ls} > 0$, W is maximal if $L_{ls} = 0$, which corresponds to the case of ordered configurations of rectangular cells such as those shown in Fig. 5.12C.

The various works of adhesion clearly are related to the densities of binding sites c_l and c_s for the l and s sides respectively, and (since an adhesive bond contains two adhesion molecules) can be expressed as $w_{ll} = c_l c_l$, $w_{ss} = c_s c_s$, and $w_{sl} = c_l c_s$. (To make these equations dimensionally correct, a factor with units of energy times length should be incorporated into each one.) Using these forms we obtain $\Delta_{ls} = \frac{1}{2}(c_l - c_s)^2 > 0$. Thus ordered arrays correspond to energetically favorable cellular patterns under rather general conditions. (Note that the above conclusions are valid for any non-equal values of c_l and c_s.)

In Fig. 5.12C two different ordered rectangular arrays, each containing the same number of cells, are shown. To analyze the difference between their stability we have to return to Eq. 5.10 and consider the last term on the right-hand side. Since the arrays contain a finite number of cells (12), N cannot be assumed to be arbitrarily large, therefore this term cannot now be ignored. For the two ordered rectangular arrays shown in Fig. 5.12C, we have $(N_l/2) \times (N_s/2) = N$, where N_l and N_s are the numbers of cells on the array boundaries along the directions of elongation and narrowing respectively. Since $L_{ls} = 0$ for both configurations, we now have to find the maximum value of $W = -\frac{1}{2}(S_l w_{ll} + S_s w_{ss}) + C$, or equivalently the minimum of $S_l w_{ll} + S_s w_{ss} = Z$ with $S_l = N_l l$ and $S_s = N_s s$, for ordered rectangular arrays. ($N_l = 8$ and 4, $N_s = 6$ and 12, respectively for the 3×4 and 6×2 ordered arrays in Fig. 5.14C.) The minimization of Z as a function of the boundary lengths then leads to the relationship $S_s/S_l = w_{ll}/w_{ss}$. (The mathematically astute reader will notice that minimization is carried out by first expressing Z in terms of N_l (or N_s) only by using the above relationship between N_l, N_s, and N, and then equating the derivative of Z with respect to N_l (or N_s) to zero.) If we now employ the fact that $l > s$ (we have not used this anywhere up to now) then we may safely assume that $w_{ll} > w_{ss}$. The above minimization procedure then leads to $S_s > S_l$, which corresponds to the 6×2 array. This resembles the array that results from convergent extension.

This model of convergent extension suggests that differential adhesion may underlie morphogenetic processes whose outcomes appear different from what would normally be expected from the classic differential adhesion hypothesis. It should be noted however, that (in contrast with the explanations given for tissue rounding-up or fragment sorting) to account for convergent extension the model of Zajac and coworkers had to combine differential adhesion with an active, "nongeneric," property of living cells: their ability to undergo planar polarization.

Neurulation

In animals with backbones, convergent extension occurs in several tissue primordia along the primary embryonic axis. As noted above, this axis is first defined by the coordinated convergence and extension of the broad somitic mesoderm and central notochord (Fig. 5.11A). Once this has occurred, the ectoderm overlying the notochord, induced by signaling molecules arising from the latter, flattens and thickens. This *neural plate* then undergoes a further set of morphogenetic movements, including convergence and extension of its own and also the elevation of two ridges to either side of the underlying notochord. The portion of the plate between the ridges sinks downward and the ridges fuse at their outermost regions. This sealing of the dorsal (back) surface of the embryo causes the neural plate to assume the form of a hollow cylinder – the neural tube – between the surface ectoderm and the notochord (Fig. 5.13). As noted at the beginning of this chapter, the neural tube is the primordium of the spinal cord of the mature animal, and this entire sequence of events is referred to as neurulation (see Colas and Schoenwolf, 2001, for a comprehensive review).

Neurulation is a process of differentiation (see Chapter 3) and pattern formation (see Chapter 7). The spinal cord arises as a distinct morphological structure that also consists of a distinct population of cells – neurons, in a particular arrangement. The generation of this new cell type is intimately tied to the change in shape and form of the neural plate. During neurulation the interplay of activators (termed "proneural" gene products) and inhibitors of neurogenesis determine precisely which cells of the dorsal ectoderm (the majority of which are capable of adopting a neuronal fate) differentiate into neurons and which adopt different fates (Brunet and Ghysen, 1999; Bertrand *et al.*, 2002).

Several of the morphogenetic phenomena in epithelial sheets described earlier are utilized during neurulation. Other neurulation-related changes are novel, though potentially explicable on the basis of familiar physical principles. The cells of the neural plate, for instance, undergo an elongation of their apical–basal axes, which can be understood by differential adhesion. The cells adhere on their basal surfaces to an underlying basal lamina. Their initially cuboidal shape indicates that the adhesion forces between the cells and the basal lamina are comparable with the homotypic forces between the cells on their lateral aspects.

Induction of the neural plate by the underlying tissues can lead to a thickening of the epithelium by one of several alternative mechanisms. The induced epithelial cells could increase the effective

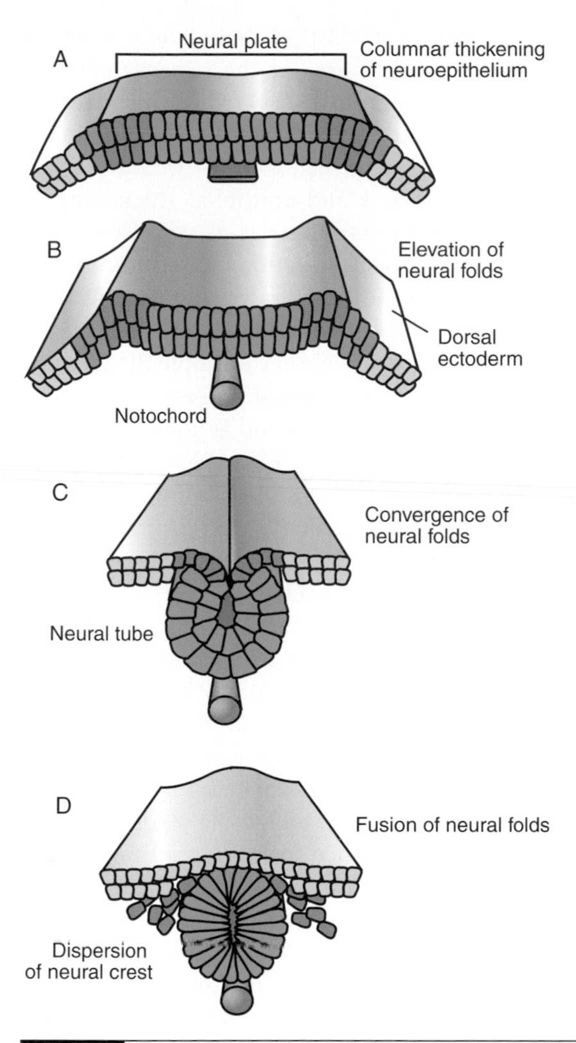

Fig. 5.13 Schematic representation, viewed as a cross-section perpendicular to the anteroposterior axis, of the successive morphogenetic changes involved in neurulation in some vertebrate species. (A) Cuboidal cells of the dorsal ectoderm along the midline of the embryo (which is elongating in the anteroposterior direction, see Fig. 5.11A) become columnar, forming the neural plate. Cells lateral to the neuroepithelium composing this plate remain cuboidal. (B) While basal portions of the central deep cells of the neural plate remain attached to the notochord (which is derived from the axial mesoderm, see Fig. 5.11A), the lateral regions of the neural plate become elevated in two parallel ridges, the neural folds. (C) The neural folds begin to converge, meeting each other at a line parallel to the notochord. As this occurs, the underlying epithelium invaginates to form a cylinder, the neural tube, which begins to pinch off from the overlying neuroepithelium. (The neural tube will subsequently develop into the spinal cord.) (D) With complete fusion of the surface ectoderm, the neural tube becomes a separate structure. Cells originally at the crests of the neural folds (the "neural crest") end up between the surface ectoderm and the neural tube. These cells detach from both structures and disperse through specific pathways as a mesenchymal population (see Chapter 6). The formation of the neural tube occurs in the fashion described along much of the central body axis of birds and mammals. In some regions of the axis of these organisms, and in other forms such as fish, the mechanics of neural tube formation is somewhat different (Colas and Schoenwolf, 2001; Lowery and Sive, 2004). (After Wallingford and Harland, 2002.)

number of cell adhesion molecules (CAMs) on their lateral surfaces and thus increase their lateral surface area at the expense of their common interface with the basal lamina. Alternatively, they may decrease the effective number of substratum adhesion molecules (SAMs) on their basal surfaces. Either mechanism (or both) would lead to elongation along the apical–basal axis and epithelial thickening, a phenomenon utilized during later organogenesis as well (see Chapter 8) and known generally as "placode formation." CAMs and SAMs, such as cadherins and integrins, are differentially regulated as neurulation begins (Levi *et al.*, 1991; Espeseth *et al.*, 1995; Joos *et al.*, 1995; Lallier *et al.*, 1996) but it is not yet clear whether a simple differential adhesion mechanism, or some more complicated process (e.g., one involving polarized microtubule assembly, Colas and Schoenwolf, 2001), drives neural plate formation.

The morphogenetic events that follow neural plate formation are: (i) convergent extension, which is mechanically independent of the convergent extension of the underlying mesodermal tissues (Keller *et al.*, 1992a, b) but utilizes similar cellular mechanisms; (ii) bending or buckling of the neural plate, which involves elevation of the neural ridges and formation of the trough-like neural groove; and (iii) closure of the neural groove (Colas and Schoenwolf, 2001). We will focus on just one of these events here, the inward bending (invagination) of the neural plate to form a groove and eventually a detached cylinder.

The invagination of the neural plate was modeled by Odell *et al.* (1981) along the same lines as those used to model the vegetal plate in the course of sea urchin gastrulation. Since their analysis has similar strengths and limitations in each case, we will present here instead the approach followed by Kerszberg and Changeux (1998), who used "cellular automata" (CA), an increasingly widely employed method for modeling morphogenetic processes (Alber *et al.*, 2003). Cellular automata are computer programs made up of interacting "cells," each of which acts like a computer programmed with a set of rules (an "automaton"). Such automata defined by simple rules sometimes give rise to surprisingly complex structures (see, for example, Wolfram, 2002). In such cases it is difficult to ascertain whether a pattern has formed for the same reasons as in a biological system. At the other extreme, the rules can be made sufficiently complex that a given CA program is indistinguishable from the standard models of physics employing dynamical laws and field concepts, like those we have considered up to now. A convenient middle strategy is to employ CA rules which are simple enough to take advantage of the method's computational speed and the ease with which parameters can be revised, but which contain sufficient biological specificity that the patterns generated can be attributed to realistic properties (Kiskowski *et al.*, 2004). The Kerszberg–Changeux model for neurulation represents an example of this middle strategy. Since it deals with epithelial folding, like the models discussed earlier in this chapter, it can serve to highlight the differences between the CA framework and those that employ mechanical and field-based concepts.

The Kerszberg–Changeux model of neurulation

The folding neural plate resembles an elongated analog of the invaginating vegetal plate that we encountered in sea urchin gastrulation. It is therefore not surprising that some of the preceding discussion is applicable to this morphogenetic process as well (Odell *et al.*, 1981; Jacobson *et al.*, 1986). Like several other models discussed in this book, the computational model of Kerszberg and Changeux (1998) explicitly uses the interplay of genetic and generic mechanisms to describe a complex developmental process, drawing in this case on the large body of data available on neurulation (reviewed in Colas and Schoenwolf, 2001). Below we summarize the general features of the model.

1. The model simulations are carried out on a two-dimensional lattice of pixels, which represents a thin slice transverse to the embryo. Each cell occupies a certain number of pixels, some being reserved for the nucleus; this number varies as cells grow and migrate. There are numerous genes that are activated during neurulation. Kerszberg and Changeux considered some explicitly, the effect of others being incorporated through model parameters. Two extracellular signaling molecules (BMP, i.e., the Bone Morphogenetic Proteins BMP-2 and BMP-4, and Sonic hedgehog, Shh) referred to as "morphogens," with predetermined concentration gradients, i.e., "morphogenetic fields", resulting from a prior pattern-forming mechanism (see Chapter 7), act on the initially homogeneous peripheral epithelial cells. These cells are held together at all times by intercellular adherens junctions; in the simulation each cell has two.

2. Depending on the local concentrations of the morphogens, which are determined by the morphogenetic fields, two sets of autoregulatory transcription factors (constituting transcription factor networks like those discussed in Chapter 3), act as genetic switches that regulate the expression level of the genes Notch and Delta. The products of these genes are widely employed positional mediators of early-cell-type determination (see Chapter 7). Notch, a neuroectodermal or neural plate gene, is activated first and later Delta, a proneural gene, is induced. Since the products of Notch and Delta are a receptor and its ligand, respectively, through their presence on adjoining cells they eventually help to define a precise topographic assignment of individual cells to a neuronal fate.

3. A major assumption of the Kerszberg–Changeux model is that differential adhesion and cell motility during neural tube formation are coupled and are under strict genetic control: neuroepithelial and presumptive neuronal cells express a particular homophilic adhesion molecule whose membrane concentration is proportional to the activities of the regulatory genes Notch and Delta (themselves set by the morphogenetic fields). Cells change shape by movement and growth. In the simulation a cell grows by the addition of pixels to its surface at randomly chosen locations. If this results in a contact with another cell that displays a sufficient concentration of the adhesion molecule, the probability of effectively adding the pixel to the first cell at this

particular site will be increased relative to that of adding the pixel at a location that does not result in a contact. The net outcome is an "effective adhesion force" that brings and keeps cells together. Cellular automata are well suited to modeling this kinetic aspect of cell adhesivity, which is not considered explicitly in purely equilibrium analyses such as the differential adhesion hypothesis (DAH; see Chapter 4).

4. As the differentiation pattern sharpens, morphogen distributions decay and cells remain committed to their acquired fate. As they continue to move and grow and eventually divide they deform under the mechanical forces to which they are subject (by contact interactions with neighboring cells). The direction of movement is biased by the adhesive forces between cells and by their cohesive interactions with the basal lamina and extracellular matrix. Division occurs with a probability that is a function of cell size and type, and these in turn depend on the expression of Notch and Delta. The pixels previously occupied by the mother cell are split into two sets belonging to each of two daughter cells, which have nuclear transcriptional states identical to that of the mother cell.

The formation of the neural tube in the simulations is depicted in Fig. 5.14. The Kerszberg–Changeux CA model differs in a number of ways from the approaches to morphogenesis discussed earlier. It is a non-equilibrium model: the cellular pattern does not evolve towards an energy minimum (as in the Mittenthal–Mazo or Drasdo–Forgacs models) or to a state with balanced forces (as in the model of Odell and coworkers). It is a discrete model: the time evolution of the concentration profiles of morphogens and gene products is not governed by differential equations, as are the dynamical systems discussed in Chapter 3. Instead, the movement of individual cells is dictated by local rules set by the authors, which are chosen on the basis of experimental results and are intended to reflect biological reality. These rules define effective forces, the variations in the expression levels of regulatory genes (i.e., Notch, Delta) and the local concentrations of molecules that mediate cell–cell interactions (i.e., adhesion proteins).

The appealing feature of such CA is that they explicitly demonstrate how simple local rules may lead to complex global patterns. Thus, they may reveal information on the hierarchical organization of molecular circuits governing cellular processes, in particular morphogenesis. Clearly their success depends on the ability to choose the "right" local rules. Because of the flexibility (and arbitrariness) of such models they can be modified and tweaked until they work, although this is not always a satisfying way of gaining a fundamental understanding of a system's behavior and can represent a drawback to such models. For example, an energy minimization principle such as that employed by the DAH may indeed be fundamental to epithelial folding, and the kinetic approach of Kerszberg and Changeux may be incomplete in this respect. The local rules of their model can, in principle, be designed to incorporate global effects, though not without sacrificing some of the elegance and facility of the CA approach.

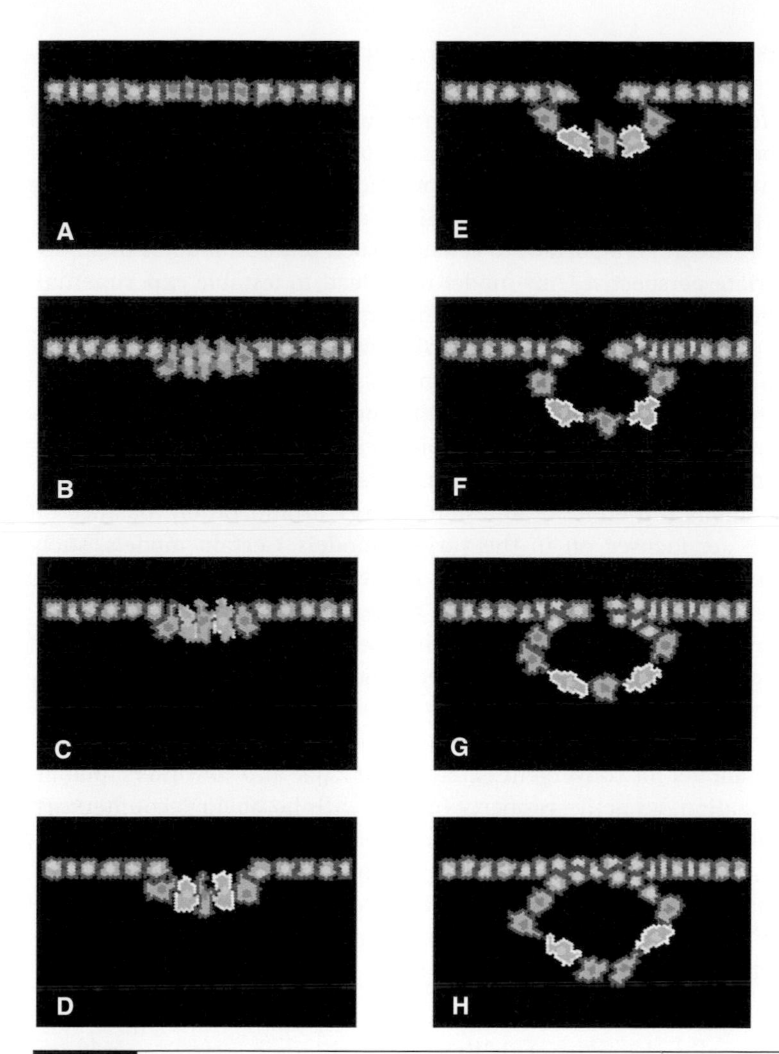

Fig. 5.14 Neural tube formation in the Kerszberg–Changeux model in a cross-sectional view similar to that of Fig. 5.13. (A) The initial epithelial sheet of cells breaks up into a central portion, the prospective neural plate, and lateral non-neural ectoderm (see Fig. 5.13), in response to the first morphogen signal. Five green blobs, nuclei expressing Notch; blue envelopes, cell membranes; small red lines, adherens junctions between adjacent cells. (B) Delineation of the neural plate in response to the second, neuralizing, morphogen signal. Three yellow blobs, nuclei expressing both Notch and its ligand Delta; central light blue envelopes, membranes displaying the Delta ligand. Two patches form, corresponding to the two ridges in the neural plate. The neural ectodermal cells begin to thicken, forming the neural plate. (C) Invagination of the neural plate due to constriction of the apical surfaces of its cells. (D) Surface epithelial cells grow over the neural cells, forming the neural folds. (E) Neural tube folding continues under the joint effects of neural cells pulling downward and epithelial cells dividing and pushing the neural folds inward. (F) Cells at the folds deform due to mechanical forces. (G) Folding is almost complete. (H) Closure of the neural tube. (From Kerszberg and Changeux, 1998, with slight modifications. Used by permission.)

Perspective

Physical models of a complex phenomenon such as epithelial folding are bound to rely on simplifying assumptions. With improved experimental input and increasing computational power, increasingly realistic models can be constructed. Since simplifications will still have to be made, it is important to consider such physical models in an appropriate perspective. Any model must lead to testable experimental predictions if it is to provide a useful explanation for a system's behavior. Most models discussed in this chapter indeed lead to specific predictions: the scaling relation in the Mittenthal–Mazo model, the eventual instability of the blastula in the Drasdo–Forgacs model, and the triggering of invagination by the excitable apical actin network in the model of Odell and coworkers. Different physical properties and phenomena (differential adhesion, energy minimization, balance of forces) are focused on in the various models. Certain models, such as that of Kerszberg and Changeux, are explicitly nonequilibrium. But even models such as those of Mittenthal and Mazo, Drasdo and Forgacs, and Odell *et al.*, which invoke equilibrium considerations (i.e., energy minimization and mechanical force balance) assume an underlying excitability of tissues – different cell types must be generated, the cytoskeleton must be mechanically excitable, and so forth. In the model for convergent extension of Zajac and coworkers, planar polarization, an active property of cells with no obvious counterpart in the nonliving world, is invoked, along with the passive physical mechanism of differential adhesion. Any comprehensive account of epithelial morphogenesis must therefore take into consideration the multiple properties of cell sheets, prominent among which are a wide variety of generic physical mechanisms.

Appendix: Linear stability analysis

To illustrate the method of linear stability analysis we consider the following dynamical system:

$$\frac{dx}{d\tau} = -a(x-y), \tag{A5.1}$$

$$\frac{dy}{d\tau} = b\left(\frac{1}{16} - xy\right)\left[\frac{(4x-3)^2}{2} + \frac{(4y-3)^2}{2} - 1\right]. \tag{A5.2}$$

These equations can be considered as the nondimensional analogues of Eqs. 5.8a, b, which arose in our discussion of the force-balance model of gastrulation earlier in this chapter. In the above equations x and y are dynamical variables (dependent on the dimensionless "time" variable τ), while a and b are positive constants. Equations A5.1 and A5.2 have three steady-state solutions, $x_1 = y_1 = 1$, $x_2 = y_2 = 1/4$, and $x_3 = y_3 = 1/2$, which make the right-hand sides vanish. We show below that the first two solutions are stable whereas the third is unstable. To accomplish this we will perform linear stability analysis around the solutions. The first step in this method is to represent the functions $x(\tau)$ and $y(\tau)$ as $x(\tau) = x_i + \varepsilon_i(\tau)$ and $y(\tau) = y_i + \eta_i(\tau)$, $i = 1, 2, 3$, where x_i, y_i denote any of the above solutions and the functions $\varepsilon_i(\tau)$ and $\eta_i(\tau)$ stand for small deviations from the solutions. Inserting these forms into Eqs. A5.1 and A5.2 and retaining terms only up to first order in $\varepsilon_i(\tau)$ and $\eta_i(\tau)$ (i.e., linearizing the equations) we obtain

$$\frac{d\varepsilon_i}{d\tau} = -a(\varepsilon_i - \eta_i), \tag{A5.3}$$

$$\frac{d\eta_i}{d\tau} = -b_i(\varepsilon_i + \eta_i), \tag{A5.4}$$

with $b_1 = \frac{15}{4}b$, $b_2 = \frac{3}{16}b$, and $b_3 = -\frac{3}{16}b$. The second step is to look for the solutions of these equations in the form

$$\varepsilon(\tau) = Ae^{-\lambda_1\tau} + Be^{-\lambda_2\tau}, \tag{A5.5}$$

$$\eta(\tau) = Ce^{-\lambda_1\tau} + De^{-\lambda_2\tau}. \tag{A5.6}$$

(Each quantity here should carry the index i, but for clarity we have omitted it.) For the time being the quantities A, B, C, D, λ_1, and λ_2 are unknown. (The mathematically sophisticated reader will recognize that λ_1 and λ_2 are the eigenvalues of the 2×2 matrix constructed from the coefficients of ε and η on the right-hand side of Eqs. A5.3 and A5.4.) To determine the stability of a given solution of the original Eqs. A5.1 and A5.2, it is sufficient to calculate λ_1 and λ_2. If λ_1 and λ_2 are both positive then the corresponding solution is stable, since as the system evolves in time both ε and η eventually vanish. To calculate λ_1 and λ_2 we insert the trial solutions given in Eqs. A5.5 and A5.6 into Eqs. A5.3 and A5.4. Performing the differentiation (remembering that $de^{-\lambda\tau}/d\tau = -\lambda e^{-\lambda\tau}$) and rearranging the equations (still omitting the index i), we arrive at

$$[\lambda_1 A - a(A - C)]e^{-\lambda_1\tau} + [\lambda_2 B - a(B - D)]e^{-\lambda_2\tau} = 0, \tag{A5.7}$$

$$[\lambda_1 C - b(A + C)]e^{-\lambda_1\tau} + [\lambda_2 D - b(B + D)]e^{-\lambda_2\tau} = 0. \tag{A5.8}$$

These equations can be satisfied only if the coefficients of the two exponential factors in each equation separately vanish. (We assume that $\lambda_1 \neq \lambda_2$, which needs to be verified once these quantities have been determined.) It is easy to see that the two equations containing λ_1 have nontrivial solutions for A and C (i.e., different from zero) if and only if λ_1 satisfies

$$\lambda_1^2 - \lambda_1(a + b) + 2ab = 0. \tag{A5.9}$$

(The above equation is equivalent to setting equal to zero the 2×2 determinant constructed from the coefficients of A and C in Eqs. A5.7 and A5.8.)

A similar analysis for the equations containing λ_2 reveals that for B and D to be nonzero, λ_2 must satisfy the same equation as λ_1

above. Solving the quadratic equation Eq. A5.9 leads to (reinserting the index i)

$$\lambda_{1,2}^{(i)} = \frac{a + b_i}{2} \pm \sqrt{\left(\frac{a + b_i}{2}\right)^2 - 2ab_i}. \tag{A5.10}$$

Here the subscripts 1 and 2 correspond respectively to the plus and minus in front of the square root. If both a and b_i are positive, as is the case for $i = 1, 2$ (see after Eq. A5.4), then it is obvious that both λ_1 and λ_2 must be positive since the second term on the right-hand side of Eq. A5.10 is smaller than the first (a and b in Eqs. A5.1 and A5.2 can be chosen so that the expression under the square root is positive for each case, $i = 1, 2, 3$). For $i = 3$, since $b_3 < 0$ we have that λ_2 is negative. Thus the steady-state solutions $x_1 = y_1 = 1$ and $x_2 = y_2 = 1/4$ are stable, whereas the solution $x_3 = y_3 = 1/2$ is unstable.

Chapter 6

Mesenchymal morphogenesis

During both gastrulation and neurulation certain tissue regions that start out as epithelial – portions of the blastula wall and the neural plate – undergo changes in physical state whereby their cells detach from one another and become more loosely associated. Tissues consisting of loosely packed cells are referred to as "mesenchymal" tissues, or *mesenchymes*. Such tissues are susceptible to a range of physical processes not seen in the epithelioid and epithelial tissues discussed in Chapters 4 and 5. In this chapter we will focus on the physics of these mesenchymal tissues.

We encountered this kind of tissue when we considered the first phase of gastrulation in the sea urchin, the formation of the primary mesenchyme, in which a population of cells separate from the vegetal plate, and from one another (except for residual attachments by processes called filopodia), and ingress into the blastocoel (Fig. 5.7). The secondary mesenchyme forms later, at the tip of the archenteron, and these newly differentiated cells help the tube-like archenteron to elongate by sending their own filopodia to sites on the inner surface of the blastocoel wall. In other forms of gastrulation, such as that occurring in birds and mammals, there is no distinction between primary and secondary mesenchyme: cells originating in the *epiblast*, the upper layer of the pregastrula, ingress through a pit (the "blastopore") and the "primitive streak" that forms behind it as the mound of tissue ("Hensen's node") surrounding the blastopore moves anteriorly along the embryo's surface (Fig. 6.1). The cells stream into the extracellular-matrix (ECM)-filled space between the epiblast and the underlying *hypoblast*, whereupon some of them are transformed into the embryo's mesenchymal middle layer, the *mesoblast*. Other cells of the epiblast, after displacing the hypoblast cells, form the embryo's epithelial endodermal layer.

In contrast with the epithelioid and epithelial tissues, in which cells are directly adherent to one another over a substantial portion of their surfaces, mesenchymes and their mature counterparts in the adult body, the connective tissues, consist of cells suspended in an ECM. Often these tissues contain more ECM than cellular material by volume. Thus there exists a set of morphogenetic processes that

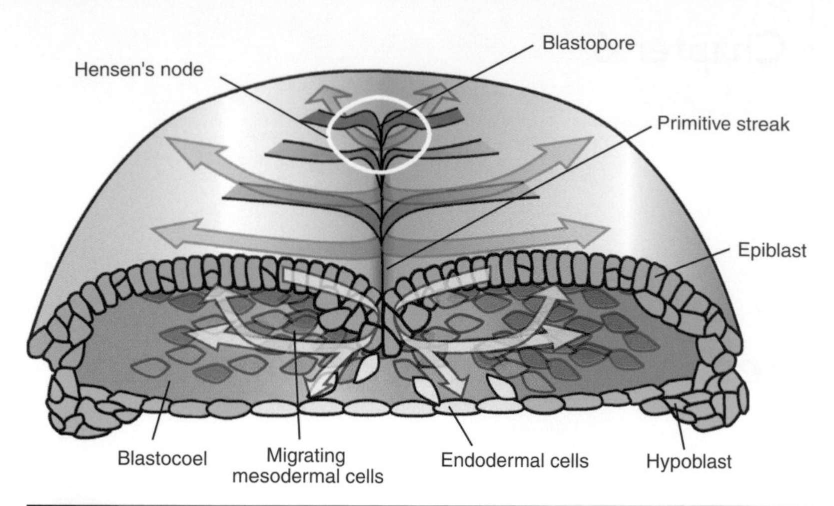

Fig. 6.1 Schematic view of a gastrulating chicken embryo. The cross-section through the anteroposterior axis is shown; the direction away from the viewer is anterior. The stage of development illustrated precedes that in the top panel of Fig. 5.13 and is roughly equivalent to the stage of frog development shown in Fig. 5.11A. The arrows show pathways of ingression and dispersion of cells entering the blastopore and primitive streak to form the mesoblast. See the main text for an additional description. Gastrulation in mammals is organized in a similar fashion.

occur in mesenchymal, but not epithelioid tissues, which depend on the physical properties of the ECM, changes in the distance between cells, the effects of cells on the organization of the ECM, and the effects of the ECM on the shape and cytoskeletal organization of cells (reviewed in Newman and Tomasek, 1996).

We will see later, in such phenomena as the separation and emergence of the limb bud from the flank or body wall of a developing vertebrate embryo (Chapter 8), that, just as in epithelioid tissues, domains of immiscibility (i.e., tissue compartments) can occur in mesenchymal tissues. This may seem surprising, since differential adhesion *per se* is not relevant to cell populations in which cells do not contact one another directly.

We will describe in this chapter "model" mesenchymes consisting of isolated ECM components mixed with cell-sized particles. Such systems have been used to demonstrate that a characteristic feature of mesenchymal-tissue ECMs – namely, the presence of large numbers of extended fibers that can arrange themselves in "paracrystalline" arrays or form random networks (Meek and Fullwood, 2001; Ushiki, 2002) – can provide the basis for mesenchymal tissue domains that behave as distinct phases. In some mature connective tissues, such as the cornea of the eye (Linsenmayer *et al.*, 1998), tendons, and bones (Cormack, 1987), highly regular arrangements do occur. These resemble "liquid crystals" (Bouligand, 1972; Gaill *et al.*, 1991; Giraud-Guille, 1996) and could potentially be analyzed in terms of the physics of these materials, which are capable of undergoing well-defined phase transitions (de Gennes and Prost, 1993). The mesenchymal tissues typically found in early embryos, however, contain ECMs with randomly arranged fibers and, as we shall see, the physical state of such a tissue can be drastically affected by the number density and aspect (length-to-diameter) ratio of such fibers, which despite their random arrangement can also form networks that undergo phase transitions.

The viscoelastic properties of ECMs are of clear importance in determining the capacity of embryonic tissues to undergo rearrangement and shape change. Equally important is the role of the ECM in providing a communication medium between cells. The cells of the epithelial and epithelioid tissues, discussed in earlier chapters, being in direct contact can signal each other by exchanging molecules through gap junctions (see Chapter 4) or by surface-bound receptor–ligand pairs ("juxtacrine" signaling; see Chapter 7). In mesenchymal and connective tissues, if cells communicate with one another then they must do so over relatively long distances, through ECMs. They can link up with each other using long, tenuous, filopodial extensions known as *cytonemes* (Ramirez-Weber and Kornberg, 1999; Bryant, 1999), which can form gap junctions at their sites of contact or thin nanometer-diameter membrane tethers (Rustom *et al.*, 2004). Mesenchymal cells can also communicate across the ECM using diffusible molecules ("paracrine" signaling), a topic discussed at length in Chapter 7. Most interestingly from the physical viewpoint is the possibility that mesenchymal cells communicate over long distances mechanically. That is to say, cells could exert a force on their local environment and, if there is a sufficient density of interconnections within this environment, other cells could experience this force. We will take up this subject and, in particular, the rigorous physical meaning of "sufficient density of interconnections," in this chapter.

In earlier chapters the physical and mathematical models we discussed were applied directly to developmental processes taking place in living systems: embryos or their tissues. In our overview of the viscoelastic and network (i.e., interconnection) properties of ECMs we will find it useful to present results from some simplified nonliving experimental models. One of these is purified type I collagen, the most abundant protein of the ECM and, for that matter, of the animal kingdom, undergoing assembly (Newman *et al.*, 1997; Forgacs *et al.*, 2003). We also describe an *in vitro* phenomenon known as "matrix-driven translocation" (MDT), which is a morphogenetic rearrangement seen in suspensions of cells or certain types of latex particles dispersed in purified ECM components (Newman *et al.*, 1985; Forgacs *et al.*, 1989). The usefulness of such an approach lies in the possibility of accounting for complex behaviors of living tissues on the basis of the interactions of a reduced number of their components. Thus the spirit of this approach is similar to that of the idealized mathematical–physical models discussed earlier.

The phenomenon of *mesenchymal condensation* is important in many events of early embryogenesis and organ formation (Hall and Miyake, 1995, 2000; Newman and Tomasek, 1996). This is usually a transient effect in development in which mesenchymal cells, initially dispersed in a matrix, move closer to one another. Condensations generally progress to other structures, such as feather germs (Chuong, 1993), cartilage or bone (Hall and Miyake, 1995; 2000), and kidney tubules (Ekblom, 1992). It is clear why condensation is characteristic of mesenchymal, but not epithelioid, tissues; the cells in the latter are already as close together as they can be. It is helpful to recognize that since "epithelial" and "mesenchymal" refer to the physical states of tissues, the conversions mentioned can go in either direction during development: epithelial to mesenchymal during gastrulation, or mesenchymal back to epithelioid tissue, transiently, as in precartilage condensations, or permanently, as in kidney tubules.

In the following we will first describe the morphogenesis of mesenchyme in the vertebrate neural crest. This will be followed by a discussion of the MDT experiment and its possible relevance to this and other mesenchymal rearrangements in the embryo. Because the MDT phenomenon suggests that mesenchymes act as coherent fluids, as a result of the network properties of their ECMs, we next present an analysis (based mainly on the concept of *percolation*) of network formation during collagen assembly. Subsequently we will describe mesenchymal condensation in several embryonic systems, with emphasis on the precartilage condensations of the developing vertebrate limb. Finally, we will bring together a number of the physical concepts from this and preceding chapters and introduce a physical model for the formation of such condensations.

Development of the neural crest

The neural crest consists of populations of cells that detach and migrate away from the dorsal ridges (referring to the embryo's back surface) of the neural tube just before, or shortly after it closes (Langille and Hall, 1993). The detachment of neural crest cells from the neural tube must involve the loss or modification of cell–cell adhesion, but which molecules and mechanisms may be involved is not yet clear. At least two Ca^{2+}-dependent cell–cell adhesion molecules, N-cadherin and E-cadherin, are lost from the neural crest cells prior to, or shortly after, detachment from the neural tube (Akitaya and Bronner-Fraser, 1992; Bronner-Fraser *et al.*, 1992).

Other suggested mechanisms for the detachment of neural crest cells from the neural tube are the degradation of adhesion molecules by secreted enzymes called proteases and the generation of tractional forces that would mechanically rip cells away from their neighbors (Erickson and Perris, 1993). Although neural crest cells are known to produce proteases that are associated with their increased motility in culture (Valinsky and Le Douarin, 1985; Erickson and Isseroff, 1989), evidence suggests that some form of epithelial-to-mesenchymal transformation, based on the acquisition of new cell-surface properties (see, e.g., Ozdamar *et al.*, 2005), is the likeliest mode of detachment of these cells (Newgreen and Minichiello, 1995; Duband *et al.*, 1995).

The detachment of cells *per se* (which we have encountered already in connection with the formation of primary mesenchyme in the sea urchin embryo, see Chapter 5) is not a mechanism that calls for novel physical explanations. But neural crest cells also translocate enormous distances (on the cell's scale), and the mechanisms of such movement, like any rearrangement of matter, do suggest that some interesting physics is at work. The cells of the neural crest give rise to a variety of widely dispersed cell types (neural crest derivatives) of the adult body. These include the nerve cells of the peripheral nervous system (i.e., other than those of the brain and spinal cord), the endocrine cells of the adrenal medulla and those of the thyroid gland that produce the hormone calcitonin, the pigment cells of the skin, cartilage and bone of the face, and the connective tissues of the teeth, several different glands, and a number of arteries. A major question for the physical understanding of development is how the cells of the neural crest convey themselves, or are conveyed, to the distant sites at which they will differentiate into these derivatives.

The unique nature of mesenchymal cells as a population free from broad cell–cell attachment permits them to rearrange and translocate by mechanisms distinct from those we have discussed for cells in epithelioid masses and epithelial sheets. For example, cells could migrate individually, moving toward sources of chemoattractants. This is referred to as *chemotaxis* (see Chapter 1). Another mode of translocation of individual cells, known as *haptotaxis*, is related to the differential adhesion mechanism we discussed for epithelia. Haptotaxis requires an external *gradient* (i.e., a nonuniform distribution) of a substratum adhesive material, typically a complex of ECM molecules. A randomly locomoting cell that binds in a reversible fashion to such a molecular substratum will eventually wind up at the site that minimizes its energy of adhesion, either locally or, if the cell's inherent motility is sufficiently strong to "kick" it past a local minimum, globally. A specific physical model of haptotaxis that characterizes the conditions under which cells may redistribute in relation to adhesive gradients was constructed by Dickinson and Tranquillo (1993).

Although chemotaxis and haptotaxis are plausible mechanisms for the translocation of mesenchymal cells during embryonic development, and haptotaxis indeed appears to be employed during the formation of mesenchymal condensations (see below), there does not appear to be a significant role for these mechanisms in the migration of the neural crest. One main pathway of neural crest dispersal in the "trunk" (main-body) region of the embryo takes them from their origin at the dorsal neural tube into lateral extracellular spaces between the ectoderm and the somites (Fig. 6.2, Path 1). The somites are blocks of mesodermal tissue that form parallel to, and to either side of, the notochord (see Chapter 7). When labeled test cells were grafted onto sites lateral to the neural tube, they migrated medially (i.e. toward the neural tube) between the ectoderm and somites and ventrally (i.e. toward the belly of the embryo – the opposite of dorsally) along blood vessels between the somites. These exogenous (i.e., transplanted)

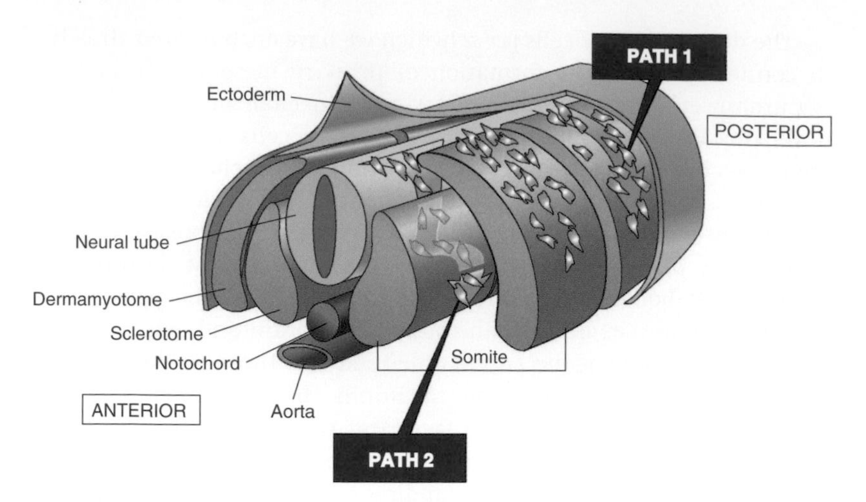

Fig. 6.2 Pathways of dispersal of trunk neural crest cells in a vertebrate embryo. The stage of development shown corresponds to the bottom panel of Fig. 5.13. In Path 1, cells move through the ECM located between the ectoderm and the somites, whereas in Path 2 they move through the anterior region of the sclerotome (the cartilage-forming region of a somite) and enter the ECM between the sclerotome and dermamyotome (the connective tissue and muscle-forming region of the somite). Path-1 neural crest cells differentiate into pigment cells of the skin, while path-2 cells differentiate into neurons, the medullary (interior) cells of the adrenal gland, and Schwann cells, the insulating cells of the nervous system. The aorta is the main artery leading away from the heart (see Chapter 8).

neural crest cells thus moved in a direction opposite to that of the endogenous (i.e., normally present) trunk neural crest cells. The other trunk neural crest pathway extends ventrally along the space that separates the portion of each somite that will form bone from the portion that will form connective tissue and muscle (the "intrasomitic space") (Fig. 6.2, Path 2). When neural crest cells were grafted onto this pathway they migrated rapidly, within 2 hr, in two directions: dorsally, eventually contacting the ventrally moving stream of host neural crest cells, as well as laterally. These experiments indicate that neither a preestablished chemotactic nor an adhesive (haptotactic) gradient exists in the embryo, since the grafted neural crest cells will move in the reverse direction from normal along these pathways, toward the dorsal neural tube (Erickson, 1985; see also Erickson, 1988).

Such puzzling results have focused attention on the ECM through which the neural crest cells move, which, along with the cells, constitutes a material that could potentially exhibit fluid properties. For example, the connective tissues of the vertebrate head are largely the product of the cephalic (also termed cranial) neural crest. These cells arise from the dorsal-most regions of the future brain and migrate laterally along several pathways (Noden, 1984). Cephalic neural crest migration in the embryo of the axolotl (an amphibian) follows a "microfinger" morphology consisting of several parallel streams (Hörstadius and Sellman, 1946) (Fig. 6.3A). Previously Noden had suggested that the invading sheet of neural crest cells might be

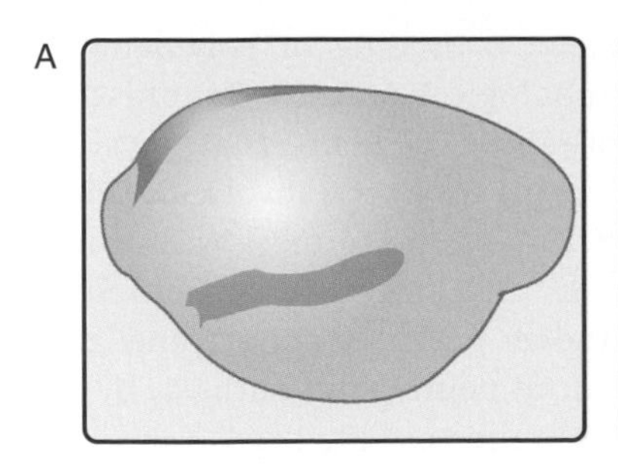
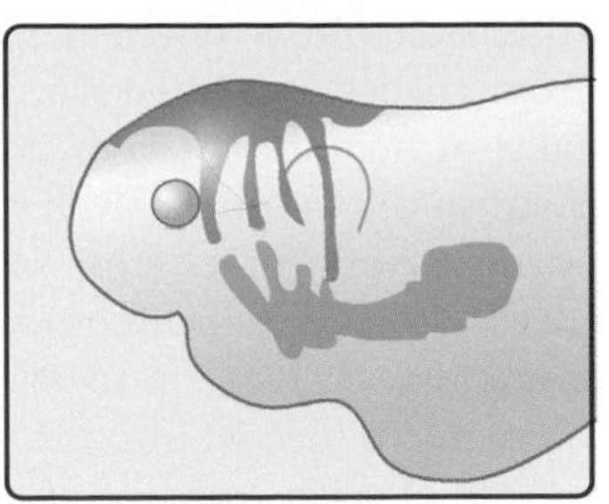

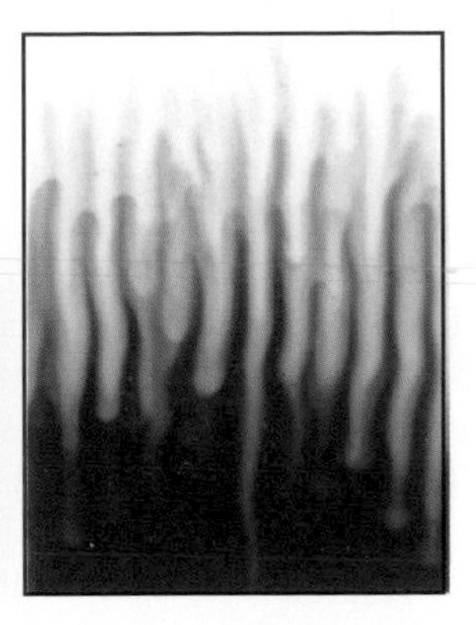

Fig. 6.3 (A) Patterns of dispersion of normal and heterotopically grafted (i.e., to an abnormal site) neural crest cells in the axolotl embryo. On the left, the right neural ridge of the head has been stained with neutral red. The left ridge has been excised, stained with Nile blue, and implanted lower down on the same side. On the right, streams of neural crest cells from the right neural ridge (red) move downwards (the normal migration pathway for this cell population) to where they meet streams of cells (blue) migrating from the graft in the dorsal direction. The whole early embryo and the portion shown of the later embryo are each about 2 mm in width. (B) The time evolution of structured flows in a polymer solution containing dextran and polyvinyl pyrrolidone (PVP). PVP in the lower layer was coupled to a dark blue dye that appears black in the figure. On the left, the initial layered-fluid preparation. On the right, the fluid configuration after 40 min. Microfingers of width on the order of 500 μm have formed by the oppositely directed vertical flow of the polymer solution. Note the striking similarity between the microfingering patterns in A and B. (A, based on Hörstadius and Sellman, 1946; B, courtesy of Dr Wayne D. Comper; see Comper *et al.*, 1987. The entire figure is based on Newman and Comper, 1990.)

divided into streams by anatomical obstructions encountered in their path (Noden, 1988). However, neural crest cells transplanted into ventral regions of axolotl embryos break into microfingers of similar dimension as they migrate dorsally through a region of the embryo, distant from the endogenous streams (Hörstadius and Sellman, 1946), that is unlikely to contain an equivalent set of obstacles. This led Comper *et al.* (1987) to suggest that such microfingering flows through the ECM might be based on a physical transport mechanism that had been observed in nonuniform solutions of polymers (Preston *et al.*, 1980). Such convective flows, which take the form of interdigitating microfingers (Fig. 6.3B), were produced in solutions of ECM proteoglycans (see Chapter 4) (Harper *et al.*, 1984) or collagen (Ghosh and Comper, 1988).

In this example, a physical system, consisting of polymers derived from tissues, exhibited a morphological outcome reminiscent of streams of neural crest cells. More direct evidence that physical, as opposed to cell-motility-driven, flow processes are involved in neural crest dispersion comes from experiments performed by Bronner-Fraser and coworkers (Bronner-Fraser, 1982, 1984, 1985; Coulombe and Bronner-Fraser, 1984). In these studies, cell-sized polystyrene latex beads were microinjected into the ventral neural crest pathway of the trunk region of avian embryos. It was found that the beads were able to translocate along this pathway, accumulating in the vicinity of endogenous neural crest derivatives (Bronner-Fraser, 1982). This occurred even if the host neural crest cells were ablated by a laser, indicating that the beads were not being passively conveyed by migrating cells (Coulombe and Bronner-Fraser, 1984).

Although nonmigratory, nonneural crest cells also translocated along the ventral trunk pathway when microinjected (Bronner-Fraser, 1982), latex beads coated with the ECM proteins fibronectin or laminin (both negatively charged) were restricted from entering the pathway. Charge, however, was not the determining factor: uncoated beads, and beads coated with type I collagen, are also negatively charged and they both translocated ventrally after injection (Bronner-Fraser, 1984). And while beads coated with polylysine, which renders them positively charged, were not capable of being translocated along the ventral neural crest pathway, polytyrosine-coated beads, with no surface charge at all, were translocated (Bronner-Fraser, 1984).

These intriguing experiments strongly suggest that the movement of actual cells along neural crest pathways does not depend on their intrinsic motility or on any obvious specific feature of their surfaces, just as it does not depend on chemotactic or haptotactic gradients (as we have seen above). Such evidence inevitably raises the possibility that some convective mechanism that is physically based (in the sense of not involving particular "living" properties) and may be termed "ECM streaming," is involved in neural crest translocation.

The extracellular matrix: networks and phase transformations

As discussed above, connective tissues of multicellular organisms (of which mesenchymes are examples) are composed of cells surrounded by complex extracellular matrices. ECMs consist of proteins, nitrogen-containing polysaccharides known as glycosaminoglycans, hybrid molecules known as proteoglycans, and, in certain mature connective tissues (bone and tooth), minerals, all in a highly hydrated state (see Comper, 1996, for reviews). The most abundant protein of ECMs, type I collagen, is a rod-like triple helical protein that undergoes assembly into macromolecular fibrils, which in turn associate to form fibers and fiber bundles (Veis and George, 1994) (Fig. 6.4). The neural crest pathways contain various ECMs, including type I

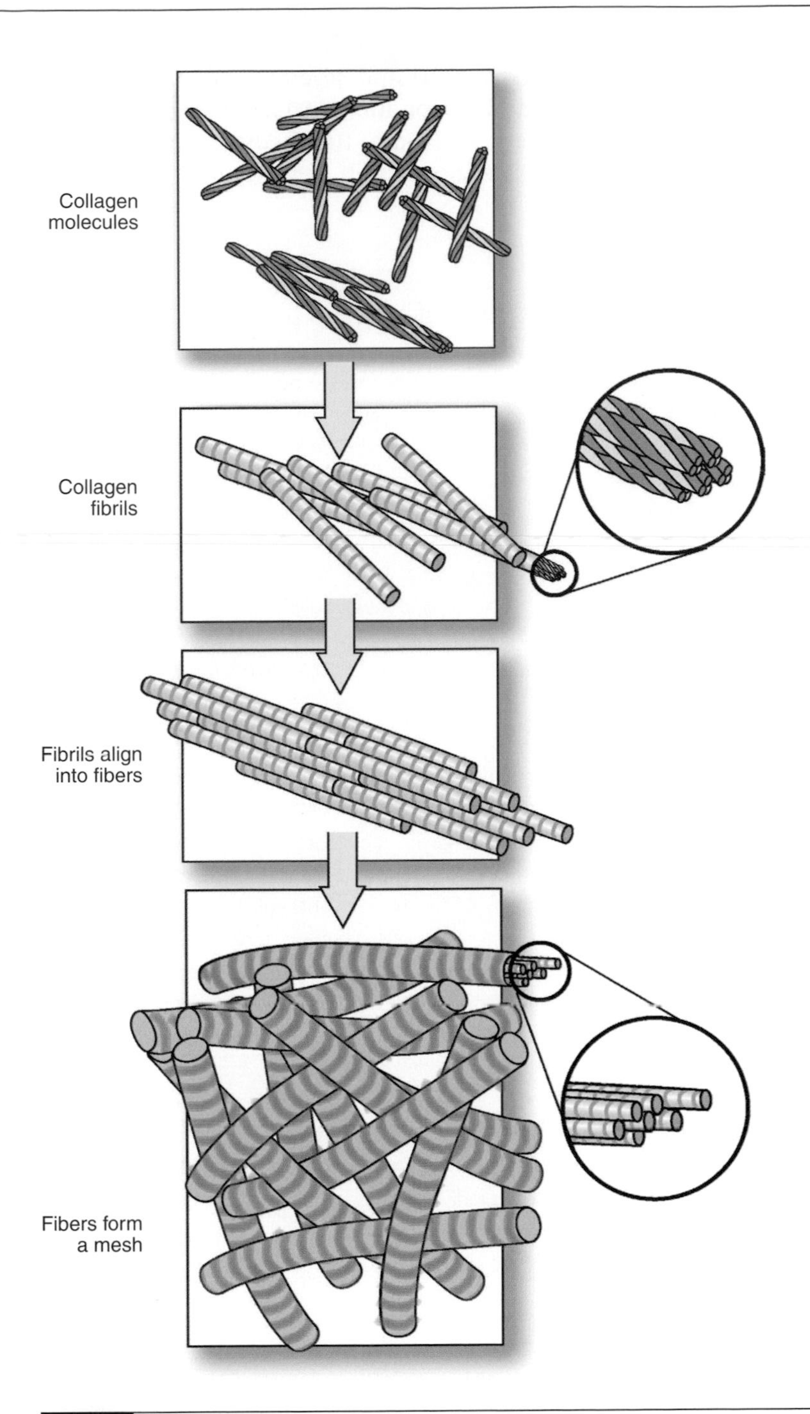

Fig. 6.4 Assembly of a type I collagen lattice, or fibrous mesh. From the top, collagen molecules (heterotrimeric triple-helical protein rods, ~300 nm in length and 1.5 nm in diameter) assemble in an end-to-end and side-by-side fashion into collagen fibrils, which in turn assemble, again in an end-to-end and side-by-side fashion, into collagen fibers. The fibers, which are of the order of a few μm in width and of indeterminate length, entangle during their assembly to form a mesh. The collagen molecule contains two identical (brown) and one distinct (beige) polypeptide chains. The collagen fibrils and fibers appear banded when viewed through an electron microscope because of the paracrystalline arrangement of the molecules and fibrils, respectively.

collagen (McCarthy and Hay, 1991) and several other fiber-forming macromolecules (Perris and Perissonotto, 2000). Fibrillogenesis – the assembly of collagen fibers – is both thermodynamically spontaneous and endothermic (Parkinson *et al.*, 1995; Kadler *et al.*, 1996; see Box 6.1).

Box 6.1 | Thermodynamics of collagen assembly

Thermodynamic spontaneity means that the process in question proceeds, like the consumption of a log by a fire, by releasing, rather than absorbing, chemical free energy. Unlike a burning log, however, assembling collagen does not release this energy in the form of heat. Heat, in fact, is absorbed during collagen fibrillogenesis (hence, it is "endothermic"). So in what form is sufficient free energy released to make up for the absorbed heat and render the entire process spontaneous?

The first and second laws of thermodynamics, applied to systems at constant temperature and pressure (i.e., most chemical and biological systems) specify that the change in the chemical (Gibbs) free energy for typical processes is

$$\Delta G = \Delta H - T\,\Delta S. \tag{B6.1a}$$

By convention, a negative value of G denotes a loss of free energy by the system and thus corresponds to thermodynamic spontaneity. H is the enthalpy, or the total heat content, of the system. Thus a loss of enthalpy, as with the burning log, will lower the system's free energy. Besides heat, the other major form of chemical free energy change is $-T\,\Delta S$, where T is the absolute temperature and ΔS is the total change in entropy of the system. The entropy is related to the organizational properties of a system; roughly speaking, the more organized or ordered the system is, the lower its entropy. (The entropy associated with a group of individuals lined up in single file is lower than when these individuals occupy random locations.) In effect, despite the fact that heat is *absorbed* during collagen fibrillogenesis, and that the collagen fibers themselves assume a *more organized* state than the (random) collection of protein monomers that give rise to them, the total entropy nonetheless undergoes a sharp increase because of the disorder induced among water molecules that were initially bound to the thousands of monomers that become assembled into each individual fiber. The consequent change in the free energy term $-T\,\Delta S$ is responsible for a negative value of ΔG and thus the thermodynamic spontaneity of the collagen assembly process.

Matrix-driven translocation

An experimental system that demonstrates the potential of composite materials made up of cells and their surrounding matrices to rearrange by physical means, thus leading to translocation of the cells, is shown schematically in Fig. 6.5 (Newman *et al.*, 1985). Here a droplet of soluble type I collagen is deposited on the right adjacent to a second such droplet, which is also populated with a small, but critical, number of living cells, or cell-sized polystyrene beads (the

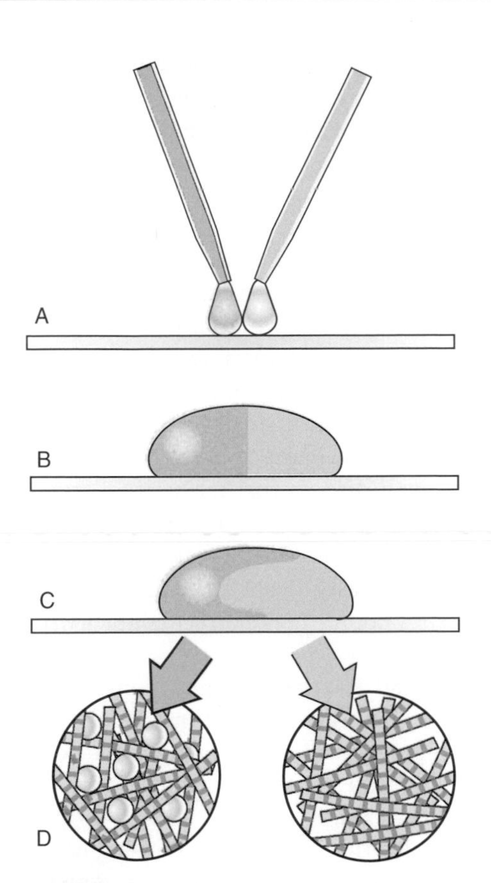

Fig. 6.5 Schematic illustration of matrix-driven translocation. (A) Two droplets of soluble type I collagen, one (on the left) containing and the other (on the right) lacking cells or cell-sized polystyrene latex beads, are deposited contiguously on the surface of a Petri dish. (B) Representation of a higher-magnification view of the two droplets, seen from the side, shortly after their fusion. (C) The translocation effect – a reconfiguration of the interface between the two droplets during the collagen assembly process. (D) A high-magnification view of the growing collagen fibrils shown in C. The fibers on the left are interacting with beads. For illustrative purposes the ratio of beads to fibers depicted in D is greater than that in the experiment. (Based on Newman, 1998b.)

same type as used in Bronner-Fraser's experiments, described above). Surprisingly, when the number density of the beads and the concentration of collagen have particular values (see below) an interface forms between the two droplets, indicating the presence of finite interfacial tension despite the fact that the compositions of the two droplets are the same, except for the presence of particles in one of them. It should be noted that these particles (cells or beads) constitute only a fraction of a percent by volume of the composite material in these experiments.

Over the next few minutes the droplet containing the particles spreads over and partially engulfs the droplet lacking particles (Newman *et al.*, 1985; Forgacs *et al.*, 1989). At higher collagen concentrations the relative movement of the two phases, referred to as "matrix-driven translocation" (MDT) only occurs when the ECM protein fibronectin, or its amino-terminal domain (comprising about 13 percent of the entire protein), is present in the droplet lacking particles (Newman *et al.*, 1985; 1987).

Matrix-driven translocation has been interpreted as follows (Newman *et al.*, 1997, 2004; Newman, 1998b): when the assembling collagen fibrils in the collagen solution reach a critical length, they can form a network that pervades the entire volume in which they are present – the droplets in this case. But when the system is perturbed by the presence of cells or polystyrene beads the network forms with different organizational properties, hence the two droplets constitute separate "phases," the bead-lacking drop being more cohesive than the bead-containing drop, as we shall discuss further below (Forgacs *et al.*, 1991; Forgacs and Newman, 1994). These "model mesenchymes" can thus behave like immiscible liquids, just as epithelioid tissues do under the differential adhesion hypothesis, despite the fact that the cells or beads in these model tissues do not make direct contact with one another.

If this picture is correct, the final relative configurations of the droplets when brought in contact with one another in the MDT experiment will be dictated by the principles of thermodynamic equilibrium: the less cohesive phase should envelop the more cohesive one. Surface tension measurements indicate that the collagen droplet lacking beads is indeed more cohesive than the one containing beads (Forgacs *et al.*, 1994). Note that the translocation caused by the phase rearrangement in MDT does not depend on individual cell motility – most persuasively, MDT occurs equally well with cells or beads – but rather is a collective property of these model mesenchymes (Fig. 6.5).

The coherent transport of mesenchymal cells occurs in neural crest dispersal, in some types of gastrulation (Wakely and England, 1977; Harrisson, 1989; Sanders, 1991), and later during organogenesis, as in the invasion of the acellular primary stroma of the cornea by mesenchymal cells from the periphery (Fitch *et al.*, 1998). The MDT experiment highlights an important physical property of mesenchymes and other connective tissues: the potential of distinct phases to be formed as the result of ECM network formation and cell–ECM interactions. In the next section we will explore the physical basis of such network formation.

Percolation, scaling and networks

Structure in physical materials is usually thought of in terms of organization. The solid form of water is more organized than the liquid form, a fact that is evident in the geometric, i.e., crystalline, arrangement of the water molecules in ice. Water vapor is an obviously disorganized arrangement of water molecules, more so even than the liquid state. Some materials can exhibit several organized states as well as disorganized ones: elemental carbon, for example, can take the form of graphite, diamond, buckminsterfullerene ("buckyballs"), and nanotubes. The common form of carbon, found in soot or charcoal, is amorphous.

Connective-tissue matrices may exhibit a variety of organizational states. Although they are much more complex materials than water or carbon, their physical properties are usually dominated by their fibrous components: collagens and the long glycosaminoglycan molecule known as hyaluronan. The type I collagen fibrils and larger-scale fibers in the tendons and ligaments of mature vertebrate organisms, for example, are densely packed and all oriented in a single direction, maximizing the tensile strength of these tissues. The same protein, type I collagen, is also the most abundant fibrous component of bone; but here it is typically organized into sheets of tissue in which the fibrils are oriented in a single direction within each sheet and the direction changes abruptly between adjacent sheets. This is the same property that gives plywood, which is made up of multiple thin layers of wood (laminated so that the wood grain changes direction from layer to layer), its enormous resistance to bending. The cornea of the vertebrate eye is also composed of type I collagen, but in this case the fibrils are orthogonal to one another in sheets, an arrangement which leads to a tissue that is both tough and transparent (Cormack, 1987).

Most adult connective tissues, nonetheless, are "irregular;" their matrices do not exhibit the paracrystalline structures found in tendon, bone, or cornea. In these tissues, and in the embryonic mesenchymes that give rise to them, ECM fibrils and fibers have no preferred orientation and are organized as random networks. Intuition might suggest that random structures by definition have no organization. In the case of networks, however, this intuition is faulty. Different states of randomness may be found within a given physical system, and there can even be physically well-defined transitions between these states. Because transformations in the state of randomly organized networks of fibrous molecules can potentially account for the coherent motion of developing mesenchymal tissues (it has been invoked to provide an explanation for MDT, for example; see above), as well as signal transduction across the cytoskeleton (Forgacs, 1995), it is useful to understand certain aspects of the physics of network formation.

The concept of *percolation* has been widely applied to the understanding of network properties in disciplines as diverse as physical, chemical, biological, engineering, and social sciences, including economics (Sahimi, 1994). It refers to a transition whereby on increasing the concentration of certain structural elements randomly distributed in a given system an interconnected network of these elements, a so-called "spanning cluster," is formed, which extends from one end of the system to the other. As an illustration of the percolation transition, consider the land-line telephone network connecting Los Angeles (LA) and New York (NY), as shown schematically in Fig. 6.6. This network consists of a multitude of cables or optical fibers

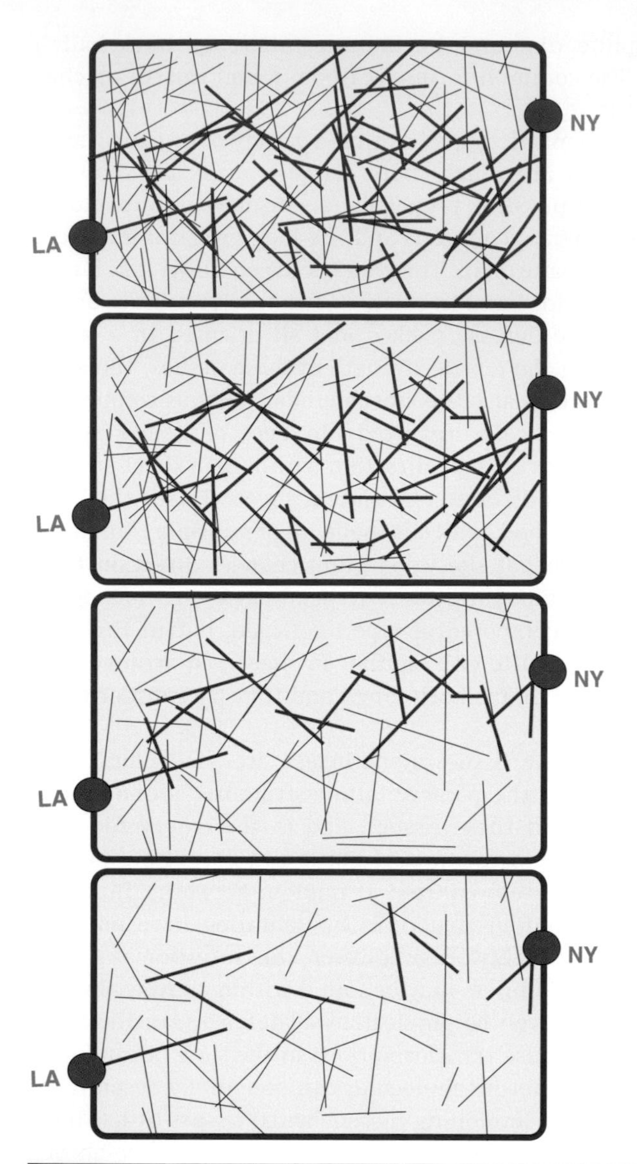

Fig. 6.6 Schematic representation of the percolation transition. Top panel: Depiction of the intact land-line telephone network between Los Angeles (LA) and New York (NY). The straight segments may denote cables or optical fibers. The heavy brown lines indicate an interconnected percolation network that extends from one end (i.e. LA) of the system to the other (i.e., NY). Middle panels: If the connections are gradually destroyed, contact between LA and NY will be maintained as long as a single connected path exists between the two cities, that is, the number of connections is above the percolation threshold. Bottom panel: Below the percolation threshold no connecting path exists and no communication between the two cities is possible, although local calls (depending only on small clusters) may still go through.

("elements") of varying finite lengths, is interconnected, and has the appearance of being random. It also has redundancy: the signal from LA can arrive in NY in a number of ways. If this network is subject to a series of disasters, more and more of the links being randomly destroyed, connection between the two cities would still be maintained up to a critical number of finite elements (although the time for the signal to arrive will typically increase). The telephone service is definitely disrupted below a threshold where an interconnected cluster of connections extending from LA to NY no longer exists, although many finite clusters of cables may still be present.

As the above example illustrates, the percolation transition is accompanied by important changes in the connectivity of the system, a well-defined topological characteristic. This defines the percolation transition as a phase transition, in the course of which the system transforms from the non-interconnected phase onto the interconnected one. The transition takes place at a critical point or "percolation threshold." In other systems that exhibit percolation transitions, changes in topology may lead to drastic alterations in physical properties. For example, above a critical concentration, elementary conducting metallic islands, randomly distributed in an insulating matrix, interconnect and the macroscopic conductance of the system becomes finite: it is capable of transmitting electric signals (Clerk *et al.*, 1990). In another example, the elastic subunits in an otherwise inelastic amorphous medium interconnect above the threshold concentration. As a consequence, the system develops macroscopic elastic properties and responds to mechanical signals (Nakayama *et al.*, 1994).

During fibrillogenesis, type I collagen, in connective tissues and in the MDT experiment, also undergoes such a transition, referred to as the "gelation transition," which is the basis of the phase behavior discussed above. We will now describe the physics of such transitions in terms of the percolation model (de Gennes, 1976a, b).

This model stipulates that gelation (the process in which an originally liquid system – a sol – with finite viscosity and no elastic modulus transforms into a different type of material – a gel – with infinite viscosity and finite elasticity) in a filamentous macromolecular system such as a collagen matrix is due to the gradual interconnection of growing fibers. The state of the network can thus be characterized by p, the number of connections formed between fibers, normalized in such a way that when all the fibers are connected $p = 1$. For sufficiently small p only isolated small clusters of connected fibers exist. However, at a threshold value $p = p_c$, which defines the sol–gel transition, the interconnections between isolated clusters lead to a continuous ("spanning") network. Percolation theory predicts power-law behavior for macroscopic physical properties such as the elastic modulus or viscosity in the vicinity of the gelation point. Such behavior is characteristic of scale-free systems (see Box 2).

In particular, the power law for the static elastic modulus or Young's modulus, E, (defined in Eq. 1.7) is

$$E \propto (p - p_c)^f, \tag{6.1}$$

which is valid for $p > p_c$. The corresponding power law for the zero-shear viscosity, η (defined in Eq. 1.2), is

$$\eta \propto (p_c - p)^{-k}, \tag{6.2}$$

Box 6.2 | Power-law behavior and scale-free networks

The power-law behavior of a physical quantity signals that the system it characterizes is scale-free. Scale-free networks, of which percolation networks represent one example, are ubiquitous across many disciplines (Barabási, 2002). The characteristic power-law behaviors discovered and observed in these networks reflect the fundamental self-organizing property of the underlying systems.

Why do power laws imply scale-free character? To explore this, let ξ represent the size of a system in terms of length, and let a physical quantity f characterizing the system be a power law function of ξ with characteristic exponent s:

$$f = C\xi^s. \tag{B6.2a}$$

Here C is a constant (i.e., it is length- or scale-independent). Length, and thus ξ, can be measured in various units: microns, feet, meters, etc. Let us change the unit of length. For example we can initially use a measuring tape in inches and then replace it with a tape in centimeters. If $\xi = A_1 l_1 = A_2 l_2$, where l_1 and l_2 are the two different units of length and A_1 and A_2 express the magnitude of ξ in terms of these units (ξ = 2.54 cm = 1 inch) then the change of units leads to

$$f = CA_1^s l_1^s = C_1 l_1^s = CA_2^s l_2^s = C_2 l_2^s. \tag{B6.2b}$$

Here the $C_i = CA_i^s$ ($i = 1, 2$) are still constants and thus f has the same power-law dependence on length no matter what tape is used to make the length measurements or, in other words, at what scale the system is studied. Note that if, for example, f depends exponentially on ξ then changing measuring tapes leads to $f = C(e^{l_1})^{A_1} = C(e^{l_2})^{A_2}$ and the functional dependence on length is modified (i.e., the power of the exponential would change). This example demonstrates an important implication of power-law behavior: the system "looks" similar under any magnification, i.e., it is scale-free. This is literally the case for fractals (see Chapter 8); these are systems (the convoluted shoreline of Norway is a celebrated example) whose local geometry is identical no matter what strength of magnifying glass is used to look at them (Mandelbrot, 1983).

which holds for $p < p_c$. The values of f and k (which are positive quantities and are called "critical exponents") are either measured or determined theoretically; in the latter case they will depend on the specific model used to describe gelation (Brinker and Scherer, 1990; Sahimi, 1994).

In the course of gelation the system exhibits complex viscoelastic behavior (see Chapter 1). Its viscosity and elasticity are both evident when it is probed at various time scales. This is typically accomplished using a viscometer with a cone or plate immersed in the system and oscillating with frequency ω. The frequency-dependent response of the evolving gel is measured in terms of viscoelastic moduli, the storage modulus $G'(\omega)$, and the loss modulus $G''(\omega)$ (Fung, 1993). These quantities are related to the Young's modulus and the viscosity in the limit of vanishing frequencies via $E = \lim_{\omega\to 0} G'(\omega)$ and $\eta = \lim_{\omega\to 0}[G''(\omega)/\omega]$. Thus E and η can be determined by measuring $G'(\omega)$ and $G''(\omega)$ at progressively smaller frequencies and extrapolating to $\omega \to 0$.

Gelation in collagen solutions, as discussed here, is a time-dependent phenomenon driven by the gradual interconnection and entanglement of growing fibers. (Sol–gel transitions can also be driven solely by changes in temperature, using existing macromolecules, as in the denatured counterpart of collagen, gelatin (Djabourov *et al.*, 1988, 1993). The analysis of gelation in terms of percolation theory as described above is based implicitly on the assumption that the extent of bond formation between fibers, characterized by the parameter p, is linear in time, i.e., $p \sim t$; thus $p_c \sim t_g$, t_g being the gelation time. This assumption does not appear to be well justified in the case of rapidly gelling collagen. Instead of relying on Eqs. 6.1 and 6.2, we will follow the more general method of Durand *et al.* (1987), which analyzes gelation by considering directly the time evolution of $G'(t;\omega)$ and $G''(t;\omega)$.

According to Durand *et al.* (1987), who were following de Gennes' (1976a, b) original proposal based on the concept of percolation, at the gelation transition the frequency-dependent viscoelastic moduli exhibit power-law behavior:

$$G'(\omega) = A\omega^{\Delta}\cos(\pi\Delta/2), \tag{6.3a}$$

$$G''(\omega) = A\omega^{\Delta}\sin(\pi\Delta/2). \tag{6.3b}$$

Equations 6.3 thus predict identical scaling laws for $G'(\omega)$ and $G''(\omega)$ at the gel point: $G'(\omega) \propto G''(\omega) \propto \omega^{\Delta}$ (A and Δ in Eqs. 6.3a, b are constants). Furthermore, it also follows that at the gel point the critical loss angle δ_c (δ is defined as $\tan\delta = G''/G'$) is related to the exponent Δ by

$$\delta_c = \frac{\pi}{2}\Delta. \tag{6.4}$$

The theory of de Gennes relates Δ to the critical exponents f and k introduced in Eqs. 6.1 and 6.2 via $\Delta = f/(f+k)$. Thus the value of δ_c does not depend on the frequency, a result that can also be obtained independently of the percolation model (Chambon and Winter, 1987; Martin *et al.*, 1989; Rubinstein *et al.*, 1989).

Equations 6.3 and 6.4 provide two independent methods of locating the sol–gel transition in time. In the first method the frequency dependence of the viscoelastic moduli at various times is measured. According to Eqs. 6.3, at the gel point, $t = t_g$, the plots of $G'(t_g;\omega)$ and $G''(t_g;\omega)$ versus $\log\omega$ should yield straight lines with identical slopes for the two functions. In the second method the experimental results obtained for the viscoelastic moduli are used to calculate the loss angle. According to Eq. 6.4, at t_g the plot of this quantity as a function of frequency should yield a straight line parallel to the horizontal axis. Moreover, a consistency check on the two methods relates the slope obtained in the first method to the value of the critical loss angle obtained in the second method.

Detailed measurements performed on the solutions of assembling collagen used in the MDT experiments have confirmed that at a well-defined point in time the storage and the loss modulus indeed exhibit power law behavior with the same exponent, and the loss angle is independent of frequency (Forgacs *et al.*, 2003; Newman *et al.*, 2004).

From this behavior and from Eqs. 6.3, it can be inferred that collagenous ECMs are capable of undergoing a percolation phase transition. This transition still occurs, moreover, in the presence of cell-sized particles in sufficient number to cause the viscoelastic properties of the assembling matrices to differ sharply from the particle-free matrices (Newman *et al.*, 2004).

Network concepts have important implications for mesenchymal morphogenesis. As we have seen above, translocating mesenchymal cells typically move during both gastrulation and neural crest dispersal as if they were mechanically coherent media (Nakatsuji *et al.*, 1986; Newgreen, 1989). This kind of behavior will be facilitated if the cells are not simply borne along in a flowing liquid but are physically linked to one another (however transiently) across the extracellular medium, allowing them to move like a flock of birds or a swarm of bees. The formation of a percolating cluster of ECM fibers is a way of ensuring that the mechanical activities of individual cells are conveyed to their nearest neighbors and also to those farther away.

Like epithelioid tissues, discussed in Chapter 5, mesenchymal tissues can undergo what appears to be phase separation. This may occur under physiological conditions, as in the formation of the limb bud primordia along the flank of vertebrate embryos (Heintzelman *et al.*, 1978; Tanaka and Tickle, 2004), during limb regeneration (Crawford and Stocum, 1988), or in artificial situations in which, for example, cell masses derived from developing fore and hind limbs are placed in contact with one another (Downie and Newman, 1994) (Fig. 6.7; see also Chapter 8). The network concept helps us to understand how mesenchymal tissues, the cells of which have no direct contact with one another, may behave as distinct phases.

Finally, networks, even if randomly constructed, can serve as communication media (Barabási, 2002). In tissues, these communication networks can operate within (Forgacs, 1995; Shafrir *et al.*, 2000; Shafrir and Forgacs, 2002) and between cells. Because cells can attach to (Gullberg and Lundgren-Akerlund, 2002; Miranti and Brugge, 2002) and exert force on (Roy *et al.*, 1999; Zahalak *et al.*, 2000; Freyman *et al.*, 2001) their microenvironments, if the microenvironment is mechanically linked, like collagen above the percolation transition, the forces can be transmitted over macroscopic distances. Similar considerations apply to the non-mechanical signaling among mesenchymal cells mediated by the cytonemes (Ramirez-Weber and Kornberg, 1999; Bryant, 1999) and tethered nanotubes (Rustom *et al.*, 2004), described at the beginning of this chapter. The networks formed by these connections from cell to cell across the ECM, like the fiber-based mechanical ones described above, can be spanning or non-spanning. As such, the nature of information transfer (i.e., whether it is global or local) in cytoneme- or nanotube-based mesenchymal networks may be analyzed by percolation models analogous to those described above.

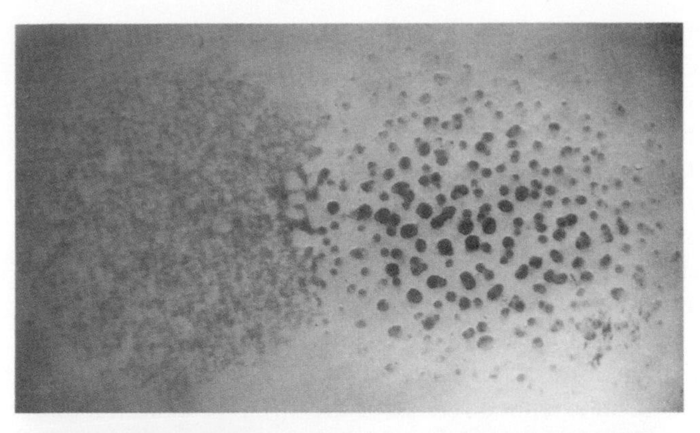

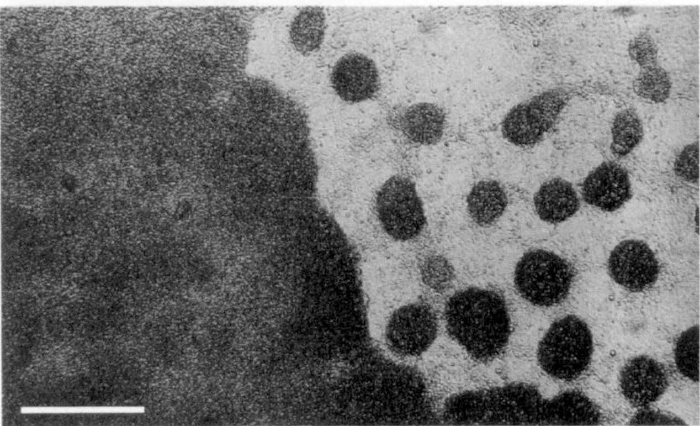

Fig. 6.7 Phase separation in mesenchymal tissues. High-density cultures of precartilage mesenchymal cells from embryonic chicken wing and leg bud were grown contiguously on a Petri dish. The upper panel shows the culture pair (wing on left, leg on right) after six days of growth, fixed and stained for cartilage with Alcian blue (the darker areas). The cartilage is uniformly dispersed throughout the wing culture but forms isolated nodules separated by noncartilage mesenchyme (the less stained areas) in the leg culture. The interface is convex from the leg side, indicating that leg tissue is more cohesive than wing tissue, a result also obtained by different assays (Heintzelman *et al.*, 1978). The lower panel shows a higher-magnification phase-contrast-microscope image of a similar pair of cultures, again after six days' growth. In this case both cultures are living and unstained. There is a clear interface between the wing culture (on the left) and the leg culture (on the right). No mixing appears to have occurred across the interface. The total horizontal distance in the top panel is ~10 mm. The bar in the lower panel is 0.5 mm long. (Reprinted from Downie and Newman, 1994, with permission.)

Mesenchymal condensation

As mentioned earlier, condensation occurs in a mesenchymal tissue when cells suspended in an ECM move closer together at particular sites. The distances traversed during this process are small – usually less than a cell diameter. Cell condensation usually occurs in two phases (Fig. 6.8). First, cells accumulate in regions of prospective condensation, which are rich in one or more adhesive ECM glycoproteins, such as fibronectin, giving rise to mesenchymal aggregates; second, cells undergo epithelialization, producing cell-surface adhesive molecules such as N-CAM and the various cadherins, molecules (with certain exceptions; see Sinionneau *et al.*, 1995) not normally expressed by mesenchymal cells (Hall and Miyake, 1995; 2000). Once they are in proximity, it is physically straightforward for cells to form direct adhesive associations. Explaining the aggregational phase of condensation in physical terms represents more of a challenge.

A variety of cellular mechanisms has been suggested for the initial stage of condensation formation. The best evidence supports a scenario based on an extended version of the differential adhesion hypothesis (DAH: Steinberg, 1978; 1998; see Chapter 4). In this interpretation, random cell movements occurring in a tissue mass in which there are local patches of increased adhesivity drive cells into higher density aggregates (Frenz *et al.*, 1989a, b; Newman and Tomasek, 1996). When test particles coated with the glycosaminoglycan heparin were mixed with limb mesenchyme cells in culture, they accumulated at sites of cell condensation in a fashion that depended on interactions of the particle surface with fibronectin. This indicated that passive movement, as the beads are buffeted by the surrounding cells, in conjunction with adhesive gradients (i.e., haptotaxis) is sufficient to translocate such particles (Frenz *et al.*, 1989a). Limb mesenchymal cells express heparin-like surface molecules (Gould *et al.*, 1992) and are thus subject to the same haptotactic forces (Frenz *et al.*, 1989b).

Cell condensation is an example of cellular pattern formation, i.e., regulated changes in cell arrangement. This is a subject we will study in more detail in Chapter 7. Interest in pattern formation has given rise to a large number of models for these processes. These models

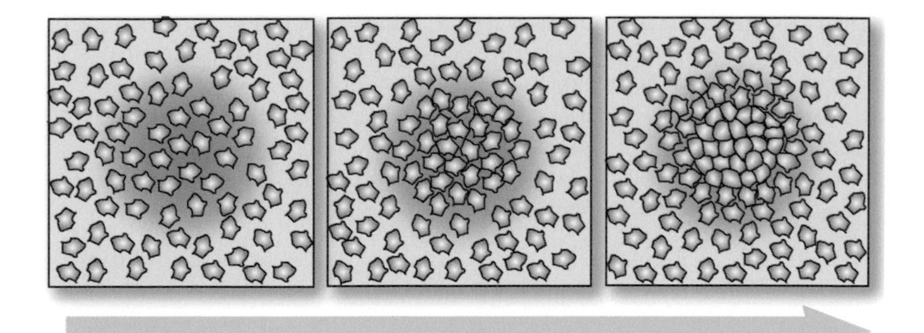

Fig. 6.8 Schematic representation of mesenchymal condensation. In the left-hand panel a population of mesenchymal cells is depicted as being scattered throughout an extracellular matrix (beige) that contains a region rich in condensation-promoting ECM molecules such as fibronectin (brown). In the middle panel cells accumulate preferentially and thus attain an elevated density in the fibronectin-rich region. In the right-hand panel, cells in high-density foci establish broad cell–cell contacts, mediated by newly expressed cell adhesion molecules (CAMs), and thus epithelialize. (After Newman and Müller, 2000.)

can be classified as continuous or discrete. In the former category, space and time are considered continuous and pattern evolution is described in terms of differential equations governing quantities (typically densities) that themselves vary continuously in space and time (Murray, 2002). We used the continuous approach in Chapter 3 to model cell differentiation and in Chapter 4 to describe gastrulation. In discrete models, such as the cellular automaton model of neurulation discussed in Chapter 5, variations in space and time occur in finite steps and thus patterns are formulated along grids or lattices. Continuous and discrete models differ from each other mostly in their initial formulations; while one or the other may be advantageous for certain applications, some models draw on features from both frameworks.

To describe cell condensation we introduce the discrete cellular Potts model (CPM), which has been applied successfully to a number of biological phenomena including differential adhesion (Graner and Glazier, 1992; Glazier and Graner, 1993; Mombach *et al.*, 1995; Jiang, Y., *et al.*, 1998). Glazier and coworkers applied this model to the initial phases of mesenchymal condensation (Zeng *et al.*, 2003). The reference data for simulations using this model were obtained from *in vitro* experiments performed in planar culture dishes. The model was therefore formulated in two dimensions.

The cellular Potts model and its application to mesenchymal condensation

The two-dimensional CPM is a representation of a collection of N cells distributed on the sites of a square lattice. With each lattice site, identified by a pair of indices (i, j), one associates a cell variable $\sigma_\tau(i, j)$. Here the index τ takes the value 1 or 0 depending on whether the lattice site is occupied by a biological cell or by its acellular environment (i.e., the culture medium). We set $\sigma_0(i, j) = 0$ for all lattice sites; $\sigma_1(i, j)$ can take on N discrete values (1, 2, ..., N) (see Fig. 6.9). A biological cell is defined by a group of neighboring lattice sites with the same value of σ_1. In the figure, all the sites in a given cell have the same color.

Cells interact with each other. We denote the energy cost of forming a unit of contact area between neighboring cells by $J\,[\sigma_\tau(i, j), \sigma_{\tau'}(i', j')]$. (Note that J is defined only for cells that contact each other.) For biological cells, J reflects their surface adhesiveness; this depends, typically, on the number of membrane-bound adhesion molecules. Since only surface energies between distinct cells are included, there is no energy cost incurred if the lattice sites belong to the same cell: $J = 0$ for $\sigma_\tau(i, j) = \sigma_{\tau'}(i', j')$. As a result of their interaction and migration cells change their shape; this corresponds to the deformation of domains with the same σ_τ (Fig. 6.9) and can be incorporated by updating the value of the cell variable at each lattice site.

Mesenchymal condensation is due to both cell–ECM and cell–cell interactions. In the simple model described above it is assumed that

Fig. 6.9 Schematic illustration of the cellular Potts model in two dimensions. A cell is represented by a set of squares (six in the figure) of the same color, each corresponding to the same value of the variable σ_{l} (as defined in the text). The two panels in the figure show two possible configurations of a cellular pattern with 13 cells arising in the course of the Monte Carlo simulation. The uncolored region represents the tissue culture medium.

J decreases linearly with the cell's integrated exposure to fibronectin (i.e., smaller values of J correspond to stronger, and thus energetically more favorable, adhesion; see Eq. 6.5). Biologically this could mean that binding to fibronectin causes cells to produce increased amounts of cell adhesion molecules such as N-cadherin. Such adhesive "crosstalk" (Marsden and DeSimone, 2003; Montero and Heisenberg, 2003) leads to positive feedback: cells encountering high levels of fibronectin tend to stay longer in those regions, thus adding the fibronectin they produce to the local ECM. By the above-stated assumption these same cells then produce more N-cadherin, making them more adhesive. The linear size of cell clusters at these sites grows continuously until the cells are unable to undergo further movement.

With these ingredients, the CPM represents the total energy of the condensing pattern as

$$E = \sum_{(i,j),(i',j')} J\left[\sigma_\tau(i,j), \sigma_{\tau'}(i',j')\right] + \lambda \sum_{\sigma} \left(A_\sigma - A_{\text{target},\sigma}\right)^2 + \mu \sum_{(i,j)} C_{\text{F}}(i,j). \tag{6.5}$$

Here the first term on the right-hand side denotes the interaction energy between cells as described above. The summation is extended only over pairs of lattice sites belonging to neighboring cells. The second term constrains the cells' surface area and the sum is over all cells. λ is the compressibility of the cell's material (larger λ values correspond to less compressible cells). A_σ is the actual cell area and $A_{\text{target},\sigma}$ is the area of the cell in the absence of compression. The third term describes the effect of preferential attachment of cells to fibronectin and here the sum includes only sites that lie within biological cells (i.e., those with $\tau = 1$). $C_{\text{F}}(i, j)$ is the concentration of fibronectin at site (i, j) and μ is the unit strength of fibronectin binding.

It is clear from this analysis that the term containing μ is responsible for the initial (i.e., aggregation) phase of condensation, whereas the term containing J is responsible for the second, epithelialization, phase.

The evolution of the cellular pattern is followed using stochastic Monte Carlo simulations in the same manner as described in Chapter 2 in connection with the Drasdo–Forgacs model. One starts with a random distribution of cells within a circular area that corresponds to the size of the "micromass" (small-diameter, high-density) culture, which is approximately 3 mm. The length scale in the computational representation is set by the pixel size (i.e., one elementary square in Fig. 6.9). Each cell occupies a certain number of pixels (six in Fig. 6.9) corresponding to its size (the linear size of a typical limb bud cell is 15 μm). At each step a lattice site is selected randomly and its cellular

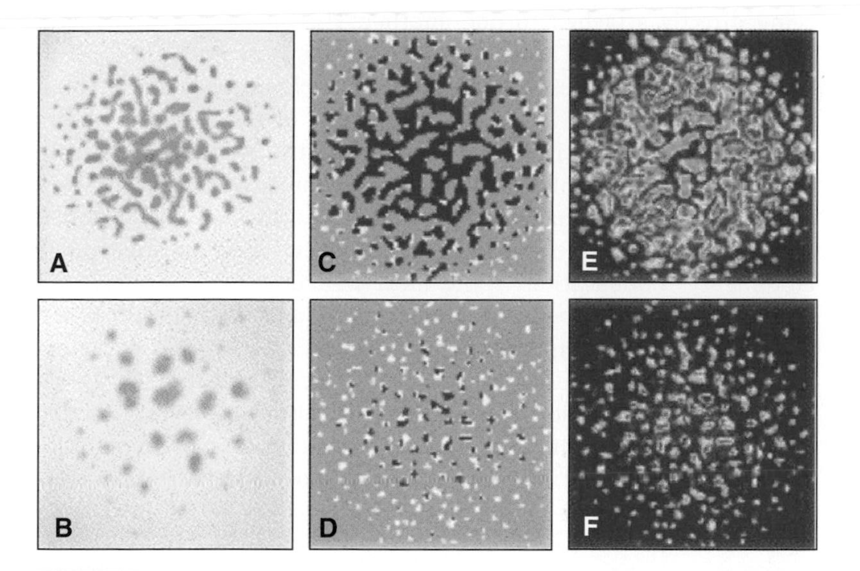

Fig. 6.10 Patterns of cartilage nodules that formed in cultures plated at two different densities, with, for comparison, simulations based on the cellular Potts model (Eq. 6.5). (A, B) Leg-cell cultures were grown for six days and stained for cartilage with Alcian blue, as described in Downie and Newman (1994) (see also Fig. 6.7 above). The initial plating densities were higher in A than in B, but in both cases they were at confluency or above, i.e., sufficient to cover the 5 mm diameter circular culture "spot" completely. The unstained regions between the nodules in A and B contain cells that failed to undergo condensation and therefore did not progress to cartilage. (C, D) Simulations of the *in vitro* condensation process. Unlike living cells in high-density cultures, cells in the two-dimensional simulations cannot readily move past one another unless there is space between them. The simulations were therefore performed at subconfluent densities of 40% and 20% coverage. As in the experiments, the simulated cells form stripes and spots at high density (C) and only spots at low density (D). The simulated cells are initially white and become blue when the local fibronectin concentration exceeds a threshold chosen to correspond to Alcian blue staining. (E, F) The distribution of fibronectin in the simulations. The fibronectin concentration ranges from red (highest) to light blue (lowest). High levels of fibronectin colocalize with the cell clusters in C and D. (Based on Zeng *et al.*, 2003; figure courtesy of W. Zeng.)

variable $\sigma_\tau(i, j)$ is changed to $\sigma_\tau(i', j')$, the value of one of its neighbors, also selected randomly. The new configuration is accepted with a certain probability P depending on the gain or loss in energy ΔE, where E is defined in Eq. 6.5, caused by this change: $P = \exp(-\Delta E / F_T)$ if $\Delta E > 0$ and $P = 1$ if $\Delta E \leq 0$. Here F_T is the cytoskeletally driven fluctuation energy discussed in connection with the Drasdo–Forgacs model near the end of Chapter 2. The system evolves towards the minimum energy state. The results of the simulations and the corresponding experiments are shown in Fig. 6.10. (For more details, see Zeng *et al.*, 2003).

Perspective

Mesenchymal cells differ from epithelial and epithelioid cells by not being directly attached to one another. This makes them susceptible to physically based morphogenetic mechanisms not applicable to other tissue types. Mesenchyme-specific mechanisms include those based on the distinctive physical properties of the ECM, its ability to flow in microfingering patterns or its ability to support the self-organization of assembling fibers, for example. Because cells in a mesenchymal tissue may be at variable distances from one another, they can also undergo local rearrangements based on short-range movement. Condensation is the most well described of these. Despite the differences between mesenchymal and epithelioid tissues, certain physical descriptions and mechanisms emerge in common: the rheological properties of tissues (describable by parameters such as viscosity and elasticity), the ability of tissues to form compartment boundaries by undergoing phase separation, and the applicability of energetic considerations, which underlies the DAH and its extensions.

Chapter 7

Pattern formation: segmentation, axes, and asymmetry

In previous chapters we found that, by virtue of their internal dynamics, cells could assume distinct states and thereby follow alternative developmental pathways. This process ultimately generated various types of terminally differentiated cells (Chapter 3). We also found that tissues made up of multiple cells, linked together directly (Chapters 4 and 5) or via an extracellular matrix (Chapter 6), could undergo alterations in shape and form (morphogenesis), leading to the development of new structures. Although developing tissues need not contain different cell types in order to undergo morphogenesis, differentiation often sets this process in motion. For example, sorting and tissue engulfment, which together comprise one of the classes of morphogenetic phenomena discussed earlier, require at least two populations of cells that are differentiated with respect to their cell-surface adhesivity.

Cell sorting, however, is unusual among the morphogenetic processes dependent on differentiation in that the initial differentiation event need not be spatially controlled. Recall that the random mixing of two cell types and the fusing of fragments of the tissues from which they were derived ultimately achieved the same morphological outcome (Fig. 4.5). Thus, if differentiation in a developing cell mass occurs in a spatially random manner, so that two cell types with differing adhesive properties come to be dispersed in a salt-and-pepper fashion, the final configuration of the eventually phase-separated tissues will always be the same. A case in point is *Hydra*, a diploblastic, i.e., two-germ-layered, organism (see Chapter 5), which starts out as a single-layered blastula. The formation of the second germ layer (i.e., gastrulation) involves the ingression of individual endodermal cells that apparently arise in a dispersed fashion in the original layer (Martin *et al.*, 1997).

While this scenario may be informative in revealing the physical consequences of sorting and the resulting boundaries of immiscibility, the emergence of distinct cell types in developing tissues is rarely spatially random. Indeed, the sorting of a random population of differentiated cells into well-organized layers will only occur if the cells are differentiated from one another with respect to adhesivity. If cells were to differentiate randomly with respect to other functions, for example, contractility (as in muscle cells) or electrical excitability (as in nerve cells), then sorting into homotypic layers would not automatically take place. In general, therefore, the acquisition of spatial patterns of differentiated cells in the developing embryo must be regulated by mechanisms other than spontaneous sorting. The physical nature of these mechanisms of spatial pattern formation is the subject of this chapter.

We encountered examples of non-sorting pattern formation when we discussed gastrulation and neurulation in Chapter 5. Recall that in each case a spatially heterogeneous distribution of cell states was needed to set things in motion. In the Drasdo–Forgacs model for gastrulation this took the form of a preexisting nonuniformity in the distribution of certain gene products, on the basis of which cells near the vegetal pole acquired nonzero spontaneous curvature. In the Kerszberg–Changeux model for neurulation it took the form of predetermined gradients of the diffusible molecules BMP and Shh. Although these models are simplified representations of the biological reality, embryos of most species exhibit spatial nonuniformities of cell state from the earliest stages of development. This is true even of mammalian embryos, where it was long thought not to be the case (see Gardner, 2001).

The term "gradient," as used in developmental biology, designates continuously varying spatial nonuniformities either in cell state (typically characterized by the cells' expression levels of a particular gene) or in the molecular microenvironment of cells that may initially be in identical states. As an example of the first meaning, early during gastrulation in the chicken the *caudal* genes (which specify a class of transcription factors) exhibit sequential activation in the newly formed neural plate and sequential extinction in axial midline structures along the anterior–posterior axis (Marom *et al.*, 1997). The resulting distribution of the *caudal* products along the embryo is indicative of spatially varying cell states. Such gradients in gene expression are *outcomes* of pattern formation.

As noted, in addition to its use in describing a distribution of cell states, the term "gradient" is also used to describe distributions of molecules in the microenvironment of cells, in particular distributions of signaling molecules that induce changes in cell state during development. Thus extracellular gradients of the *Xenopus nodal-related* (Xnr) factors, which are secreted by the embryonic ectoderm, mediate mesoderm induction during the blastula stage (Agius *et al.*, 2000). The factors that constitute such gradients are referred to as "morphogens" (a term coined by the mathematician A. M. Turing; Turing, 1952). There is considerable evidence that diffusible morphogens play important roles throughout development (this topic is reviewed in Green, 2002).

As we have seen, a new cell pattern may arise by the rearrangement of preexisting cells of different types or by the generation of new cell types. *Morphogenetic* pattern formation (or more commonly, "morphogenesis") refers to mechanisms that generate new patterns without generating new cell types (Salazar-Ciudad *et al.*, 2003). Pattern-forming mechanisms in which new cell types are generated can be divided into two subcategories, those that generate new cell types without employing cell–cell signaling, termed *cell autonomous* mechanisms, and those that do so by using cell–cell signaling, termed *inductive* mechanisms (Salazar-Ciudad *et al.*, 2003). As we will see, each type of biological pattern formation mechanism can employ a variety of distinct physical processes. The following sections will describe examples of each category of pattern formation listed above and illustrate the different ways in which physical mechanisms – adhesion, diffusion, oscillatory and other dynamical systems properties – enter into our understanding of these phenomena.

Basic mechanisms of cell pattern formation

While most changes in cell pattern during development involve both morphogenesis (i.e., changes in tissue form and shape; see Chapters 4, 5, and 6) and cell differentiation (i.e., stable changes in cell state; see Chapter 3), it is useful to focus at first on mechanisms in which only one or the other occurs (Salazar-Ciudad *et al.*, 2003) (Fig. 7.1A, B).

Mechanisms that do not generate new cell states

As indicated above, *morphogenetic* pattern-forming mechanisms (Fig. 7.1A) are defined by their capacity to change the relative arrangement of cells over space without affecting their states. We have considered already one such morphogenetic mechanism, *differential adhesion*, which generates pattern by the sorting of preexisting cell types into homotypic islands.

Other morphogenetic mechanisms that lead to changes in cell pattern without changes in state include *oriented cell division* or *directed mitosis*, in which a tissue elongates in a preferential direction because of the nonrandom alignment of its cells' mitotic spindles (Gong *et al.*, 2004) and *migration*. Migration can be directionally random, or it can be random but speeded up by an ambient chemical cue ("chemokinesis"), or it can have a preferred direction in relation to a chemical gradient (chemotaxis) or an insoluble substratum gradient (haptotaxis). Note that these cues do not change the cells' differentiated states, only their relative positions. *Convergent extension*, a reshaping of tissue masses during gastrulation, which involves the mutual intercalation of planar polarized cells without changing their state, is also such an example. The *contraction* of individual cells, mediated by actin–myosin complexes, can have dramatic morphogenetic effects on neighboring cells and on the tissue as a whole (Beloussov, 1998). Contraction is propagated in epithelial tissues by direct physical attachment between cells (see, for example, Hutson *et al.*, 2003), and in mesenchymal tissues by the extracellular matrix. *Apoptosis*, the programmed loss of cells, changes the pattern of the cells that remain behind.

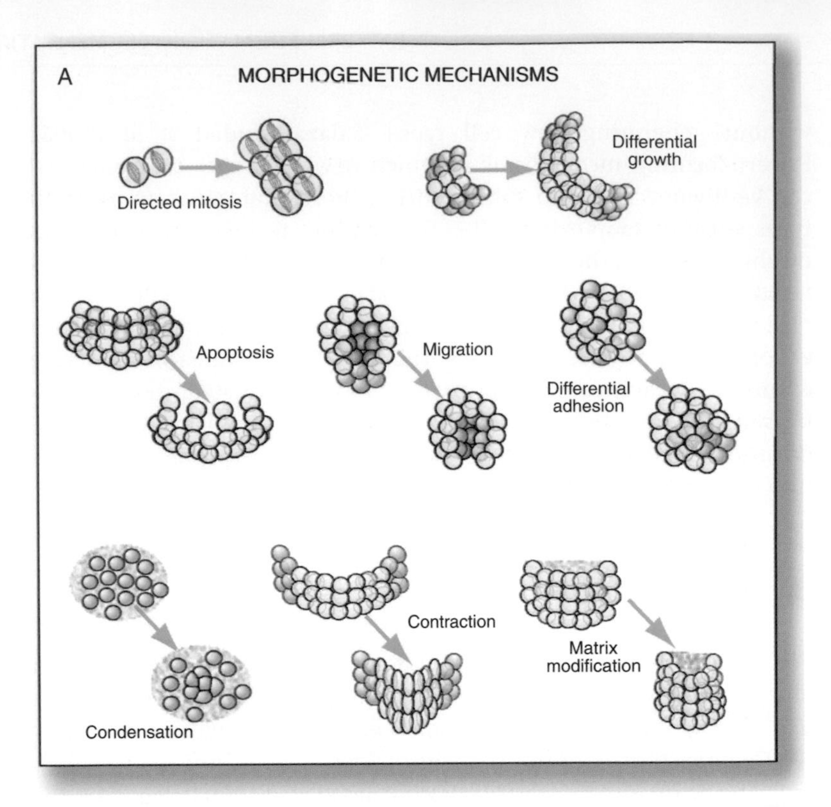

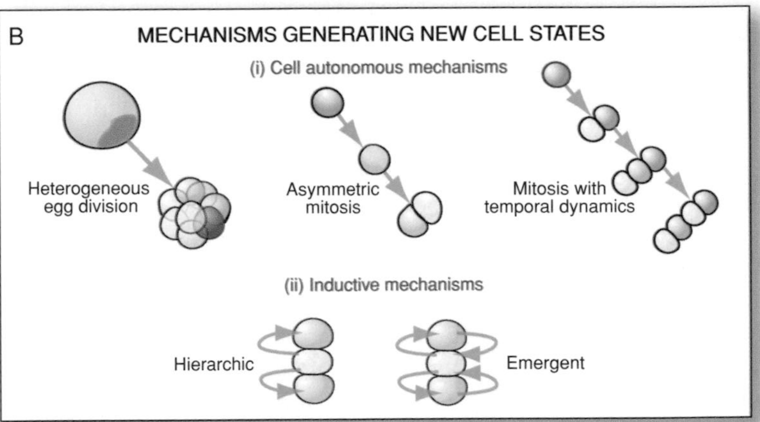

Fig. 7.1 Schematic representation of the basic pattern-forming mechanisms. Panel A shows morphogenetic mechanisms, which bring cells into new spatial relationships with one another without generating new cell types. *Directed mitosis*: consistently oriented mitotic spindles can reshape a growing tissue. *Differential growth*: cells dividing at a higher rate (yellow) can alter tissue shape. *Apoptosis*: the programmed death of specific cells (blue) can transform an established pattern into a different one. *Migration*: a pattern is altered by the directed migration of cells (blue) to a new location. *Differential adhesion*: a change in pattern can result from the grouping of cells with different adhesive properties (the blue cells are more cohesive than the yellow cells). *Condensation*: some mesenchymal cells in a matrix become tightly associated with one another. *Contraction*: the contraction of a subset of cells (yellow) can cause the buckling of a tissue. *Matrix modification*: the swelling, new deposition, loss, or molecular modification of an ECM can cause budding and other shape changes in a tissue.

(Cont.)

Mechanisms that generate new cell states

There are two major categories of basic pattern-forming mechanisms that produce patterns by cell differentiation, i.e., by generating new cell states (Fig. 7.1B): those that do not depend on cell–cell signaling (*cell autonomous* mechanisms) and those that do (*inductive* mechanisms) (Salazar-Ciudad *et al.*, 2003). *Cell autonomous mechanisms* all generate new cell states by employing cell division. In these mechanisms, mitosis occurs under conditions that are nonuniform either in space or in time, causing the daughter cells to exhibit distinct cell states. In general these cell states may be transient or stable, but here we consider a new developmental pattern to have formed only if transient differences in state are converted into stable differences, i.e., new cell types. As we saw in Chapter 3, this usually occurs by the establishment of persistent changes at the transcriptional level.

Let us first consider how cell division can be *spatially* nonuniform with respect to cell state. Certain eggs have materials such as yolk or informational RNAs or proteins transported into them in a spatially asymmetric fashion during oogenesis (Wikramanayake *et al.*, 1998; Zhang *et al.*, 1998). When the zygote resulting from this egg undergoes its first division, a cellular pattern is automatically formed: one blastomere will contain greater amounts of the stored material than the other. In such cases, the initiation of pattern formation is a passive consequence of an earlier cellular event. Such differences in cell state may then be converted into differences in cell type (see, for example, the review by Chan and Etkin, 2001). Another variant of this cell autonomous patterning mechanism is the preferential transport, on the basis of intracellular polarity, of materials to one of the daughter cells during cell division (asymmetric mitosis, Fig. 7.1B). This mechanism may employ transport along the cytoskeleton by molecular motors, a general cellular process that has been studied by physical

Panel B shows pattern-forming mechanisms that involve the generation of new cell states. (i) Cell-autonomous mechanisms produce pattern changes by generating new cell types without employing cell–cell signaling. *Division of a heterogeneous egg*: the egg cytoplasm contains spatially separated sets of different molecules (indicated by different colors) resulting in distinct blastomeres after cleavage. *Asymmetric mitosis:* molecules are differentially transported into the two halves of a dividing cell, resulting in different daughter cells. *Internal temporal dynamics coupled to mitosis*: cells which have levels of molecular determinants that oscillate with a period different from that of the cell cycle can produce periodic spatial patterns. In the example shown, the green cell (a stem cell) divides to produce unequal progeny, another green cell and a cell that is yellow or blue, depending on the state of the regulatory cycle; the latter is out of synchrony with the cell cycle. This leads to the alternating production of yellow and blue cells (see Newman, 1993 for additional details). (ii) Inductive mechanisms produce changes in pattern by employing cell–cell signaling to generate new cell types. *Hierarchic induction*: the inducing cell (yellow) affects neighboring cells but the induced cells (blue) do not influence the production of the inducing signal. *Emergent induction*: the inducing cell affects neighboring cells, which in turn signal back, influencing the inducing signal. (Based on Salazar-Ciudad *et al.*, 2003, with additions.)

methods (see Howard, 2001). Cell division may also occur under conditions in which the cell state is nonuniform in *time*. This requires that something other than the cell cycle acts as a clock. As we saw in Chapter 3, the cyclical production of biomolecules can result from biochemical networks with inherent oscillatory dynamics. According to the model of Borisuk and Tyson (1998) the cell cycle itself is based on such a dynamic oscillation. If a regulatory molecule unrelated to the cell cycle also exhibits oscillatory dynamics then it can act as a clock, and a cell pattern can emerge from the interplay of the two oscillations (Newman, 1993; Holtzendorff *et al.*, 2004). Consider, for example, a population of dividing cells in which one of the daughter cells eventually becomes converted into one of several new cell types, and the other daughter remains uncommitted. Such populations are called "stem cells." By itself, stem cell character does not imply any particular pattern or arrangement of cells. But now if we assume that the fate of the differentiating daughter cell is determined by the particular level of a periodically varying regulatory molecule it contains when it is "born," under a certain choice of parameters a spatially periodic pattern of cells will be generated (mitosis with temporal dynamics, Fig 7.1B).

Inductive mechanisms, the second category of basic pattern-forming mechanisms that generate specific cellular configurations by producing new cell types, do so by employing cell-to-cell signaling (Fig. 7.1B); cell division may or may not be involved. Signaling can be directly from one cell to an adjacent cell (*juxtacrine* signaling) or via morphogens and other signals that are transmissible across several cell diameters (*paracrine* signaling). The most ubiquitous form of juxtacrine signaling during development, the Notch–Delta system, employs just one cell-surface-bound receptor–ligand pair. The Notch and Delta proteins, and by implication this signaling system, are considered to be evolutionarily ancient because of the *conservation* (i.e., the limited extent of change) of their sequences over the more than half-billion years since insects diverged from vertebrates. This system is used in a wide variety of embryonic contexts to set up cell patterns, including, as we have seen previously, neurogenesis in the neural plate (Tiedemann, 1998; see also the discussion of the Kerszberg–Changeux model in Chapter 5) and in the peripheral nervous system (Wakamatsu *et al.*, 2000; Morrison *et al.*, 2000), somitogenesis (Jiang, Y. J., *et al.*, 2000; Oates and Ho, 2002; see below), left–right asymmetry (Raya *et al.*, 2003, 2004; see below), muscle cell differentiation (Umbhauer *et al.*, 2001), and blood vessel formation (Uyttendaele *et al.*, 2001).

Paracrine-type inductive mechanisms are also widespread during development (Green, 2002). Typically, these employ morphogens, such as members of the transforming growth factor, i.e., TGF-β, superfamily (TGF-βs, activin, Nodal, BMPs, GDFs), or the fibroblast growth factor FGF family (FGF-2, -3, -4, etc.). In Chapter 1 we discussed the reasons why free diffusion inside cells can only occur on scales that are small compared with the cell's linear dimensions. The situation is quite different, however, in multicellular embryos and in the cell aggregates that form organs. Morphogens can diffuse in the extracellular space and form gradients by interacting with cell surface receptors. The best-documented case of the use of this mechanism is the induction of mesoderm by *Xenopus* Xnr factors, mentioned in the introduction to this chapter. Morphogens can also spread by *cell relay* mechanisms. Here the effect of the morphogen spreads as if by diffusion, but what is actually propagating across the domain is an activity passed from cell to cell by signaling rather than by any substance (Reilly and Melton, 1996).

In addition, morphogens can act on cells in an *autocrine* fashion (i.e., on the same cells that produce them). If the morphogen happens also to stimulate its own production (as is the case with the TGF-βs; Van Obberghen-Schilling *et al.*, 1988), a positively autoregulatory chemical reaction loop will result. Such positive feedback situations become "explosive" unless something is present to curtail them. In any realistic biological system, therefore, an "inhibitory" morphogen, i.e., one that suppresses the production of the "activating" or positively autoregulatory morphogen, will also be present. In his 1952 paper in which he introduced the term "morphogen," Turing considered the situation when an activating morphogen also induces the production of the inhibiting morphogen. This implies that the activator and inhibitor both emanate from the same sites. Turing then assumed that the two molecules diffused away from their (common) sites of production at different rates. Such "reaction–diffusion" systems (which could be purely chemical, as well as biological) were demonstrated mathematically to have inherent pattern-forming capability (Turing, 1952).

With certain parameter choices reaction–diffusion systems can thus produce morphogen patterns that are significantly more complex than the monotonic gradients that would result from simple morphogen diffusion (Turing, 1952; Harrison, 1993; Meinhardt and Gierer, 2000). The cell patterns induced by these morphogen profiles will then be correspondingly complex. It has been suggested by various workers that reaction–diffusion mechanisms underlie the generation of planar cell polarity (discussed in Chapter 5 in relation to convergent extension; Amonlirdviman *et al.*, 2005), axis formation and the generation of left–right asymmetry in the early embryo (Meinhardt, 2001; see below), the formation of pigment stripes in the skin of fish (Kondo and Asai, 1995), feather bud formation in the skin of birds (Jiang, T., et al., 1999), and skeletal pattern formation in the vertebrate limb (Newman and Frisch, 1979; Miura and Shiota, 2000a; Hentschel *et al.*, 2004; see Chapter 8).

Inductive pattern-forming mechanisms can themselves be subdivided into two broad categories, *hierarchical* and *emergent* (Salazar-Ciudad *et al.*, 2000, 2001a, b) (Fig. 7.1B). In hierarchical patterning mechanisms, transmission of the inductive signal, whether juxtacrine or paracrine, is unidirectional. In emergent patterning mechanisms, there are reciprocal inductive interactions between cells. Both types of mechanism are prevalent during development; indeed, most episodes of pattern formation during embryogenesis contain hierarchical and emergent steps. In Chapter 10 we discuss the implication of these different mechanisms for the evolution of embryonic pattern.

In the remainder of this chapter we will describe specific biological examples of, and physical models for, developmental patterning by cell-autonomous and juxtacrine mechanisms and by autocrine and/or paracrine gradient and reaction–diffusion mechanisms. The developmental events we will consider are segmentation, the formation of periodically arranged, clustered, epithelial-cell subpopulations, spatially regulated mesoderm induction, and axis formation and the associated generation of left–right asymmetry. These events all occur during early development but not in every species and not always in the same order. Mesoderm induction occurs prior to gastrulation in all triploblasts (organisms with three germ layers, see Chapter 5) whereas segments or epithelial clusters may form both before and after gastrulation, depending on the species. Axis formation and the generation of left–right asymmetry are postgastrulation determinants of form, but the sequence of these events differs in different species. Here we will follow an order of presentation that is motivated by the logic of the physical mechanisms involved.

Segmentation

A wide variety of animal types, ranging across groups as diverse as insects, annelids (e.g., earthworms), and vertebrates, undergo *segmentation* early in development, whereby the embryo, or a major portion of it, becomes subdivided into a series of tissue modules. These modules typically appear similar to each other when initially formed; later they may follow distinct developmental fates and the original segmental organization can be all but obscured in the adult form. Somite formation (or "somitogenesis") is a segmentation process in vertebrate embryos in which the presomitic "paraxial" mesoderm (i.e., the mesoderm directly to either side of the notochord) becomes organized into parallel blocks of tissue. The somites are transient structures that eventually give rise to mature tissues and structures, some of which retain the segmental or modular organization (the vertebrae – hollow cylinders of bone that surround and protect the spinal cord – and the ribs), and some of which do not (the dermis of the dorsal skin, the muscles of the back, body wall, and limbs) (Gossler and Hrabe de Angelis, 1998; Keller, 2000; Pourquié, 2001).

Somitogenesis takes place in a sequential fashion. The first somite begins forming as a mesenchymal condensation (see Chapter 6) in the anterior region of the trunk (i.e., the main body region). Each new somite forms just posterior to the previous one, budding off from the rostral (towards the nose) portion of the unsegmented paraxial mesoderm (Fig. 7.2). As they mature, the somites epithelialize, that is, become epithelioid tissues, the *sclerotome*, which forms the vertebrae and the ribs, and the *dermamyotome*, which forms the dermis and the muscles. Eventually, 50 (chick), 65 (mouse), or as many as 500 (certain snakes) of these segments will form.

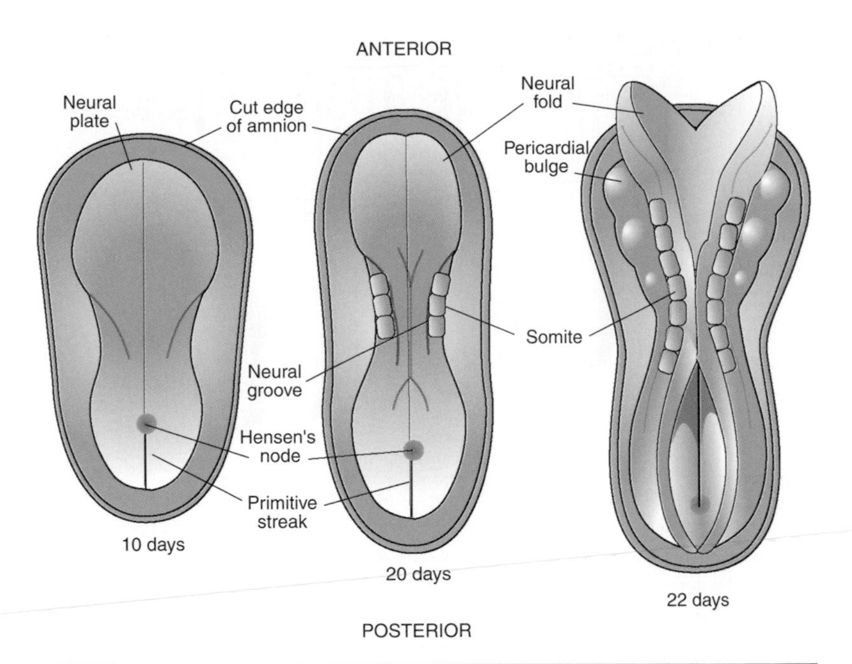

Fig. 7.2 Somitogenesis in the human embryo, looking down on the dorsal surface. On the left, the 19-day embryo. Gastrulation has advanced, but the primitive streak is still visible at the posterior end of the embryo (see Fig. 6.1 for a different view of a chicken embryo, which has similar features at this early stage). The neural plate and neural groove have formed but no somites have emerged by this stage, which is roughly equivalent to that shown in Fig. 5.13B. In the center, the 20-day embryo. The amnion is a fluid-filled membrane, composed of embryo-derived cells, that forms a protective covering over a portion of the (mammalian and avian) embryo. Three pairs of somites have formed, beginning at the anterior end of the future trunk. The neural folds have begun to converge (see Fig. 5.13C). On the right, the 22-day embryo. Several additional pairs of somites have formed, each new pair having been added posterior to the previous one. The neural folds have begun to fuse, as have the paired heart tubes (see Fig. 8.1), which reside within the pericardial bulge. (After Langman, 1981.)

In the late nineteenth century the biologist William Bateson speculated that the formation of repetitive blocks of tissue, such as the somites of vertebrates or the segments of earthworms, might be produced by an oscillatory process inherent to developing tissues (Bateson, 1894). More recently, Pourquié and coworkers made the significant observation that *c-hairy1*, an avian homologue of the *Drosophila* gene *hairy*, is expressed in the paraxial mesoderm of avian embryos in cyclic waves whose temporal periodicity corresponds to the formation time of one somite (Palmeirim *et al.*, 1997; Pourquié, 2003). The *c-hairy1* mRNA and its protein product, a transcription factor, are expressed in a temporally periodic fashion in individual cells, but since the phase of the oscillator is different at different points along the embryo's axis, the areas of maximal expression sweep along the axis in a periodic fashion (Fig. 7.3).

Other studies indicate that somite boundaries form when cells that have left a posterior growth zone move sufficiently far from a source of FGF8 in the tailbud (the posterior tip of the embryo)

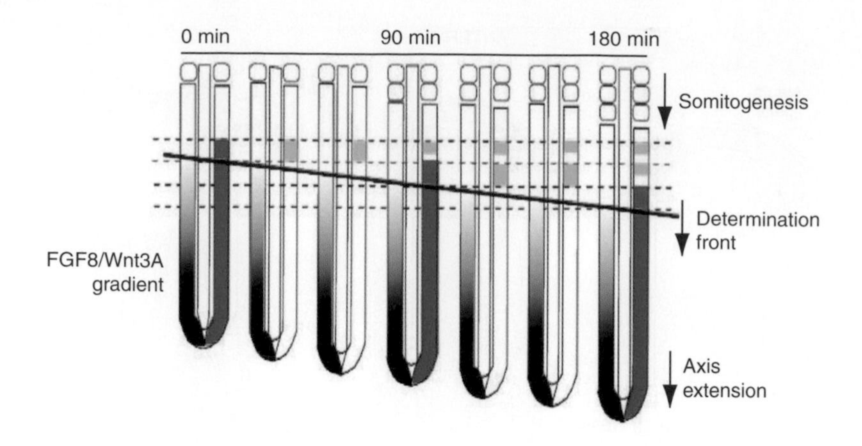

Fig. 7.3 The model proposed by Pourquié for segment formation in vertebrates, based on mouse and chick data. A gradient of FGF8 (see the main text), shown in black, regresses posteriorly during somitogenesis. The anterior boundary of the gradient defines the determination front, which corresponds to the position of the wavefront (the thick black line). (A coordinately expressed gradient of Wnt3A plays a similar role; Aulehla *et al.*, 2003). The oscillatory (i.e., waxing and waning with the developmental stage) expression of *c-hairy1* and related genes is shown in red. The expression of genes of the Mesp family, which encode transcription factors involved in somite boundary formation, is shown in blue-green. (Reprinted, with permission, from Pourquié, 2003.)

(Dubrulle *et al.*, 2001). The FGF8 gradient (along with the coordinated activity of Wnt3A; Aulehla *et al.*, 2003) thus acts as a "gate," which, when its low end coincides with a particular phase of the segmentation clock, results in the formation of a boundary (Pourquié, 2003) (Fig. 7.3). The general features of this mechanism, called the "clock and wavefront" model, were predicted two decades before there was any direct evidence for a somitic oscillator (Cooke and Zeeman, 1976).

The Lewis model of somitogenesis

In all the vertebrates studied, certain genes are differentially expressed in the anterior and posterior portions of each *somitomere*, the domains of tissue in the presomitic mesoderm (PSM) that will eventually give rise to the somites. These always include genes of the Notch–Delta pathway or their modulators (Jen *et al.*, 1999; Holley *et al.*, 2000, 2002; Oates and Ho, 2002; Dale *et al.*, 2003). As noted earlier in this chapter, the Notch–Delta system is the most common form of juxtacrine signaling during early development. Notch is a surface-bound receptor that transmits signals received from outside the cell into the cell's interior. Members of the Delta family of proteins are ligands of Notch. Delta is not secreted, but remains bound to the cells that produce it. The Notch–Delta interaction is just the first step in the generation of a patterned arrangement of cells; the downstream effects of Notch activation or lack of it include transcriptional alterations that lead to new cell types (Artavanis-Tsakonas *et al.*, 1999) (see Chapter 3).

In the mouse (Jouve *et al.*, 2002) and in zebrafish (Jiang *et al.*, 2000), mutations that inactivate the Notch pathway disrupt somitogenesis.

During somitogenesis in zebrafish the Notch ligand DeltaC oscillates in the PSM with the same period as a pair of transcription factors, encoded by the genes *her1* and *her7*, which are related to chicken *c-hairy1* (see above). Lewis (2003) has devised a model for somitogenesis in the zebrafish based on a mechanism for biochemical oscillation proposed independently by himself and Monk (2003). Lewis hypothesized that *her1* and *her7* constitute an autoregulatory transcription factor of the sort discussed in Chapter 3 and, furthermore, that DeltaC is a downstream effector of this oscillation. The two genes negatively regulate their own expression (Holley *et al.*, 2002; Oates and Ho, 2002) and are positively regulated by Notch signaling (Takke and Campos-Ortega, 1999; Oates and Ho, 2002). Even though, in mutants in which Notch signaling is disrupted, somitogenesis breaks down (Holley *et al.*, 2002), the resulting pattern of *deltaC* expression in the PSM is not what would be expected if the oscillator had halted in all the cells. Rather, there is a salt and pepper distribution of DeltaC, suggesting that the oscillator has become asynchronous. This has led Lewis to propose that juxtacrine signaling by the Notch pathway, usually considered to act in the determination of cell fate (see below), in this case acts to keep cells in the segment-generating growth zone in synchrony (Lewis, 2003). Such synchrony is an experimental finding (Stern and Bellairs, 1984; Primmett *et al.*, 1989) that enters into most models of segmentation.

Lewis provided a simple mechanism for the oscillatory expression of the *her1* and *her7* genes, which we briefly summarize here. The model is based on the assumption that there exists a feedback loop in which the Her1 and Her7 proteins bind directly to the regulatory DNA of their own genes to inhibit transcription. Also incorporated into the model is the recognition that there is always a delay T_{m} between the initiation of transcription and the initiation of translation (since it takes time for the mRNA molecule to translocate into the cytoplasm), as well as a delay T_{p} between the initiation of translation and the emergence of a complete functional protein molecule.

These ingredients are put into mathematical language in the following way. For a given autoregulatory gene, let $m(t)$ be the number of mRNA molecules in a cell at time t and let $p(t)$ be the number of the corresponding protein molecules. The rates of change of m and p are then assumed to obey the following equations:

$$\frac{\mathrm{d}p(t)}{\mathrm{d}t} = am(t - T_{\mathrm{p}}) - bp(t), \tag{7.1}$$

$$\frac{\mathrm{d}m(t)}{\mathrm{d}t} = f\,(p(t - T_{\mathrm{m}})) - cm(t). \tag{7.2}$$

Here the constants b and c are the decay rates of the protein and its mRNA, respectively, a is the rate of production of new protein molecules, and $f(p)$ is the rate of production of new mRNA molecules.

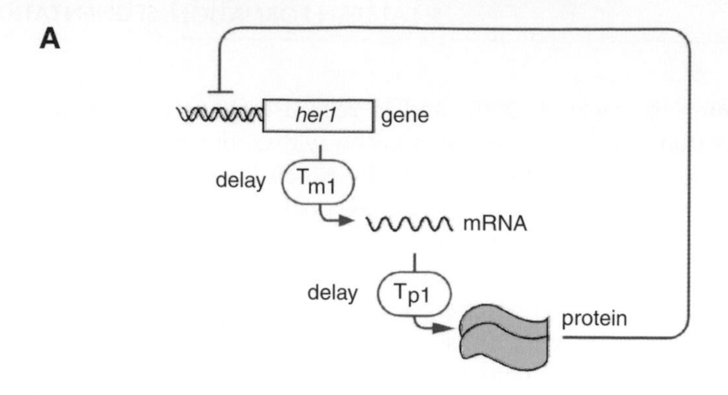

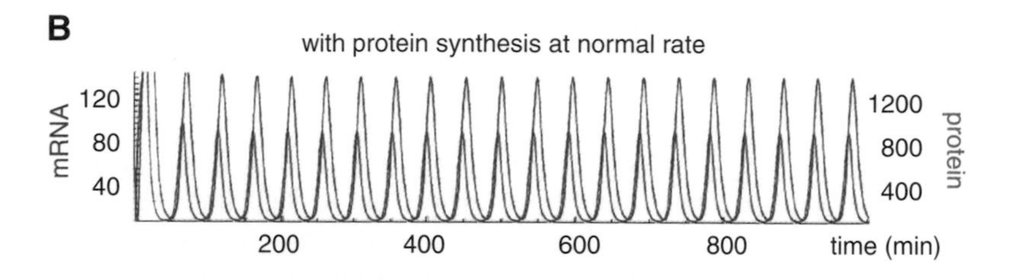

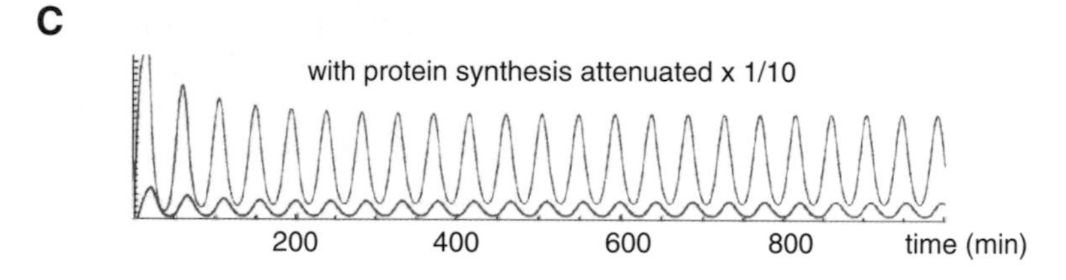

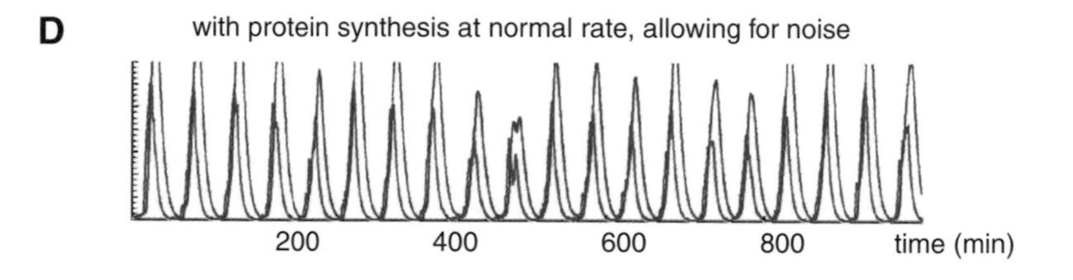

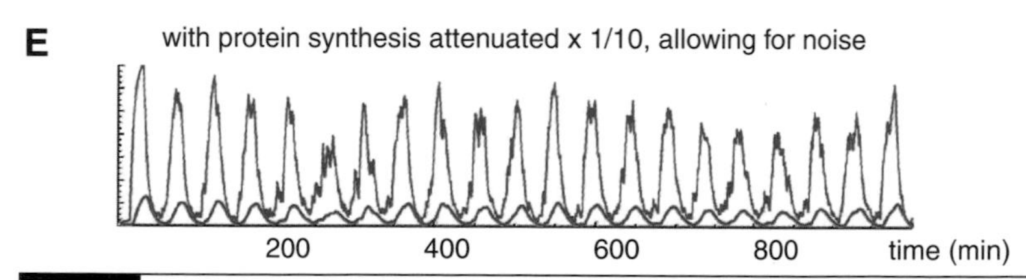

Fig. 7.4 Cell-autonomous gene-expression oscillator for zebrafish somitogenesis. (A) Molecular control circuitry for a single gene, *her1*, whose protein product acts as a homodimer to inhibit *her1* expression. In the case of a pair of genes (i.e., *her1* and *her7*) the analogous circuit would contain an additional branch with coupling between the two branches. (B) Computed behavior for the system in A (defined by Eqs. 7.1 and 7.2), in terms of the numbers of mRNA molecules per cell in red and protein molecules per cell in blue. Parameter values were chosen appropriate for the *her1* homodimer oscillator (see the form of the function $f(p)$ in the text) on the basis of experimental results: $a = 4.5$ protein molecules per mRNA molecule per minute; $b = c = 0.23$ molecules

(Cont.)

The function $f(p)$ is assumed to be a decreasing function of the amount of protein. (The form used by Lewis and Monk is

$$f(p) = \frac{k}{1 + p^2/p_0^2},$$

with constants k and p_0, to represent the action of an inhibitory protein, assumed to be a dimer. The results of simulations turned out to be quite insensitive to the specific form of $f(p)$.)

The above time-delay differential equations were numerically solved for Her1 and He7 (for which Lewis, drawing on experimental results, was able to estimate the values of all the model parameters in Eq. 7.1 and 7.2). The solutions indeed exhibit sustained oscillations in the concentration of Her1 and Her7, with predicted periods close to the observed ones (Fig. 7.4). The important conclusions from this analysis are that no oscillations are possible if delay is not incorporated (i.e., if $T_m = T_p = 0$) and that the oscillators are quite insensitive to the rate of protein synthesis (i.e., to the value of a, Eq. 7.1). Furthermore, Lewis showed that incorporation into the model of the inherently noisy nature of gene expression (by adding stochastic effects to the deterministic equations 7.1 and 7.2) reinforces continued oscillations (Fig. 7.4). (Without noise, oscillations are eventually damped, which would upset normal somite formation beyond the first few.)

The sequence of blocks of tissue generated by the clock and wavefront mechanism become somites only after a physical boundary forms at their interfaces. In the quail, the expression of the gene *Lunatic fringe*, a modulator of *Notch* signaling, at the position −1 (i.e., one somite length caudal to the most recently formed intersomitic morphological boundary) initiates fissure formation at the boundary between the PSM and what will become the posterior half of the next-forming somite (Sato *et al.*, 2002).

As will be seen below, the generation of left–right asymmetry in the vertebrate embryo utilizes many of the signaling molecules employed in somitogenesis, e.g., FGFs, the Wnt and Notch pathways. Dynamical pattern-forming systems using shared diffusible components would interfere with each other in the absence of special ways

per minute, corresponding to protein and mRNA half-lives of 3 minutes; $k = 33$ mRNA per diploid cell per minute, corresponding to 1000 transcripts per hour per gene copy in the absence of inhibition; $p_0 = 40$ molecules, corresponding to a critical concentration of around 10^{-9} M (moles per litre) in a 5 μm diameter cell nucleus; $T_m \approx 20.8$ min; $T_p \approx 2.8$ min. (C) Decreasing the rate of protein synthesis (to $a = 0.45$) causes little or no effect on the period of oscillation. All the other parameters are the same as in B. (D) The computed behavior for the system in A when the noisy nature of gene expression is taken into account. To model stochastic effects an extra independent parameter is introduced, the rate constant k_{off} for the dissociation of the repressor protein (i.e., Her1) from its binding site on the regulatory DNA of its own gene (i.e., *her1*). Results are shown for $k_{off} = 1\ \text{min}^{-1}$, corresponding to a mean lifetime of 1 min of the repressor bound state. (E) The same as in D except for the rate of protein synthesis, which is as in C. Parameter values not mentioned explicitly in C–E are the same as in B. (Reprinted from Lewis, 2003, with permission from Elsevier.)

of protecting against this. It is significant, therefore, that mammals, birds, and fish all employ the diffusible lipid molecule retinoic acid, in a mechanism that resists the tendency of somites on opposite sides of the body to go out of register as they are forming and left–right asymmetry is simultaneously being established (Vermot *et al.*, 2005; Vermot and Pourquié, 2005; Kowakami *et al.*, 2005).

Epithelial patterning by juxtacrine signaling

As discussed above, juxtacrine signaling is one of the inductive mechanisms that can lead to pattern formation by the production of new cell types. In juxtacrine communication, signaling molecules anchored in the cell membrane bind to and activate receptors on the surface of immediately neighboring cells, thus typically leading to patterning on the scale of a few cells.

In the discussion of the segmentation model of Lewis (2003), Notch signaling was employed to establish synchrony among neighboring cells of a single type. More typically, the Notch–Delta system is used to specify different fates (i.e., the commitment to specific types) in adjacent cells of a developmentally equivalent population. An example of this is the differentiation of gonadal cells in the nematode *Caenorhabditis elegans*. The predifferentiated cells can differentiate into either an anchor cell (AC) or a ventral uterine precursor cell (VU). The choice between AC or VU, however, depends on the interaction between the Notch-type receptor, LIN-12, of one cell with its Delta-like ligand, LAG-2, on the adjacent cell. Activation of LIN-12 in one cell forces that cell to adopt the VU fate, whereas cells in which the action of LIN-12 is suppressed adopt the AC fate (Greenwald, 1998).

In other cases, nonequivalent precursor cells communicate through Notch–Delta signaling to progress to the next stage of differentiation. When this occurs, intrinsic or extrinsic factors confer a bias to one of two neighbors, and this is then sharpened or consolidated by Notch–Delta interactions. This was the hypothesized role of Notch and Delta in the Kerszberg–Changeux model for neurulation discussed in Chapter 5. It is also the role played by this juxtacrine pair in the induction of mesoderm at the beginning of sea urchin gastrulation. Specifically, between the eighth and ninth cleavage in the sea urchin *Strongylocentrotus purpuratus* the micromeres (see Chapter 5; Fig. 5.7) begin to express Delta, which triggers the specification of the adjoining inner ring of cells (bearing a Notch receptor and referred to as Veg2 cells; see Fig. 5.7), to a non-skeletogenic mesodermal fate (Sherwood and McClay, 2001). In amphibians (Green, 2002; see below) and mammals (Beddington and Robertson, 1999), the equivalent gastrulation-related inductive steps appear to occur via paracrine, rather than juxtacrine, mechanisms.

Initial evidence suggested that Notch's function was mainly to repress the acquisition of a new state of differentiation by cells that carry the Notch protein (Artavanis-Tsakonas *et al.*, 1999): the Notch receptor-bearing cell is inhibited from assuming a differentiated state, while the Delta ligand-bearing cell is free to do so. Consequently, the Notch–Delta pathway was thought to act only via lateral inhibition. More recently, however, this simple model has come into conflict with evidence for longer-range effects mediated by Notch-dependent positive differentiation signals (Rook and Xu, 1998; Pourquié, 2000).

A model constructed by Collier *et al.* (1996) for Notch–Delta signaling, which made use only of lateral inhibition, gave rise to pattern over a range of one or two cells. Here we summarize the model of Sherratt and coworkers (Owen *et al.*, 2000; Wearing *et al.*, 2000; Savill and Sherratt, 2003), which predicts more realistic patterns via juxtacrine communication on a longer scale on the basis of a positive feedback mechanism (Owen *et al.*, 1999; see also Webb and Owen, 2004). The model, like some of those presented in Chapters 3 and 5, employs "linear stability analysis" to predict transitions between alternative dynamical states. (See the Appendix to Chapter 5 for an example of how this works).

Juxtacrine signaling: the model of Sherratt and coworkers

Juxtacrine signaling can lead to pattern formation via either lateral inhibition or lateral induction (i.e., activation). (Inhibition and activation are the main outcomes of what we referred to as inductive signaling earlier in this chapter. Often, however, the term "induction" is used synonymously with activation, and we will occasionally follow this usage in what follows.) Experimental evidence suggests that the Notch–Delta ligand–receptor interaction (see the model for neurulation in Chapter 5) can produce both inhibition (Haddon *et al.*, 1998) and induction (Huppert *et al.*, 1997). The model of Sherratt *et al.* (Owen and Sherratt, 1998; Owen *et al.*, 2000; Wearing *et al.*, 2000) provides a general mechanism for lateral induction through juxtacrine communication in an epithelial sheet, which the authors propose as a prototype for the Notch–Delta system. Embryonic epithelial patterns that may arise in this fashion include the veins in the Drosophila wing, the feather germs (primordia) in avian skin, and stem cell clusters in the epidermis (Owen *et al.*, 2000; Savill and Sherratt, 2003). Below we summarize the general features of the model.

1. Consider a two-dimensional cellular sheet, a portion of which is depicted in Fig. 7.5. Let $a_{ij}(t)$, $f_{ij}(t)$, and $b_{ij}(t)$ denote respectively the number of ligand molecules, free receptors, and bound receptors on the surface of the cell at the intersection of row i and column j (Owen *et al.*, 2000).

2. Sherratt and coworkers used a simple kinetic scheme to model the time variation of $a_{ij}(t)$, $f_{ij}(t)$, and $b_{ij}(t)$. Ligands and free receptors can reversibly associate to produce bound ligand–receptor complexes (at a rate k_a) or decay (with respective decay constants d_a and d_f). Bound ligand–receptor complexes can dissociate (at a rate k_D and thus produce new ligands and receptors at the same rate) or be internalized (at a rate k_I). The distinguishing feature of the model is a positive feedback mechanism: apart from the dissociation of the bound complexes,

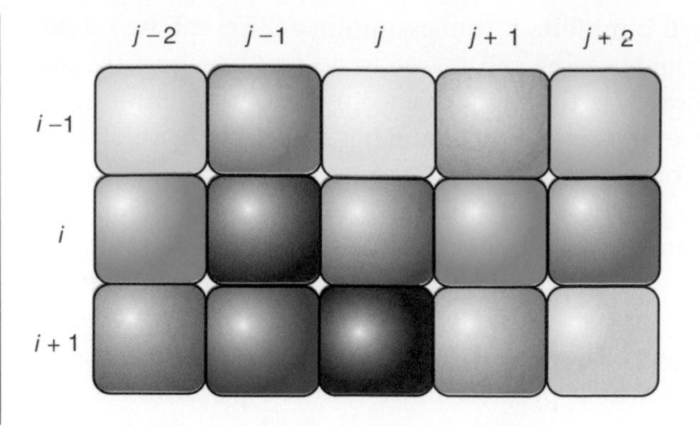

Fig. 7.5 Part of the cellular grid used in the simulation of Sherratt and coworkers (Owen *et al.*, 2000). Each square represents a single cell in an epithelial sheet. Neighboring cells are connected via Notch–Delta ligand–receptor pairs. The rows and columns of cells are labeled by the indices i and j, respectively. The position of a cell is uniquely determined by the pair ij.

there is an additional source of ligands and receptors with a rate of production that increases with the level of occupied receptors (governed by the functions $P_a(b_{ij})$ and $P_f(b_{ij})$, which are both increasing functions of b_{ij}). In mathematical form,

$$\frac{da_{ij}}{dt} = -k_a a_{ij}\langle f_{ij}\rangle + k_D\langle b_{ij}\rangle - d_a a_{ij} + P_a(b_{ij}),$$

$$\frac{df_{ij}}{dt} = -k_a\langle a_{ij}\rangle f_{ij} + k_D b_{ij} - d_f f_{ij} + P_f(b_{ij}), \tag{7.3}$$

$$\frac{db_{ij}}{dt} = k_a\langle a_{ij}\rangle f_{ij} - k_D b_{ij} - k_I b_{ij}.$$

For the values of the rate constants in the above equations, Sherratt and coworkers used the experimental data for epidermal growth factor and its receptor (Waters *et al.*, 1990). For the specific forms of the functions $P_a(b_{ij})$ and $P_f(b_{ij})$ see Owen *et al.* (2000). The symbol $\langle\ \rangle$ in Eqs. 7.3 denotes an average over nearest neighbor cells and is the reflection of juxtacrine signaling. For example,

$$\langle a_{ij}\rangle \equiv \tfrac{1}{4}(a_{i,j-1} + a_{i,j+1} + a_{i-1,j} + a_{i+1,j}). \tag{7.4}$$

The terms on the right-hand side represent the total number of ligand molecules on the surfaces of cells (in two dimensions) adjacent to the cell at the position (i, j) in Fig. 7.5. It is assumed that the ligands are uniformly distributed along the four lateral faces of a cell attached to its neighbors.

3. Once a homogeneous (i.e., spatially uniform) steady-state solution, denoted by a_h, f_h, b_h, of the system given by Eqs. 7.3 is determined (it is easy to show that at least one such nontrivial solution exists), linear stability analysis around this solution, along the lines described in the Appendix to Chapter 5, can be performed (for details see Wearing *et al.*, 2000). The authors introduced the expressions $a_{ij} = a_h + \tilde{a}_{ij}$, $f_{ij} = f_h + \tilde{f}_{ij}$, $b_{ij} = b_h + \tilde{b}_{ij}$, where the quantities with wavy lines are assumed to be small perturbations and in general inhomogeneous (i.e., varying over space). Inserting these forms into Eqs. 7.3, using the fact that a_h, f_h, b_h satisfy the original equations with the left-hand sides set to zero, and retaining only terms linear in the small quantities (i.e., linearizing the equations around the homogeneous solutions), a set of equations is generated for the spatially

varying quantities $\tilde{a}_{ij}$, $\tilde{f}_{ij}$, $\tilde{b}_{ij}$, whose solutions are sought in the form $\tilde{a}_{ij} = \tilde{a}e^{\lambda t}\cos kr$ or $\tilde{a}_{ij} = \tilde{a}e^{\lambda t}\sin kr$ (and similarly for $\tilde{f}_{ij}$ and $\tilde{b}_{ij}$; $r = (i, j)$ is a compact notation for the cell position). Here $\tilde{a}$ is a constant, λ is the temporal growth rate and k is a wavenumber. (Actually the solutions can be written in the complex-number form $\tilde{a}_{ij} = \tilde{a}e^{\lambda t + ikr}$, where $\mathrm{i} = \sqrt{-1}$; in view of the relationship $e^{ikr} = \cos kr + \mathrm{i}\sin kr$, this is just a convenient way to incorporate simultaneously the forms containing $\cos kr$ and $\sin kr$.)

For spatial pattern formation, it is required that the homogeneous solution a_h, f_h, b_h be (i) stable under homogeneous perturbations and (ii) unstable under inhomogeneous perturbations. The first case corresponds to $k = 0$, $\lambda < 0$, whereas for the second case $k \neq 0$ and $\lambda > 0$. For a complex system of this sort, linear stability analysis typically identifies solutions for several possible wavenumbers, which are called "modes." Furthermore, for each mode there is a functional relationship between k and the corresponding λ. The fastest-growing mode will be the one that eventually prevails, so it is this mode that is identified with the developing pattern. Which of the modes is the fastest growing depends on the parameters in Eqs. 7.3.

In principle, Eqs. 7.3 determine all possible particular patterns of the dynamical system, without any stability analysis. However, finding the inhomogeneous solutions of such a complex system presents a formidable mathematical challenge. Linear stability analysis is a systematic way to explore the wide range of possibilities. Once a particular inhomogeneous solution describing a pattern is identified by linear stability analysis, one may attempt to locate the region in parameter space where this solution is stable using numerical methods, as was done by Sherratt and coworkers (Owen *et al.*, 2000). (Note that linear stability analysis can only establish the possibility of a particular spatial pattern, because the amplitude ($e^{\lambda t}$) of the mode that characterizes the pattern grows in time and thus the linear stability analysis itself breaks down eventually.) The system defined by Eqs. 7.3 contains five parameters in addition to the two functions $P_a(b_{ij})$ and $P_f(b_{ij})$, and thus such an analysis is highly nontrivial.

In Fig. 7.6 we reproduce a pattern generated by the model of Sherratt *et al.* that corresponds to specific forms of the functions $P_a(b_{ij})$ and $P_f(b_{ij})$.

Mesoderm induction by diffusion gradients

The nonuniform distribution of any chemical substance, whatever the mechanism of its formation, can clearly provide spatial information to cells. Let us assume, for example, that a population of cells is capable of assuming more than one differentiated fate – ectoderm vs. mesoderm, for example – or differentially adhesive versions of the same cell type confronting each other across a compartmental boundary. We have seen above how cell-autonomous or juxtacrine mechanisms are capable of generating such spatially patterned configurations. Exposing an array of cells to varying amounts of an external signaling molecule is another, even simpler, way to make them behave differently from one another. And molecular diffusion is probably the most physically straightforward and reliable way of producing a range of levels of a substance across a spatial domain.

For at least a century, embryologists have considered models for pattern formation and its regulation that employ diffusion gradients. Only in the last decade, however, has convincing evidence been produced that this physical mechanism is utilized in early development. The prime evidence comes from studies of mesoderm induction, a key event preceding gastrulation (see Chapter 5) in the frog, *Xenopus*. As noted above, important early steps in mesoderm induction in sea urchins occur by a Notch–Delta-based juxtacrine mechanism. In contrast, analogous events in *Xenopus* occur through paracrine inductive signaling by secreted factors. "Animal caps" (prospective ectoderm) of *Xenopus* blastulas, when explanted alone, become epidermis and express all the molecular and morphologic features of skin. The prospective fates of the vegetal pole explants are the endoderm of the gut, as well as extra-embryonic tissue. Nieuwkoop juxtaposed these two types of explants and demonstrated that the vegetal poles induced the animal caps to become muscle instead of skin (Nieuwkoop, 1969).

The following kinds of study strongly suggested that a paracrine mechanism was at work (see the review by Green, 2002). Various soluble factors of the TGF-β superfamily (e.g., activin, BMPs, Nodal) and the FGF family could substitute for the vegetal pole cells (i.e., the "vegetal pole organizer") in the experiments of Nieuwkoop. Interference with receptors for these factors in animal cap cells, or the administration of protein inhibitors of their activities, such as Noggin or Chordin, attenuated induction. Different doses of activins and FGFs elicited different cell-type-characteristic patterns of gene expression in animal cap cells. Finally, both TGF-β (McDowell *et al.*, 2001) and FGFs (Christen and Slack, 1999) can diffuse over several cell diameters.

None of this proves beyond question that the simple diffusion of morphogens between and among cells, rather than some other, cell-dependent, mechanism, actually establishes the gradients in question. Kerszberg and Wolpert (1998) for example, asserted that the capture of morphogens by receptors impedes diffusion to the extent that stable gradients can never arise by this mechanism. They proposed that morphogens are instead transported across tissues by a "bucket brigade" mechanism in which a receptor-bound morphogen on one cell moves by being "handed off" to receptors on an adjacent cell. This could be accomplished, for example, by repeated cycles of exocytosis and endocytosis, referred to as "planar transcytosis" (Freeman, 2002).

While arguments can be made in either direction (i.e., for diffusion or for the bucket-brigade mechanism) on the basis of simple models and thought experiments, sometimes the best way of answering a question of this sort is to set up a mathematical model with a realistic degree of complexity (one that would defy intuitive examination or straightforward analytical solution) and solve it numerically. This is what Lander and coworkers (2002) did, to address the

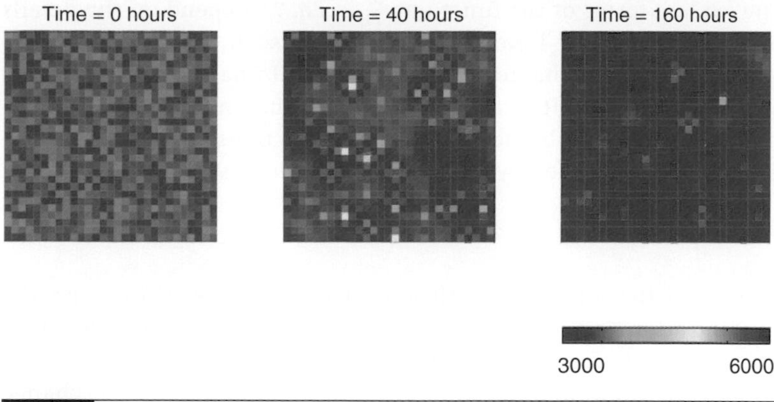

Fig. 7.6 An extended quasi-periodic spatial pattern that results from lateral induction in the model of Sherratt and coworkers (Owen *et al.*, 2000), defined by Eqs. 7.3. Pattern development is induced by small random perturbations about the homogeneous equilibrium applied throughout a 30×30 grid of cells. Only the density of bound-ligand–receptor complexes (molecules per cell, see the color scale) is shown. The final pattern has a characteristic wavelength (with some irregularities), which depends crucially on the strength of feedback, as quantified by the functions $P_a(b_{ij})$ and $P_f(b_{ij})$: the stronger the feedback the longer the characteristic wavelength. The simulation results correspond to the following parameter set in Eqs. 7.3: $k_a = 3 \times 10^{-4}$ molecules^{-1} min^{-1}, $k_D = 0.12$ min^{-1}, $k_I = 0.019$ min^{-1}, $d_a = 6 \times 10^{-3}$ min^{-1}, $d_f = 0.03$ min^{-1}. The feedback functions are given for all cell positions (i, j) by $P_a(b) = C_1 b/(C_2 + b)$ and $P_f(b) = C_3 + C_4^3 b^3/(C_5^3 + b^3)$, with $C_1 = 110$, $C_2 = 2500$, $C_3 = 90$, $C_4 = 7.4$, and $C_5 = 5450$. The periodicity of the pattern in the figure is in qualitative agreement with the results of linear stability analysis. (Reprinted from Owen *et al.*, 2000, with permission from Elsevier.)

question whether the simple extracellular diffusion of morphogens combined with receptor binding can plausibly set up developmental gradients. The biological system that they modeled was the *Drosophila* wing imaginal disc (see Chapter 5), where quantitative estimates of the spreading of morphogens are available (Entchev *et al.*, 2000; Teleman and Cohen, 2000). However, their conclusions may also apply to other systems, such as mesoderm induction in *Xenopus* (see above) and vertebrate limb development (see Chapter 8). Since the model of Lander *et al.* (2002), as well as others (for axis formation and left–right asymmetry), discussed later in this chapter, is based on generalized reaction–diffusion systems, first we briefly outline the standard mathematical analysis used to study such systems.

Reaction–diffusion systems

The rate of change in the concentrations c_i, $i = 1, 2, \ldots, N$, of N interacting molecular species is determined by their reaction kinetics and expressed in terms of ordinary differential equations

$$\frac{dc_i}{dt} = F_i(c_1, c_2, \ldots, c_N). \tag{7.5}$$

The explicit forms of the functions F_i in Eq. 7.5 depend on the details of the reactions. We have seen examples of such equations in Chapter 3 (cf. Eqs. 3.1), Chapter 4 (cf. Eqs. 4.3) and Chapter 5 (cf. Eqs. 5.1). Spatial inhomogeneities also cause time variations in the concentrations even in the absence of chemical reactions. If these inhomogeneities are governed by diffusion then, in one spatial dimension,

$$\frac{\partial c_i}{\partial t} = D_i \frac{\partial^2 c_i}{\partial x^2}. \tag{7.6}$$

Here D_i is the diffusion coefficient of the ith species. (Because the concentrations now depend continuously on both x and t we need to use symbols for partial differentiation, as explained in Box 1.1.) In general, both diffusion and reactions contribute to the change in concentration at any point in the reaction domain, and the time dependence of the c_i is governed by reaction–diffusion equations:

$$\frac{\partial c_i}{\partial t} = D_i \frac{\partial^2 c_i}{\partial x^2} + F_i(c_1, c_2, \ldots, c_N). \tag{7.7}$$

Generation of morphogen gradients: the analysis of Lander and coworkers

Lander and coworkers (Lander *et al.*, 2002) applied the above formalism in their attempt to resolve the controversy concerning whether morphogen gradients are set up by diffusion or by more elaborate mechanisms (e.g., the bucket-brigade mechanism; Kerszberg and Wolpert, 1998). They considered the generation of a gradient of the protein Decapentaplegic (Dpp), a morphogen expressed in the *Drosophila* wing imaginal discs. The authors chose this system because Dpp had previously been visualized in living embryos by introducing a genetic construct, in which Dpp was fused to green fluorescent protein (Entchev *et al.*, 2000; Teleman and Cohen, 2000). They could thus compare their model's predictions with experimental results directly. Below we summarize the assumptions of the model and the main conclusions of its analysis.

1. The wing disc represents an essentially two-dimensional system in which the morphogen sources consist of a linear array of cells in the center of the disc. Since Lander *et al.* were primarily interested in the formation of gradients perpendicular to this array (along the anteroposterior axis in the case of Dpp), they formulated their model in one dimension (Fig. 7.7). They assumed that morphogens are introduced at a constant rate v at one end of the system and absorbed at a distance 100 μm away (x_{max}), representing a distance of about 40 cell diameters.

2. As morphogen molecules diffuse in the intercellular space, they interact with receptors on the cells' surfaces and establish ligand–receptor complexes, which form and decay with respective rates k_{on} and k_{off}. The model also allowed for the internalization and subsequent degradation of the morphogen–receptor complexes at a rate k_{deg}.

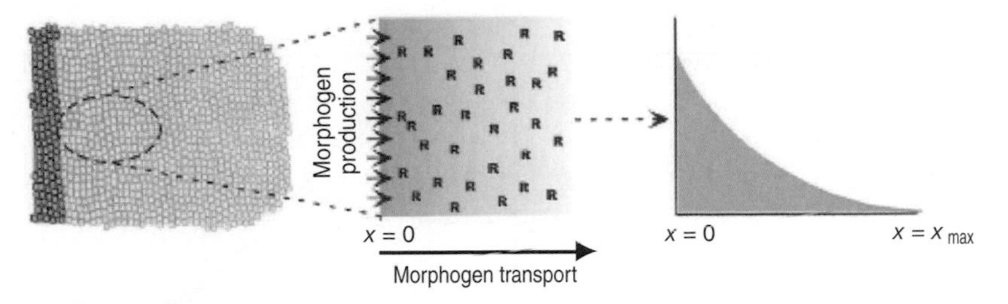

Fig. 7.7 Representation of morphogen fields in the model of Lander *et al.* (2002). In the left-hand panel a stripe of cells produces a morphogen (orange) that spreads between cells over a distance of approximately 40 cell bodies (light grey). This situation mimics the Dpp gradient in the wing imaginal discs of the *Drosophila* larvae. The middle panel illustrates the proposed mechanism of gradient generation: morphogen molecules emanating from a linear source encounter and bind to a homogeneous distribution of receptors (R) in the adjacent two-dimensional space. On the right, this situation is simplified to a one-dimensional model with constant morphogen production at $x = 0$, absorption at $x = x_{max}$, and an initially uniform receptor concentration throughout. (Reprinted from Lander *et al.*, 2002, with permission from Elsevier.)

3. If R_{tot} is the total number of receptor molecules per cell and A and B are, respectively, the fractional (normalized to R_{tot}) concentration of free and receptor-bound morphogens, then the evolution of the morphogen pattern (subject to the above conditions) is governed by the following equations:

$$\frac{\partial A}{\partial t} = D\frac{\partial^2 A}{\partial x^2} - k_{on}R_{tot}A(1-B) + k_{off}B, \tag{7.8a}$$

$$\frac{\partial B}{\partial t} = k_{on}R_{tot}A(1-B) - k_{off}B - k_{deg}B. \tag{7.8b}$$

In the most general reaction–diffusion systems (cf. Eq. 7.7) every molecular species is capable of diffusing (see also below). But since the diffusive term is missing in Eq. 7.8b, the above equations represent a special type of reaction–diffusion system, in which freely diffusing molecules (i.e., morphogens) react with immobilized ones (i.e., receptors). All the parameters in the above equations, including the morphogen diffusion coefficient D, can be estimated reasonably accurately from experimental data.

4. For parameter values compatible with the experimental results, biologically "useful" situations with steady-state gradients of receptor occupancy are generated by the system defined by Eqs. 7.8a, b. For the gradient to be biologically useful it must be able to distribute spatial information over the entire field of cells and thus not be too steep. One situation is illustrated in Fig. 7.8A, where the solution of Eqs. 7.8a, b for B is plotted for a particular set of model parameters (see figure legend). This solution is not biologically useful, since the variation in B, and thus the ability of the morphogen to produce pattern, is at any time restricted to a negligibly narrow spatial region (i.e., the gradient is too steep). A different set of model parameters, however, produces a morphogen gradient that is biologically useful

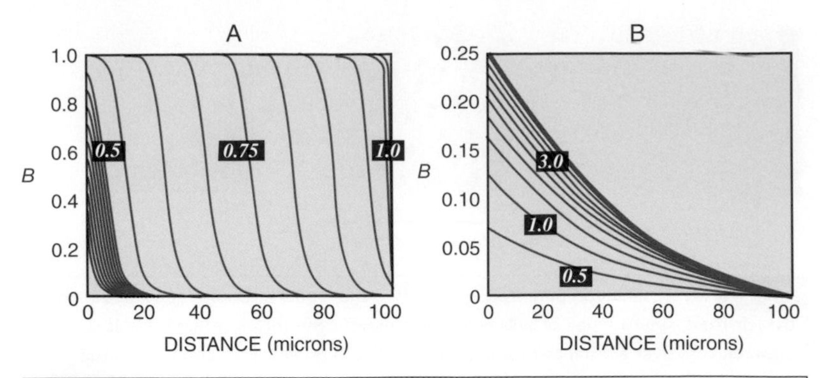

Fig. 7.8 Morphogen gradients generated in the model of Lander *et al.* (2002). Equations 7.8 were solved with the initial conditions (at $t = 0$) $B = 0$ for all x and $A = 0$ for all values of x different from zero and the boundary conditions $A = B = 0$ at $x = x_{max}$ and a nonzero flux, $\partial A/\partial t = v/R_{tot}$, at $x = 0$. (A) A biologically ineffective gradient of the fractional receptor occupancy B, for parameter values $D = 10^{-7}\text{cm}^2\ \text{s}^{-1}$, $k_{deg} = 2 \times 10^{-4}\ \text{s}^{-1}$, $v/R_{tot} = 5 \times 10^{-4}\ \text{s}^{-1}$, $k_{on}R_{tot} = 1.32\ \text{s}^{-1}$, $k_{off} = 10^{-6}\ \text{s}^{-1}$. (B) A biologically "useful" gradient of the fractional receptor occupancy for the parameter values listed above, with the exception of $v/R_{tot} = 5 \times 10^{-5}\ \text{s}^{-1}$ and $k_{on}R_{tot} = 0.01\ \text{s}^{-1}$. The time interval between successive curves is 300 s in panel A and 1800 s in panel B. The cumulative time represented by selected curves is shown in hours. The curves in B, unlike those in A, approach steady-state receptor occupancy. (After Lander *et al.*, 2002.)

(Fig. 7.8B). For this choice of parameters, steady-state receptor occupancy occurs in about 4 hr, and the resulting gradient extends over the entire field. Not surprisingly, such solutions arise only when the rate of morphogen production v is slower than receptor turnover, $v < R_{tot}\, k_{deg}$.

Further generalization of the model (allowing for ligand–receptor complexes internalized by the cells to signal from within endocytic compartments, followed either by their return to the cell surface or destruction) does not change the conclusion that the extracellular diffusion of morphogens can effectively set up gradients which can eventually determine cell fate (Lander *et al.*, 2002). The same study also provided strong quantitative evidence that other mechanisms of morphogen transport (e.g., bucket-brigade or transcytosis) would require a series of cell-biological events to occur at implausibly fast rates. But since endocytosis plays an additional role in the long-range movement of Dpp that is not addressed in the pure extracellular diffusion model (Kruse *et al.*, 2004), the question of how the Dpp gradient is actually established cannot be considered to be entirely settled.

Once the gradient of a signaling molecule is established (by whatever means), the next step is for cells to respond to it. There are various ways in which this can happen. The simplest way in which cells can interpret a spatially graded signal is through unidirectional hierarchical induction (Fig. 7.1B(ii)), where the gradient serves as positional information and the cells respond to it according to a corresponding set of internal thresholds (Wolpert, 1969). A more complex and general way in which cells respond to a nonuniform signaling environment is by an emergent system of positive and negative feedback interactions with neighboring cells (Fig. 7.1B(ii)). In such cases the shape of the resulting gradient need not be monotonic, as it would with simple diffusion, and achievement of the induced cell states is more of an active process than for a positional information mechanism. Both of these categories of gradient-response, or inductive, mechanisms presuppose that changes in tissue geometry and topology via morphogenetic mechanisms (Fig. 7.1A) occur after all new cell states have been induced. An even more complex scenario occurs when tissue reshaping occurs simultaneously with the generation of new cell states, resulting in continual rearrangement of the signaling environment (Salazar-Ciudad and Jernvall, 2002). There are numerous examples of such "morphodynamic" mechanisms (Salazar-Ciudad *et al.*, 2003), but we will not consider them here. Instead, we will conclude this chapter with several examples of emergent inductive pattern formation in the establishment of the vertebrate body plan.

Control of axis formation and left–right asymmetry

The prefertilization amphibian egg is radially symmetric about the animal–vegetal axis. After fertilization (see Chapter 9), radial symmetry is broken during the first cleavage division, when the cortical cytoplasm actively rotates 30° in the direction of the animal pole while the deeper, internal, cytoplasm, containing dense yolky cells, remains oriented by gravity (Fig. 7.9). The site where the sperm enters

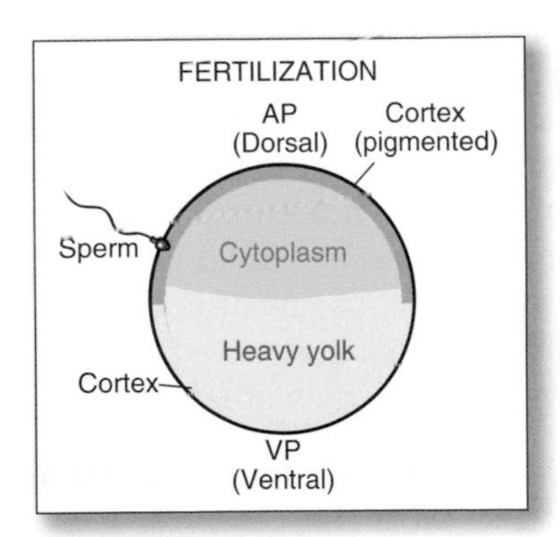

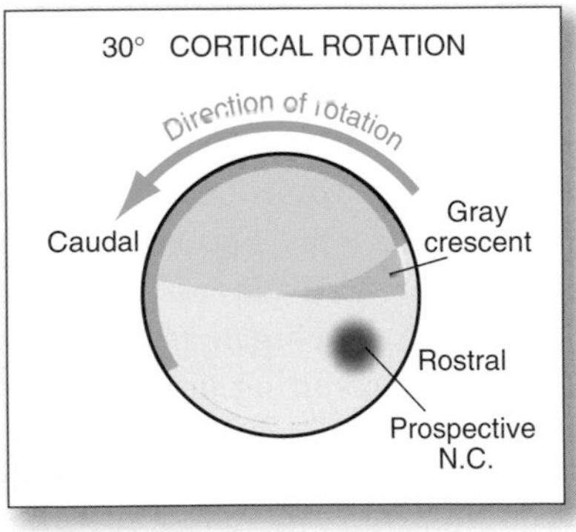

Fig. 7.9 Cortical rotation in amphibian embryos. The egg, before (on the left), and after (on the right) the 30°postfertilization rotation of the cortical cytoplasm (medium and light gray) relative to the inner cytoplasm (blue). The cortex rotates in such a way that the sperm entry point moves vegetally during the first cell cycle. The gray crescent seen in the right-hand panel is formed by the overlapping of the pigmented animal pole cytoplasm with non-pigmented vegetal pole cytoplasm. The animal pole (AP) and vegetal pole (VP), which is close to the future site of the Nieuwkoop center, "Prospective N.C.," and the prospective dorsal, ventral, rostral (anterior), and caudal (posterior) regions of the animal are shown. (Based on Elinson and Rowning, 1988, with changes.)

has traditionally been designated as the future ventral (belly) surface of the embryo, the opposite side as the future dorsum (back), and the axis connecting them as the dorsal–ventral (or dorsoventral) axis.

Improved marking techniques that permit the following of individual cells and their progeny in living embryos (Lane and Smith, 1999; Lane and Sheets, 2000) have led to revisions of earlier established *Xenopus* "fate maps." (A fate map is an assignment of the cells of an experimentally unperturbed early embryo, e.g., a blastula or gastrula, to the regions or specific cells of the fully developed organism to which they will give rise). In particular, it was suggested that the longstanding designation of the dorsoventral axis contained serious inconsistencies. This, in turn, led Lane and Sheets (2002) to redefine the principal axes in the *Xenopus* embryo. They proposed to reassign the dorsoventral axis to the animal–vegetal aspect of the embryo. The traditional dorsoventral axis is now designated as the rostrocaudal (nose to tail) or anteroposterior axis (Figs. 7.9 and 7.10; see also Kumano and Smith, 2002, and Gerhart, 2002). This reassignment of axes harmonises the terminology used to describe *Xenopus* development with that for other vertebrates. In particular, the assignment of the rostrocaudal axis is now consistent with the general head-to-tail development of vertebrate embryos. The Lane and Sheets designation of the axes is used in Figs 7.9 and 7.10 and in the following description.

The postfertilization differential cytoplasmic rotation creates a region on the side of the embryo opposite to the sperm entry point (i.e., the rostral side, see Fig. 7.9) called the "gray crescent." Next to this region gastrulation will be initiated, by the formation of an indentation, the "blastopore," through which the surrounding cells invaginate.

Spemann and Mangold (1924) discovered that the rostral blastopore lip (containing cells derived from the gray crescent) constitutes an *organizer*, a population of cells that directs the movement of other cells. The action of the Spemann–Mangold organizer ultimately leads to convergent extension, the formation of the notochord, and thus the body axis (see Chapter 5). When they transplanted an organizer from another gastrula into an embryo at a point some distance from the embryo's own organizer, Spemann and Mangold observed that two axes formed, resulting in conjoined twins. Since then other classes of vertebrates have been found to have similarly acting organizers.

A property shared by organizers in other developing systems (including invertebrates such as *Hydra*, one of the systems in which the organizer phenomenon was first identified; Browne, 1909) is that if an organizer is extirpated then adjacent cells differentiate into organizer cells and assume its role. This suggests that one of the functions of the organizer is to suppress nearby cells with similar potential from exercising it. This implies that the body axis is a "self-organizing" system. Another manifestation of self-organization is that if animal and vegetal cells from a *Xenopus* embryo are dissociated, separately reaggregated, and confronted with each other as amorphous cell masses then not only mesodermal tissue (see the section on mesoderm

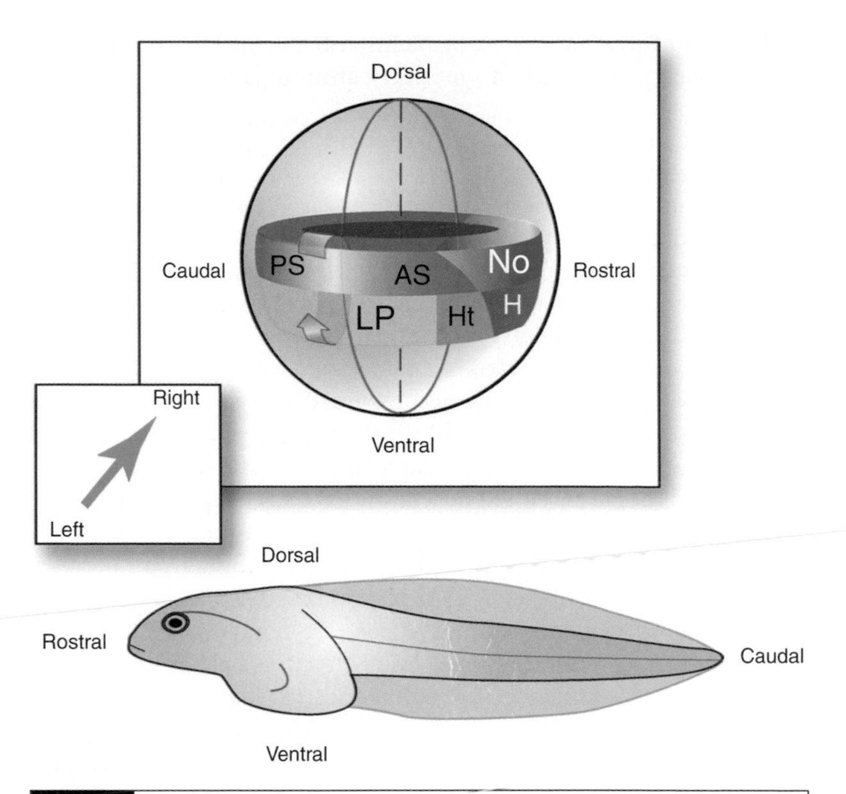

Fig. 7.10 Fate map of a *Xenopus* embryo at the late blastula stage. By this stage the three germ layers are already determined, the mesoderm (shown as an annulus) arising from the cells of the equatorial or marginal zone. The blastula is viewed from its left, as is the tadpole (larval stage frog) into which it will develop. During gastrulation the mesoderm undergoes inversion (curved arrow), accounting for the rostrocaudal axis reversal seen between the blastula and tadpole stages. The primordia of the notochord (No), the anterior and posterior somites (AS, PS), the lateral plate mesoderm (LP), the heart (Ht), and the head mesoderm (H) are indicated. The rostrocaudal (i.e., anteroposterior) and dorsoventral aspects of the blastula (which are the same as in the fertilized egg shown in Fig. 7.9) and of the tadpole are indicated. (After Gerhart, 2002, and Lane and Sheets, 2002.)

induction above) but also a recognizable notochord and somites will be induced (Nieuwkoop, 1973, 1992).

The formation of the body axis in vertebrates also exhibits another unusual feature: while it occurs in an apparently symmetrical fashion, with the left and right sides of the embryo seemingly equivalent to one another, at some point this symmetry is broken. Genes such as *nodal* and *Sonic hedgehog* start being expressed differently on the two sides of the embryo, and the whole body eventually assumes a partly asymmetric morphology, particularly with respect to internal organs such as the heart.

In his pioneering paper on reaction–diffusion systems Turing (1952) demonstrated that such systems have inherent pattern-forming potential, with a particular propensity to exhibit symmetry breaking across more than one axis. In the following, we present the application of this framework by Meinhardt to axis formation in vertebrates and the breaking of symmetry around these axes.

Meinhardt's models for axis formation and symmetry breaking

Viewed from the perspective of physics, early development, during which the organism acquires its final shape, is a series of symmetry-breaking events starting from a highly symmetrical spheroidal egg and arriving at a body with a much lower degree of symmetry. The understanding of symmetry breaking has long occupied physicists because it underlies the striking phenomena of phase transitions. Phase transitions are typically accompanied by a reduction in symmetry. A well-known example will illustrate this point. In a paramagnet, molecules are elementary magnets, each with its own random orientation. These can be aligned in an external magnetic field, but once the field is turned off the molecules continue their uncorrelated existence and the system shows no net magnetization. Upon decreasing the temperature, however, a critical point is reached below which the elementary magnets couple. If they are now aligned by an external field, they remain aligned even after the field is turned off: the paramagnet changes to a ferromagnet. The originally perfectly isotropic, featureless system, with a net magnetization of zero in every direction, gradually self-organizes into a "pattern." It becomes anisotropic, characterized by a single axis defined by the direction of its net magnetization.

It took great efforts to decipher the mechanism responsible for the paramagnet–ferromagnet and similar phase transitions. At high temperatures, the thermal energy (proportional to the absolute temperature T, see Chapter 1), which causes strong fluctuations and tends to disorient the molecules, dominates over the interaction energy between neighbors. With decreasing temperature, the thermal energy diminishes and the interaction energy can more effectively compete with it: adjacent molecules align and gradually induce others to follow suit. Short-range correlations become amplified and, as a consequence, the system responds to external influences (e.g., an imposed magnetic field) in a collective manner. The message from this example is that symmetry breaking, with its macroscopic manifestation, may be induced from a microscopic localized source (e.g., individual molecules in a paramagnet) as a consequence of competition between antagonistic driving forces.

Can our knowledge of the physical mechanisms underlying phase transitions provide insight into the generation of developmental axes? We may consider the various uncommitted cells in the early embryo to be analogous to the molecules with random magnetic orientation. Organizers, such as the vegetal pole, the "Nieuwkoop center" (Fig. 7.9), and the Spemann organizer, may be considered as analogous to the local sources whose influence, under permissive conditions, is propagated through progressive interactions between cells in such a way that a pattern, and eventually an axis, may be induced to form. While thermal fluctuations are not relevant in early development, the states of cells in a developmental "field" (a term often used in embryology in analogy with its use in physics; Gilbert, 2003) are susceptible to influences by adjacent cells and diffusible signals. Specific competing processes analogous to thermal randomization and magnetic dipole interactions may therefore be responsible for the formation of axes.

Meinhardt hypothesized (Meinhardt, 2001; see also Meinhardt and Gierer, 2000) that the expression levels of known genes involved in axis formation (e.g., *VegT*, *FGF*, *brachyury*, *β-catenin*, etc.) are outcomes and indicators of such biochemical competition. Below we summarize his reaction–diffusion-based model of symmetry breaking and axis formation in amphibians.

1. The first goal that a model of axis formation must achieve is to generate an organizer *de novo*. Biochemically, this is reflected in a pattern of high local concentrations and graded distributions of one or more signaling molecules, which can be produced by the coupling of a self-enhancing feedback loop, acting over a short range, and a competing inhibitory reaction, acting over a longer range. Meinhardt proposed a simple system acting in the xy plane (representing two-dimensional sections of the embryo) that consists of a positively autoregulatory activator (with concentration $A(x, y; t)$) and an inhibitor (with concentration $I(x, y; t)$). The activator controls the production of the inhibitor, which in turn limits the production of the activator. This process can be described by the following reaction–diffusion system (Meinhardt, 2001)

$$\frac{\partial A}{\partial t} = D_A \left(\frac{\partial^2 A}{\partial x^2} + \frac{\partial^2 A}{\partial y^2} \right) + s \frac{A^2 + I_A}{I\,(1 + s_A A^2)} - k_A A \tag{7.9a}$$

$$\frac{\partial I}{\partial t} = D_I \left(\frac{\partial^2 I}{\partial x^2} + \frac{\partial^2 I}{\partial y^2} \right) + s A^2 - k_I I + I_I. \tag{7.9b}$$

The terms proportional to A^2 specify that the feedback of the activator to its own production and that of the inhibitor is in both cases nonlinear. The factor $s > 0$ describes positive autoregulation, the capability of a molecule to induce positive feedback and regulate its own synthesis. This may occur by purely chemical means ("autocatalysis"), which is the mechanism originally considered by Turing (1952). More generally, in living tissues, positive autoregulation occurs if a cell's exposure to a factor that it has secreted causes it to make more of the same factor (Van Obberghen-Schilling *et al.*, 1988). The inhibitor slows down the production of the activator (by the $1/I$ factor in the second term in Eq. 7.9a). Both activator and inhibitor diffuse and decay with respective diffusion coefficients D_A, D_I and rate constants k_A, k_I. The small baseline inhibitor concentrations I_A and I_I can initiate activator self-enhancement or suppress its onset, respectively, at low values of A. The factor s_A, when present, leads to the saturation of positive autoregulation. Once the positive autoregulatory reaction is under way, it leads to a stable self-regulating pattern in which the activator is in dynamic equilibrium with the surrounding cloud of the inhibitor.

2. The various organizers and subsequent inductive interactions leading to symmetry breaking, axis formation and the appearance of the three germ layers in amphibians during gastrulation can all be modeled, in principle, by the reaction–diffusion system defined by Eqs. 7.9a, b or by the coupling of several such systems. The biological relevance of such reaction–diffusion models depends on whether there exist molecules that can be identified as activator–inhibitor pairs. Meinhardt's model starts with a default state, which consists of ectodermal tissue. Patch-like activation generates the first "hot spot," the vegetal pole organizer, which induces endoderm formation (a simulation is shown in Fig. 7.11A). A candidate for the diffusible activator in the corresponding self-enhancing loop for endoderm specification is the TGF-β-like factor Derriere, which activates the transcription factor VegT (Sun *et al.*, 1999). Evidence for the existence of an inhibitor is more circumstantial. VegT expression remains localized to the vegetal pole, but this is not because the surrounding cells lack the competence to produce VegT (Clements *et al.*, 1999). Subsequently, a second feedback loop forms a second hot spot in the vicinity of the first, in the endoderm. This is identified with the Nieuwkoop center, an organizing region, which appears in the rostro-vegetal quadrant of the blastula (see Fig. 7.9) at around the 32-cell stage (Gimlich, 1985, 1986). A candidate for the second self-enhancing loop is the molecular assembly of the Wnt-pathway (including β-catenin), a ubiquitous early developmental signaling cascade (Fagotto *et al.*, 1997). Meinhardt made the interesting suggestion that the inhibitor for this second loop might be the product of the first loop (i.e., the vegetal pole organizer). As a result of this local inhibitory effect, the Nieuwkoop center is displaced from the pole (see the simulation in Fig. 7.11B). With the formation of the Nieuwkoop center the spherical symmetry of the embryo is broken. In Meinhardt's model this symmetry breaking "propagates" and thus forms the basis of further symmetry breakings, in particular left–right asymmetry (see below).

3. The vegetal pole and the Nieuwkoop center are examples of the localized spot-like patterns generated by activator–inhibitor reaction–diffusion systems in Eqs. 7.9a, b (with $s_A = 0$). Such systems can also induce the full zonal separation of cells with ectodermal, endodermal, and mesodermal specification into stripes (if $s_A > 0$) and with this establish the dorsoventral axis (which, according to the new axis assignments, coincides with the animal–vegetal axis, as discussed above). This is achieved by using several competing feedback loops in such a way that in any cell only one of these loops can be active. For a realistic description of pattern formation, loops that are able to induce spots and zones must be coupled (Fig. 7.11C).

4. By secreting several diffusible factors, the Nieuwkoop center induces the formation of the Spemann–Mangold organizer (Harland and Gerhart, 1997). Interestingly, if the second feedback loop, responsible for the Nieuwkoop center, is not included in the model then two Spemann–Mangold organizers appear, symmetrically with respect to

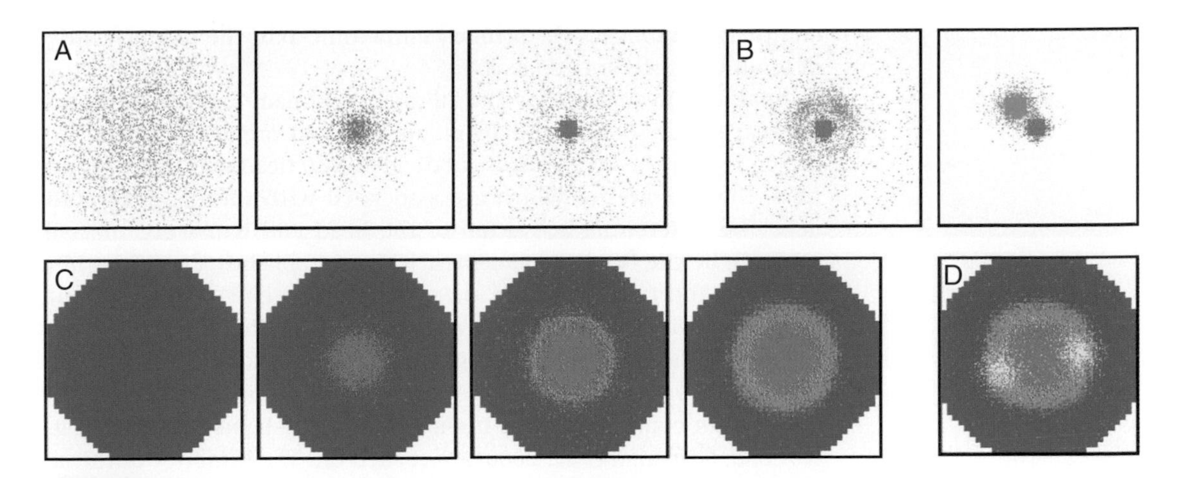

Fig. 7.11 Pattern formation in the reaction–diffusion model of Meinhardt. (A) Induction of the vegetal pole organizer. Left: The interaction of an autocatalytic activator, the TGF-β-like factor Derriere (red), and a long-ranging inhibitor (whose production it controls and which, in turn, limits the activator's production), creates an unstable state in an initially near-uniform distribution of the substances (the inhibitor is not shown). Middle and right: A small local elevation in the activator concentration above steady-state levels triggers a cascade of events governed by Eqs. 7.9, a further increase in the activator due to autocatalysis, the spread of the concomitantly produced surplus of inhibitor into the surrounding area, where it suppresses activator production (middle), and the establishment of a new stable state in which the activator maximum (at the hot spot) is in a dynamic equilibrium with the surrounding cloud of inhibitor (right). The right-hand panel also shows the near-uniform distribution of the activator (green) in the second reaction–diffusion system, discussed in B. (B) Induction of the Nieuwkoop center. Once the first hot spot has formed, it activates a second self-enhancing feedback loop. The production of the activator (green) in this reaction is inhibited by the vegetal pole organizer itself. As a consequence, the Nieuwkoop center is displaced from the pole. (C) The zonal separation of ectoderm, endoderm, and mesoderm. The competition of several positive feedback loops ensures that in one cell only one of these loops is active. As the results indicate, reaction–diffusion systems can produce not only spot-like organizers but also zones. Endoderm (red) forms, in the way shown in A, from a default ectodermal state (blue). The mesodermal zone (green; it forms by the involvement of the *FGF–brachyury* feedback loop, Schulte-Merker and Smith, 1995) develops a characteristic width from the additional self-inhibitory influence in Eqs. 7.9. (D) Induction of the Spemann–Mangold organizer. The activation of an organizing region (yellow) within the still-symmetric mesodermal zone would imply a competition over relatively long distances. In such a case, Eqs. 7.9 lead to the occurrence of two organizing regions, a biologically unacceptable result. The strong asymmetry shown in B prevents this and suggests a reason for the existence of the Nieuwkoop center. (Reprinted from Meinhardt, 2001, with permission from the University of the Basque Country Press.)

the animal–vegetal axis, and no symmetry breaking occurs (Fig. 7.11D). With the formation of the Spemann–Mangold organizer the developmental process is in full swing. The organizer triggers gastrulation, in the course of which the germ layers acquire their relative positions and the notochord forms. This long thin structure marks the midline of the embryo, and this midline itself inherits organizer function and eventually establishes the primary embryonic axis, the rostral–caudal or anterior–posterior (AP) axis.

Spatial information for the AP axis is already contained in the pregastrula, as soon as the Spemann–Mangold organizer forms. The reason is that the organizer is subdivided into head and trunk organizers, the activities of which are associated with specific gene products. The differential activation of the head and trunk organizer is achieved by the Nieuwkoop center, which establishes a gradient in a TGF-β-like morphogen. At the low concentration end of this gradient, away from the Nieuwkoop center, the genes for the trunk organizer (e.g., *Xnot*, the *Xenopus* homologue of zebrafish *floating head*; Danos and Yost, 1996) are activated. Near the Nieuwkoop center, where the morphogen concentration is high, genes of the head organizer (e.g., *goosecoid*; Gritsman *et al.*, 2000) are turned on. The inhibitor of the activator for the head organizer is the activator of the trunk organizer and vice versa. A simulation of midline formation based on Meinhardt's model is shown in Fig. 7.12.

5. Finally, the breaking of left–right symmetry can be understood as again a competition between already existing and newly developing self-enhancing loops, similarly to the formation of the Nieuwkoop center and prospective germ layer zones. The molecule that best fulfils the role of the "left" activator in Meinhardt's model is the product of the *nodal* gene, so called because it is expressed around Hensen's node, the avian and mammalian counterpart of the Spemann–Mangold organizer, as well as around the Spemann–Mangold organizer itself (Weng and Stemple, 2003). The Nodal protein, which is a diffusible, positively autoregulatory, member of the TGF-β superfamily, induces expression from the embryonic midline of another TGF-β-related molecule, Lefty, and Lefty antagonizes Nodal production (Juan and Hamada, 2001; Branford and Yost, 2002; Yamamoto *et al.*, 2004). Because Nodal and Lefty are antagonistic diffusible signals that differ in the range of their activities (Chen and Schier, 2002; Branford and Yost, 2002; Sakuma *et al.*, 2002), the ingredients for a symmetry-breaking event along the primary embryonic axis are present (Solnicka-Krezel, 2003).

Interestingly, although the major molecular players involved in symmetry breaking in the frog and the mouse appear to be the same, in the chicken, another popular experimental system for studying axis formation, things appear to be quite different. In this species, when Nodal is initially expressed in Hensen's node, it is already spatially asymmetric (Pagan-Westphal and Tabin, 1998). The asymmetry in expression thus appears at a relatively earlier stage than in the frog or mouse and is unlikely to be induced by a diffusible signal (Dathe *et al.*, 2002). In fact, *nodal*'s asymmetric expression in the chicken embryo appears to result from a juxtacrine Notch–Delta-based signaling system, such as that described above for epithelial patterning (Raya *et al.*, 2003, 2004). A molecularly realistic dynamical model for the generation of axis-associated gene expression patterns in the chicken can be found in Raya *et al.* (2004).

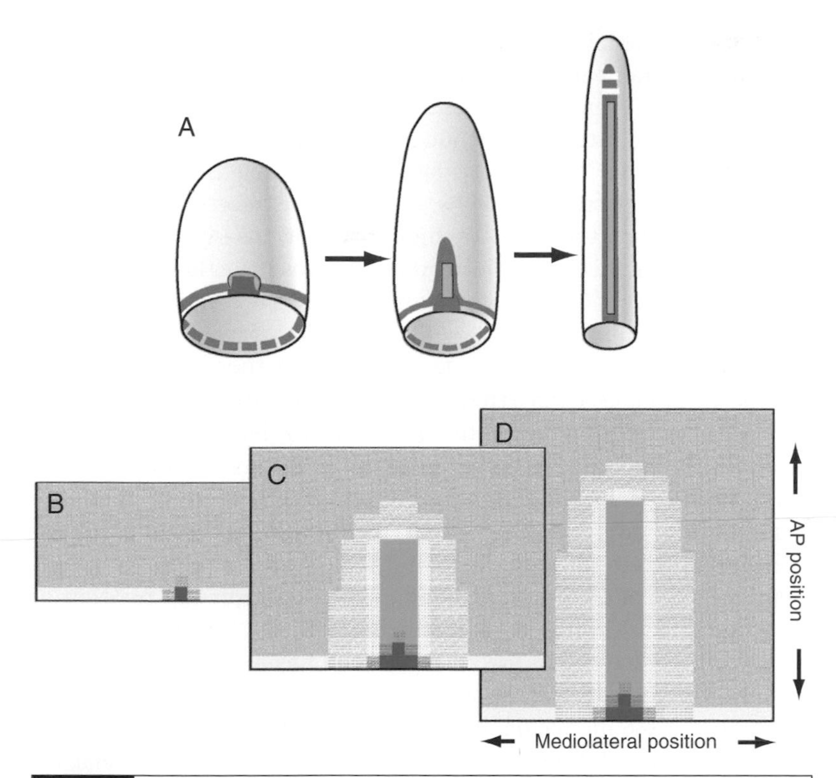

Fig. 7.12 Formation of the midline and enfolding of the anteroposterior (AP) axis according to the reaction–diffusion model of Meinhardt. (A) Schematics of the hypothesized processes involved in axis formation. Cells close to the blastopore (red annulus) move towards the Spemann–Mangold organizer (blue). The spot-like organizer redirects the cells in its vicinity: initially they are drawn to the organizer but then lose their attraction to it, extending from the posterior towards the anterior, so that they leave as a unified stripe-like band. Stem cells in the organizer region may contribute to this band to form the most central ("medial" as opposed to "lateral") element (green) of the dorsal surface. Such highly coordinated cellular motion and strong positional specification along the enfolding AP axis require the simultaneous action of several feedback loops. (B–D) A simplified simulation of the scenario just described. A reaction–diffusion system tuned to make stripes (green; compare with the zone formation in Fig. 7.11) is triggered by the organizer (blue). The organizer itself is the result of a self-enhancing system activated in a spot-like manner. (In B–D the blastopore is shown in yellow.) Repulsion between the spot system (the organizer) and the stripe system (the notochord) causes elongation of the latter. Saturation in self-enhancement (due to the factor s_A in Eq. 7.9a) ensures that the stripe system does not disintegrate into individual patches and thus establishes the midline. This, in turn, acts as a sink for a ubiquitously produced substance (pink), which could be the product of the BMP-4 gene (Dosch *et al.*, 1997); the local concentration of this substance (shades of pink) is a measure of the distance from the midline. This simulation is simplified in that the actual movement of cells toward the organizer is not considered; instead, the midline elongates by the addition of lines of new cells next to the blastopore. (A redrawn and B–D reprinted from Meinhardt, 2001, with permission from the University of the Basque Country Press.)

Embryonic axial asymmetry is reflected in left–right differences in the expression pattern of certain genes, such as *Sonic hedgehog* and *nodal* itself, and eventually in the left–right asymmetry of organs such as the heart and intestines. Although total inversion of the symmetry of internal organs (*situs inversus totalis*) usually has little adverse impact on health, partial inversions, in which the heart, for example, is predominantly on the right, are highly deleterious (Aylsworth, 2001), and perhaps for this reason vertebrate embryos have the means to ensure that 99.99% of humans, for example, have the standard left–right asymmetry.

If we entertain Meinhardt's hypothesis that the axis-forming system of vertebrates is governed by a reaction–diffusion mechanism then we are confronted with the question how left–right asymmetry is so reliably generated in these organisms. Supp and coworkers found that a mouse genetic variant called *iv*, characterized by random positioning of the internal organs, carries a mutation in the microtubule-associated protein left–right dynein (LRD) (Supp *et al.*, 1997). Dyneins form the part of the molecular motor that controls the rotation of cilia, and LRD is expressed in "monocilia" (cilia are motile extensions of the cell surface and in this case each cell has only one) in Hensen's node in the mouse. The mutation in LRD in the *iv* mutant has the result that these monocilia are immotile (Supp *et al.*, 2000).

This motivated Nonaka and coworkers to look at the relationship between these monocilia and the distribution of Nodal around Hensen's node in the mouse embryo (Nonaka *et al.*, 1998). They found that the vortical motion of the monocilia in Hensen's node is implicated in left–right symmetry breaking. The clockwise rotational movement of several hundred nodal monocilia generates leftward (i.e., anticlockwise) fluid currents (nodal flow) that bias the distribution of *nodal* and thus determine the polarity of the broken symmetry (Nonaka *et al.*, 2002). In a striking confirmation of their model for axis symmetry breaking, when Nonaka *et al.* artificially reversed the direction of nodal flow in wild-type embryos, or subjected the *iv* mutant to leftward flow, they were able to induce the normal arrangement of internal organs.

These results have a ready interpretation in terms of the reaction–diffusion mechanism for embryonic axis generation described above (Solnicka-Krezel, 2003). Reaction–diffusion mechanisms in nonliving systems are indifferent to which side is left and which side is right, the local initiators of activation typically arising by random fluctuations (see, for example, Castets *et al.*, 1990; Ouyang and Swinney, 1991). In embryonic systems, where there is clearly a premium on having pattern-forming mechanisms that generate the same results in each successive generation, molecular or structural cues may act as guides, biasing a mechanism capable of producing multiple outcomes to generate one that is biologically functional. The monocilia near Hensen's node in the mouse and the embryonic organizers of other vertebrate species (Essner *et al.*, 2002) appear to play such a role. Here the inherent pattern-forming potential of one physical system (i.e., the axis-forming reaction–diffusion process comprising the cells and various gene products described above) is channeled and restricted by its interaction with another set of physical systems (i.e., the motile monocilia). Although these systems are intrinsically different in scale, the collective behavior of the appropriately arranged monocilia (a product of evolution; see Chapter 10 for a discussion of the evolution of developmental mechanisms) tips the balance in the otherwise indifferent pattern-forming mechanism in such a way as to reliably produce a biologically functional macroscopic outcome.

Perspective

Embryos employ a strikingly wide range of physical mechanisms to establish the characteristic spatial patterns of cell differentiation. Most of these mechanisms – adhesion-based cell sorting, juxtacrine feedback loops, the establishment of gradients over the multicellular spatial scale by diffusion – are virtual inevitabilities in simple aggregates composed of cells capable of differentiation. Epithelia, containing cells in direct contact with one another, or mesenchymes, containing ECMs permeable to diffusible factors, can only employ physically based mechanisms arising from, or compatible with, their material nature. Living tissues thus have an inherent capacity to generate spatially heterogeneous developmental outcomes. When the relevant physical mechanisms are embodied in, or integrated with, fine-tuned molecular signal-response mechanisms (e.g., the Notch–Delta or Nodal–Lefty couples), they become indispensable components of reliable developmental pattern-formation mechanisms.

Chapter 8

Organogenesis

As a consequence of generating and reshaping the epithelia of their ectodermal and endodermal germ layers and (for those species that have them) the mesenchymes that make up the mesodermal germ layer and neural crest, embryos take on one of approximately three dozen stereotypical "body plans." These body plans can be asymmetric (sponges), radially symmetric (e.g., hydra and sea urchins), bilaterally symmetric (e.g., planaria, insects), or "bilaterally asymmetric" (vertebrates; see the last section of Chapter 7). While it is obvious that nothing in development, or in any other domain of biology for that matter, can occur without the participation of physical mechanisms, the high degree of structural and dynamical complexity of most living systems makes it exceedingly difficult, in general, to follow the workings of basic physical principles or appreciate their roles in generating characteristic biological phenomena. It is therefore remarkable to how great an extent (as the preceding chapters have shown) the tissue generation and reshaping processes leading to embryonic body plans can be accounted for by physical forces and properties – adhesion, viscoelasticity, dynamical multistability, network formation, positive and negative feedback dynamics, diffusion – that also govern the behaviors of nonliving condensed materials, that is, by "generic" physical mechanisms.

The ability to understand aspects of early development in generic physical terms can be attributed, in part, to the likelihood that the original multicellular organisms were simple, loosely organized, cell masses, whose forms were determined to a great extent by their inherent physical properties (see Chapter 10). Early-stage modern-day embryos, however "programmed" they may be by sophisticated genetic mechanisms of pattern formation that have evolved over the last half-billion years, are likely still to retain many of these presumed original physical determinants of form and pattern (Newman and Müller, 2000).

Attempts to use basic physical models to understand changes at later developmental stages encounter greater difficulties. The simultaneous actions of multiple morphogenetic and patterning processes over time, along with the acquisition by cells of specific differentiated properties (e.g., the contractility of muscle cells, the electrical excitability of nerve cells, the secretion of solid matrices by cartilage and bone cells) causes the developing body to become increasingly complex, both structurally and functionally. And once an active circulatory system is in place (in species that have one), and "endocrine" tissue types have been generated that are capable of transmitting tissue-specific regulatory signals through this network of branching and anastomosing (i.e., intercommunicating) conduits, the hope of accounting for coordinated changes in body shape and form by one or two generic physical mechanisms must be abandoned.

Nonetheless, in most animal species the various specialized organs develop in a semi-autonomous fashion from distinct clusters of cells ("organ rudiments") during a period after the establishment of the body plan but prior to the extensive elaboration and integration of the embryo's or fetus's anatomical and physiological systems. Organogenesis, the formation of bodily organs, therefore can make use of all the generic physical mechanisms discussed previously in relation to body-plan formation and in certain cases utilize some novel ones. A physical mechanism that generates a branched network of tubules, for example, while not a likely determinant of any known animal body plan can come into play at the level of organogenesis, during the formation of insect tracheal systems or the lungs, kidneys, and vascular systems of vertebrates. Branched tubes can also provide the basis for external appendages such as limbs, in animals with exoskeletons (e.g., insects), as we saw when we considered the Mittenthal–Mazo model in chapter 5.

We considered the physical properties of epithelial and mesenchymal tissues separately in Chapters 4, 5, and 6 and, indeed, during development of the germ layers and body plans these tissue types often act independently of one another, as, for example, in neural crest dispersion. In other cases, as happens with convergent extension in vertebrates and archenteron elongation in sea urchins, the two tissue types behave in a coordinate or parallel fashion, as if they were subject to similar forces. During organogenesis, however, *epithelial–mesenchymal interactions*, in which the specific biomechanical properties of the two components act in a synergistic fashion in producing an outcome, become a major determinant of morphogenesis.

We saw hints of this capability in the description of neurulation in Chapter 5. In that case, the mesenchymally derived notochord (by that stage a stiff rod) induced the overlying epithelial sheet to form neural folds. However, in organ rudiments not constrained to produce a structure with a central axis the mesenchymal component will typically retain its "liquid" state during morphogenesis, the tissue spreading as a coherent mass and the individual cells moving relative to one another through the ECM. As we will see below, epithelial–mesenchymal interactions under these more general conditions can lead to the formation of branched and lobulated structures, such as those seen in the pancreas and salivary glands, or of arrays of structural elements that are repetitive along more than one axis, as seen in bird feather tracts or the vertebrate limb skeleton. And while we saw examples of epithelium-to-mesenchymal transformation in mesoderm generation during gastrulation (Chapter 5) and during the formation of the neural crest (Chapter 6), organogenesis is often accompanied by a transformation of mesenchyme to epithelium, with associated physically induced morphogenetic consequences.

In this chapter we will describe several examples of organ formation, in particular, assembly of the blood vessel network (vasculogenesis), gland development (branching morphogenesis), and skeletogenesis in the vertebrate limb. As earlier, we will introduce, as far as possible, physical concepts not encountered previously along with the biological topics covered.

Development of the cardiovascular system

The cardiovascular system of vertebrate organisms is a network of tubes that defines a single, enclosed, blood-filled space. The heart, an initially tubular structure that undergoes extensive modifications during development, is interconnected with millions of tubular blood vessels, beginning with the pulmonary vein and artery and the large-caliber aorta and vena cava (two large veins), which extend directly from the heart, continuing through the branched arteries and veins of intermediate caliber, on to the small-caliber arterioles and venules and minuscule capillaries responsible for local blood supply. In the systemic circulation, oxygenated blood is pumped from the heart through the aorta and is carried by the arteries to the body's organs and tissues; oxygen-depleted blood from these peripheral sites is returned to the heart by way of the veins and ultimately the vena cava. Reoxygenation of the blood takes place by means of the pulmonary circulation, in which the pulmonary artery conveys deoxygenated blood from the heart to the lungs and the pulmonary vein returns it to the heart in oxygenated form. (Thus in this artery–vein pair alone, the normal oxygenation status is reversed.) Since the circulatory system is essentially closed (with rare exceptions, as in lymph nodes, where white blood cells of the immune system can squeeze through the walls of small blood vessels, or in the liver, where the capillary walls are discontinuous), oxygen and soluble nutrients are supplied to the tissues, and waste materials conveyed away, by diffusion across the membranes of the *endothelial* cells, which line all the vessels and comprise the major structural component of the capillary walls.

Despite the eventual continuity of the cardiovascular system's tubular network, the components that represent its extremes in size and function develop independently of one another. In the chicken, for example (see Fig. 8.1), the two lateral tubes that ultimately fuse with one another to make the primitive heart begin forming at 25 hr of development (top panel), by 29 hr (bottom panel) fusion has taken place and the heart has already begun to beat. In all vertebrate organisms the initially straight heart tube undergoes "looping," conversion into a curved tube (Fig. 8.2) directed, in the vast majority of cases, towards the right side of the embryo (though ultimately the larger half of the mature, asymmetric, heart resides in the left half of the chest (Manner, 2000; see Fig. 8.2D).

Interestingly, looping and the subsequent development into a morphologically normal heart is a self-organizing property of cardiogenic mesenchyme that does not require fusion of the parallel tubes shown in Fig. 8.2A. In embryonic mice carrying a mutation in the gene specifying the transcription factor Foxp4, no midline fusion of the heart tubes takes place. Nonetheless, each bilateral heart-forming region is capable of developing into a four-chambered heart (Li *et al.*, 2004).

Cardiac looping is the first morphological manifestation of the left–right asymmetry in body-axis formation discussed in Chapter 7, although it is preceded by molecular asymmetries. In *Xenopus* (Breckenridge *et al.*, 2001) and zebrafish (Chen *et al.*, 1997), for example, cardiac looping follows a left-predominant expression of the morphogen BMP4 in the linear heart tube. Taber and coworkers determined that the looping actually comprises two mechanical motions, a rotation and a bending to the right (Voronov and Taber, 2002). While the bending motion may be influenced by asymmetric gene expression, Taber and coworkers found, using video tracking of fluorescently labeled live myocardial (developing heart muscle) cells, that the chicken heart has no intrinsic ability to rotate (Voronov *et al.*, 2004). In a striking example of the role of physical mechanisms in morphogenesis, these investigators demonstrated that external forces, exerted by the intraembryonic *splanchnopleure* (the ventrally located layer of mesoderm and endoderm (Fig. 8.1, top panel) that eventually forms the wall of the digestive tube, and the veins associated with this layer of tissue), drive the rotation and determine the directionality of looping in the initial stages of this process.

During the same period small blood vessels, which will eventually be part of a completely interconnected network of tubes that includes the heart, are developing at sites distant from that organ (Rupp *et al.*, 2003). In the yolk sac (a membranous structure outside the embryo proper, which in the chicken embryo plays a similar role to the vegetal pole cells of the frog egg) and in the splanchnopleure, cords of cells arise that eventually become tubular by a process of pattern formation referred to as "cavitation" (Lubarsky and Krasnow, 2003). During cavitation, the cord's internal core differentiates into blood cells, thereby becoming subject to flow (driven by the beating of the newly forming heart, once connectivity is established), while its external layer remains in place, forming a vessel.

In the *somatopleure*, the dorsally located layer of mesoderm and ectoderm that eventually forms the body wall (Fig. 8.1, top panel),

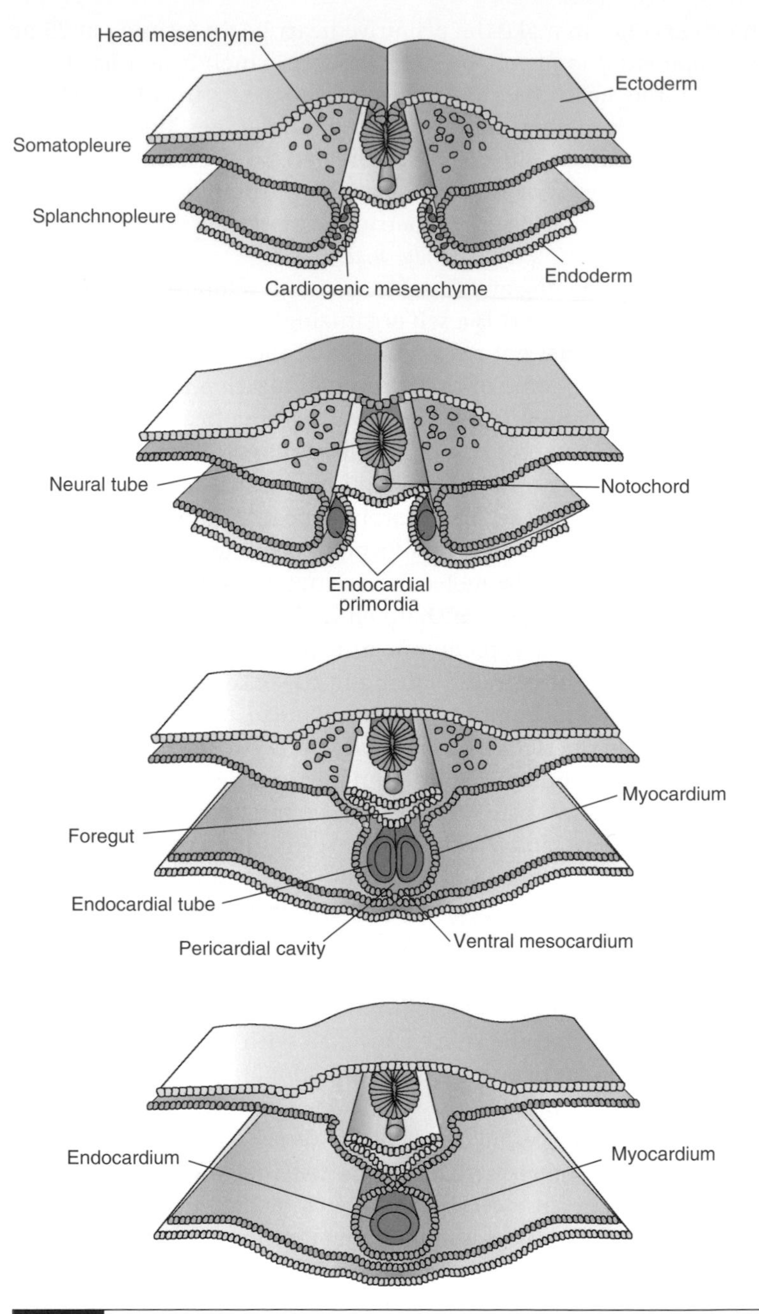

Fig. 8.1 (opposite)

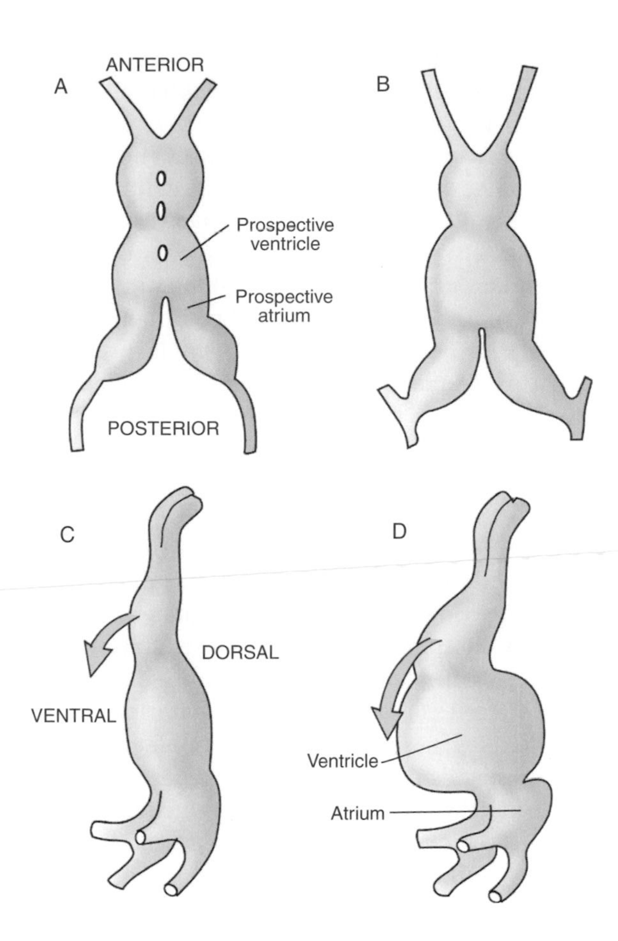

Fig. 8.2 Fusion of the two heart tubes and cardiac looping. (A, B) A ventral view of fusing heart tubes in the human embryo at developmental stages corresponding to the chicken stages shown in the lowest two panels of Fig. 8.1. (C) The fused heart tube in B is viewed from the left side at a slightly later stage. The anterior portion of the tube bends in a ventral and posterior direction (C) and, with continued development, rotates slightly to the right (D). (After Langman, 1981.)

Formation of the chick heart. Top panel: During the same period that neurulation is occurring (see Fig. 5.13) the mesoderm lateral to the somite forming segmental plate splits along the anteroposterior axis into two layers, the *parietal* and *visceral* mesoderm. The parietal layer, along with the ectoderm that lies dorsal to it, forms the somatopleure, the future body wall. The visceral layer, along with the endoderm beneath it, forms the splanchnopleure, the future wall of the digestive tube. (The body wall and digestive tube take up their definitive spatial relationship, the former surrounding the latter, after the embryonic layers undergo lateral folding and ventral fusion along the midline, not shown here). Middle panels: The heart starts out as two parallel tubes arising from the visceral mesoderm, each consisting of an inner, *endocardial*, layer, and an outer, *myocardial* layer. The space separating the two layers is filled with an extracellular matrix known as cardiac jelly. The endocardium will form the heart's inner lining, chamber walls and valves, while the myocardium will form its muscles (bottom panel). Transverse section through the anteroposterior axis at the heart-forming region of the chick embryo at (top to bottom panels) 25, 26, 28, and 29 hr. (After Gilbert, 2003.)

a population of cells arises that also forms vascular tubes, but, in contrast with the cells of the yolk sac and splanchnopleure, it does so without simultaneously giving rise to blood cells. The two vessel-forming cell populations vascularize distinct regions of the embryo (the aorta being somewhat exceptional in receiving contributions from both). These different mechanisms of *vasculogenesis* (i.e., the *de novo* formation of blood vessels) suggest the existence of two distinct cell types – the unipotential *angioblast* of the somatopleure and the bipotential *hemangioblast* of the yolk sac and splanchnopleure, although the existence of the hemangioblast as a single precursor cell type is still speculative (Pardanaud *et al.*, 1996; Pardanaud and Dieterlen-Lievre, 2000; Eichmann *et al.*, 2002).

Cardiogenesis (i.e., heart formation) and vasculogenesis (i.e., the formation of capillaries and eventually veins and arteries, Fig. 8.3) are thus both examples of tube morphogenesis (Fig. 8.4), some other examples of which we have encountered before. Recall that the vertebrate neural tube was described in Chapter 5 as forming by the raising and fusing of parallel ridges along the neural plate, a process that Lubarsky and Krasnow (2003) refer to as "wrapping." Cardiogenesis and vasculogenesis by cells of splanchnopleural origin, in contrast, occur by the formation of cell cords followed by cavitation, as described above, whereas vasculogenesis by cells of somatopleural origin occurs by still another mechanism of tube morphogenesis, referred to as "cord hollowing" (Lubarsky and Krasnow, 2003). During this process (which is similar to the lumen-forming process depicted in Fig. 4.2), the rearrangement of cells without cell loss, within an initially solid cord, yields the tubular morphology. Finally, some very small caliber blood vessels form by a process of "cell hollowing" (Lubarsky and Krasnow, 2003) in which individual cells arranged in linear chains undergo a process of cytoplasmic rearrangement and vesicle fusion such that a continuous channel is formed within the chain.

All these tube-forming mechanisms depend on the polarized nature of epithelial cells. As we saw in Chapter 4 (see Fig. 4.2), polarized epithelial cells that have non-adhesive molecules on their apical surfaces will automatically rearrange in such a way that a cavity or lumen forms, lined by the cells' non-adhesive surfaces. However, the energy barrier separating the tubular morphology from the cystic or closed sac-like morphology is probably not large, as evidenced by the relatively high frequency in humans of polycystic kidney disease, a condition in which kidney tubules and ducts frequently take the form of cysts (Wilson, 2004).

Epithelial sheets can also form tubular structures by extrusion-type processes, often referred to as evagination or invagination. We saw examples of the latter in our discussion of amphibian gastrulation (Chapter 5) and of the former when we considered the Mittenthal–Mazo application of the differential adhesion hypothesis to epithelial sheet morphogenesis and the formation of the insect leg (Chapter 4). Extrusion from a preexisting tube, referred to

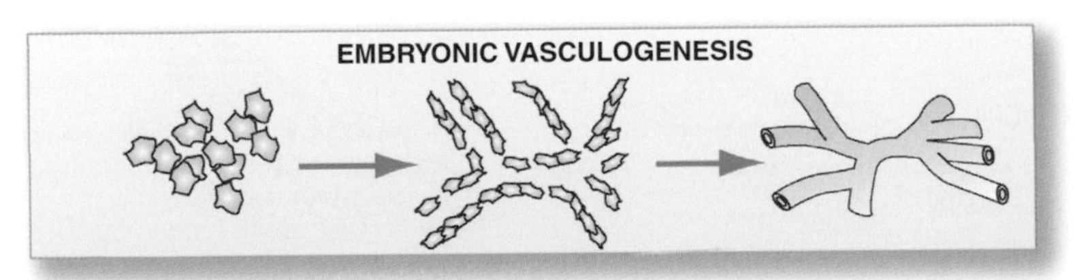

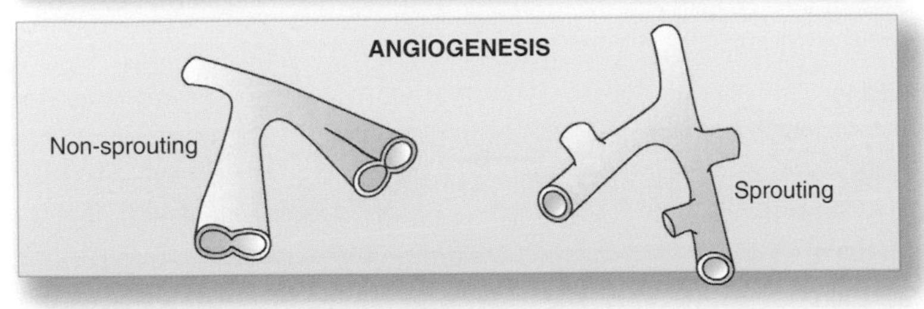

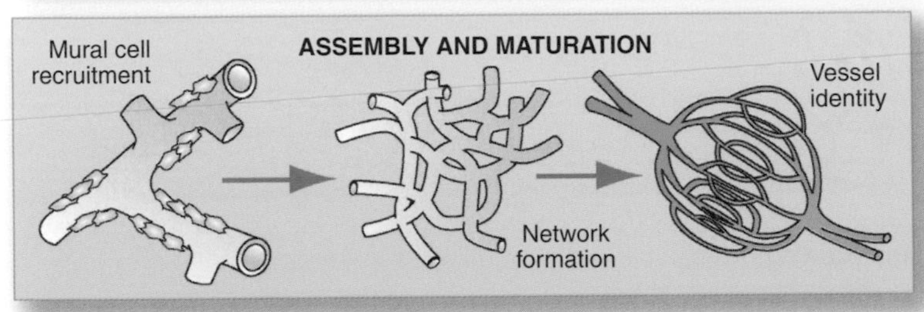

ORGANOTYPIC DIFFERENTIATION

Fig. 8.3 The hierarchy of morphogenetic events during the development of blood vessels. Top panel: The primary formation of blood vessels occurs through the mechanisms of vasculogenesis. Embryonic vasculogenesis results from the coalescence of mesodermal precursor cells (angioblasts) to form a capillary network. Upper middle panel: further development of blood vessels by angiogenesis, i.e., the formation of vessels and vascular networks from preexisting vascular structures. This can occur through classical sprouting angiogenesis with the formation of interconnected branches or anastomoses (on the right) or through the mechanisms of nonsprouting angiogenesis (on the left). Nonsprouting angiogenesis occurs by intussusceptive microvascular growth (IMG) – the focal insertion of a tissue pillar – or by the longitudinal fold-like splitting of a vessel. Sprouting angiogenesis and intussusception contribute to the increasing complexity of a growing vascular network. Lower middle panel: as the network assembles and matures, mural (i.e., outer-wall) cells, which include pericytes and smooth-muscle cells, are recruited to the surfaces of the vascular tubes, and directional blood flow ensues. Cellular and biomechanical factors appear to be involved in determining vascular identity (i.e., whether the vessels are capillaries, arteries, or veins). Bottom panel: microenvironmental cues (ECM, cell contacts, and organ-associated growth factors) regulate the organotypic differentiation of a newly formed vascular tree with distinct types of endothelia. These are not pictured, but include: "continuous" endothelia in which cells form an unbroken single layer, seen in most capillaries and larger blood vessels; "discontinuous" endothelia in which adjacent cells have gaps between them, found in the liver, for example, and "fenestrated" endothelia in which the membranes of the endothelial cells are perforated, found in the kidney and endocrine organs. (After Augustin, 2001.)

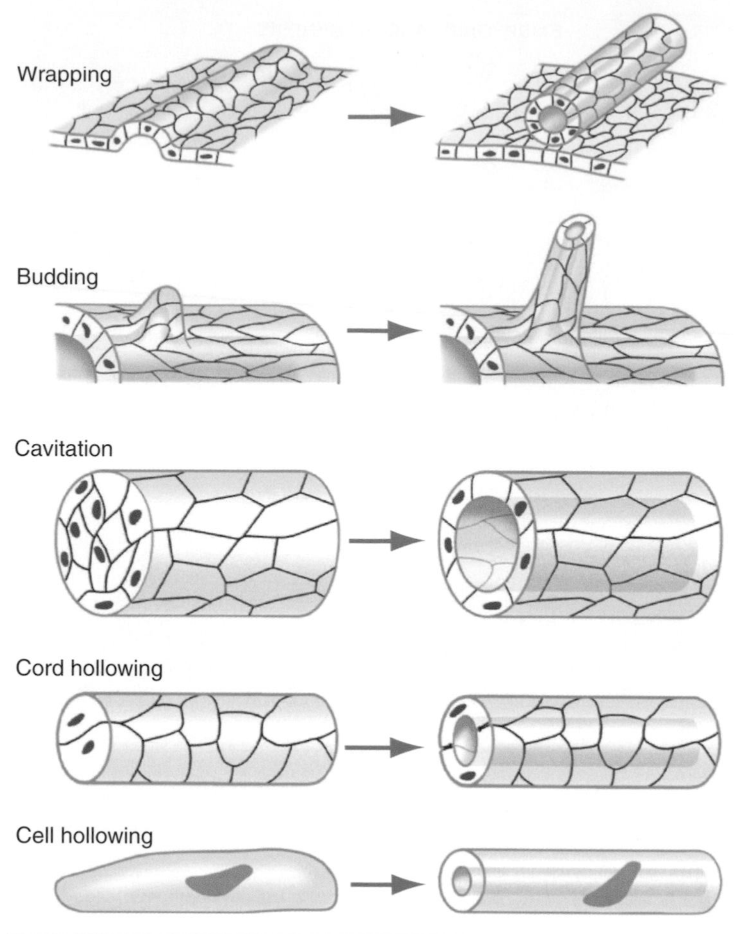

Fig. 8.4 Morphological processes of tube formation. *Wrapping*: a portion of an epithelial sheet invaginates and curls until the edges of the invaginating region meet and seal, forming a tube that runs parallel to the plane of the sheet. *Budding* (also referred to as sprouting): a group of cells in an existing epithelial tube or sheet migrates out and forms a new tube as the bud extends. The new tube is a branch of the original tube. *Cavitation*: the central cells of a solid cylindrical mass of cells are eliminated, converting the cylinder into a tube. *Cord hollowing*: a lumen is created *de novo* by the rearrangement of cells, without cell loss, into a thin cylindrical cord. *Cell hollowing*: a lumen, spanning the length of a single cell, forms by rearrangement of the cell's membrane and cytoplasm. (Reprinted from Lubarsky and Krasnow, 2003, with permission from Elsevier.)

as "budding" by Lubarsky and Krasnow (2003) and as "sprouting" by other investigators, is the main basis of *angiogenesis*, the formation of new blood vessels from preexisting ones (Patan, 2000).

Angiogenesis is responsible for creating the anastomosing network of capillaries, which nourishes all tissues in the healthy body, as well as the "neovascularization" (abnormal new capillary formation) of pathological tissues such as tumors (Verheul *et al.*, 2004) and the diabetic retina (Campochiaro, 2000). Angiogenesis can occur by "intussusceptive" microvascular growth (IMG), whereby vessels split by lumenal insertions of tissue-dividing walls, or septa, or by the longitudinal fold-like splitting of a preexisting vessel (Augustin, 2001). However, budding or sprouting angiogenesis is the most common mechanism of capillary network formation (Fig. 8.3). Capillaries in the embryo and during neovascularization are induced to sprout by exposure to members of the VEGF (vascular endothelial growth factor) family of diffusible glycoproteins secreted by target tissues or by the capillaries themselves. The VEGFs act as chemoattractants to angioblasts (Folkman, 2003).

Both initial vasculogenesis and remodeling angiogenesis appear to be self-organizing processes that construct efficient transport networks of fractional dimensionality, i.e., these networks are "fractals." It will be helpful to explore the nature of fractals as a prelude to considering the physical processes that may underlie vascular network formation.

Fractals and their biological significance

In our discussion of network formation in Chapter 6 we saw that at the percolation transition drastic changes can take place in a system's behavior. In the example of a telephone communication system, if the interconnected network of cables ceased to percolate, communication between distant cities was disrupted (see Fig. 6.6). Let us now imagine the line segments in Fig. 6.6 to be pipes instead of cables. If we impose a pressure difference between LA and NY then, once a percolating network of pipes forms, flow between distant locations will be established. We can now begin to see how the percolation formalism, first introduced to account for the properties of extracellular matrices, might also be applicable to blood vessel formation.

Another physical example of network formation via percolation is shown in Fig. 8.5, which is a schematic illustration of an amorphous material with metallic fragments randomly dispersed in an insulating matrix. If a voltage difference is set up between the two sides of this material then below the percolation transition no current can flow across the system because no communication exists between the two sides through the metallic islands (denoted by identical filled squares in Fig. 8.5). However, at a critical concentration of the metallic component a percolation transition takes place and the metallic islands link up into an interconnected network through which charges can move from one side of the system to the other. The tenuous network that emerges at the percolation transition is a *fractal* and various biological structures have been interpreted in terms of such networks (e.g., the vasculature structure and the network of airways in the lung).

To understand the concept of fractals we recall that mass = density times volume. Thus, a cubic box (having a negligible mass of its own) containing some material of density ρ, comprises a mass $M = \rho L^3$, where L is the side of the box. A box with side $2L$ will hold eight times as much material, and in general a box with side λL will have

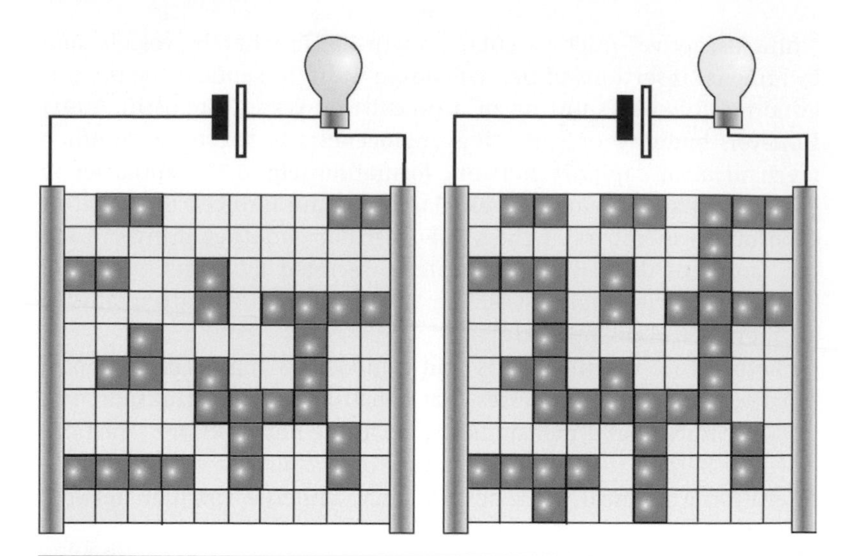

Fig. 8.5 Schematic representation of an amorphous metal. The filled squares denote the metallic component. The left-hand panel shows the situation below the percolation threshold. No continuous metallic connection exists between the two contacts on the left and right, represented by the thick bars, thus no current can flow and the bulb is not lit. The right-hand panel shows the situation above the percolation threshold. Continuous paths (two in the figure) for the charges exist, the electric circuit is closed, and the bulb is lit.

a mass $M(\lambda L) = \lambda^3 M(L)$. The power 3 appears here because the box is three-dimensional. A similar analysis with "two-dimensional" boxes (i.e., squares) would yield a power 2 and, in general, in a d-dimensional space the analogous relationship is

$$M(\lambda L) = \lambda^d M(L). \tag{8.1}$$

Equation 8.1 can be solved for the function $M(L)$ on setting $\lambda = 1/L$, to yield as expected $M(L) \propto L^d$. (Note that the factor on the left-hand side of Eq. 8.1 becomes $M(1)$, which is clearly independent of L and thus can be identified with the scale-independent density, ρ.)

The above results are well known even without resorting to equations such as Eq. 8.1. However, it is exactly the relationship in this equation that is needed to describe fractal objects, for which the Euclidian dimension d in Eq. 8.1 is replaced by the fractal dimension d_f. Since $d_f < d$ (see below), dimensional analysis indicates that the density of fractal objects, in contrast with ordinary materials, must be length dependent:

$$\rho \propto L^{d_f - d}. \tag{8.2}$$

Thus, upon increase in size, a fractal becomes less dense, which is illustrated by the Sierpinski carpet (Mandelbrot, 1983), shown in Fig. 8.6 and discussed in Box 8.1.

It is not surprising that those biological structures whose main physiological function is material exchange with their surroundings

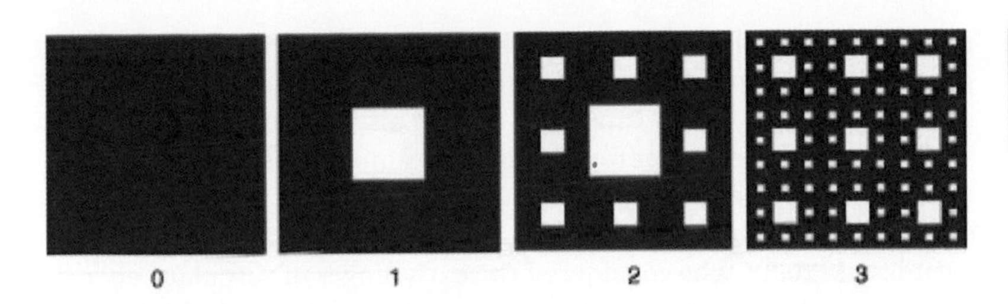

Fig. 8.6 The construction of the Sierpinski carpet. For details, see Box 8.1.

have a fractal character. The vascular system supplies organs with oxygen and disposes of waste, the respiratory airway network exchanges oxygen with carbon dioxide. The efficiency of such exchanges requires a large surface area: it is precisely here that fractal networks have an unparalleled advantage when compared to other geometries. This is quite obvious in the Sierpinski carpet, for example, in which the perimeter of the holes (i.e., the "surface area" between the percolating filled region and the surrounding material) grows without limit. We will now see that such an arrangement naturally arises in a plausible model for blood vessel development.

Box 8.1 | The Sierpinski carpet

The Sierpinski carpet is a fractal embedded in two-dimensional Euclidean space. To construct this object one starts with a filled square of side 1, Fig. 8.6, panel 0, and creates a hole by removing the middle ninth of its area as shown in panel 1. In the next step the middle ninth of each of the remaining eight small squares is removed (panel 2) and the process is iterated *ad infinitum*. Clearly, the density (the brown, filled, fraction of the total area) of the ultimate Sierpinski carpet vanishes while the total perimeter of the holes is infinite. To determine the fractal dimension of the Sierpinski carpet let us calculate its density as a function of the square's side at each iteration. For the original square with side 1, the density is 1; after the first iteration it is 8/9 (out of the nine squares with side 1/3, eight are filled, and one is empty; the side of the original square in terms of the side of the smaller squares is 3). After the second iteration the density is $(8/9)^2$ (out of the 81 squares with side 1/9, 64 are filled; the side of the original square is now 9). Finally, after the kth iteration $\rho = (8/9)^k$ and the side of the original square in terms of the smallest squares is $L = 3^k$. Using Eq. 8.2 (which, as a result of the fact that we are working with dimensionless quantities, is now an equality) the fractal dimension of the Sierpinski carpet is given by $(8/9)^k = (3^k)^{d_f - d}$, which yields (with $d = 2$) $d_f = 3 \ln 2 / \ln 3 \approx 1.893 < 2$.

Early vasculogenesis: the analysis of Bussolino and coworkers

As discussed above, the early stages of vasculogenesis involve (i) the chemotactically driven migration of endothelial cells and network formation, (ii) network remodeling, and (iii) differentiation into tubular structures. The network can best perform its biological functions if it is a fractal. However, fractal geometry is the end result of a dynamical process and *a priori* it is not evident how the interaction and migration of endothelial cells can bring about this nontrivial structure.

There have been numerous attempts to model vasculogenesis and angiogenesis in terms of percolating fractal networks. Here we concentrate on the model of Bussolino and coworkers (Gamba *et al.*, 2003; Serini *et al.*, 2003), who considered the early stages of vascular network formation. These authors performed *in vitro* experiments with human endothelial cells plated on Matrigel, an extracellular matrix preparation. The role of the chemotaxis of endothelial cells in response to a form of VEGF was analyzed, as well as the dependence of vessel assembly on cell density. To gain insight into the mechanisms driving network formation, the dynamics of the various interacting components was modeled using biologically motivated assumptions. The model calculations reproduced the experimental observations and strongly suggested that endothelial cells have the inherent capacity to self-organize into percolating networks with fractal geometry.

The experimental images in Fig. 8.7 (upper row) show the time evolution of capillary network formation, when the original plating cell density ρ_0 is above the percolation threshold ($\rho_0 \approx 100$ cells/mm^2) and

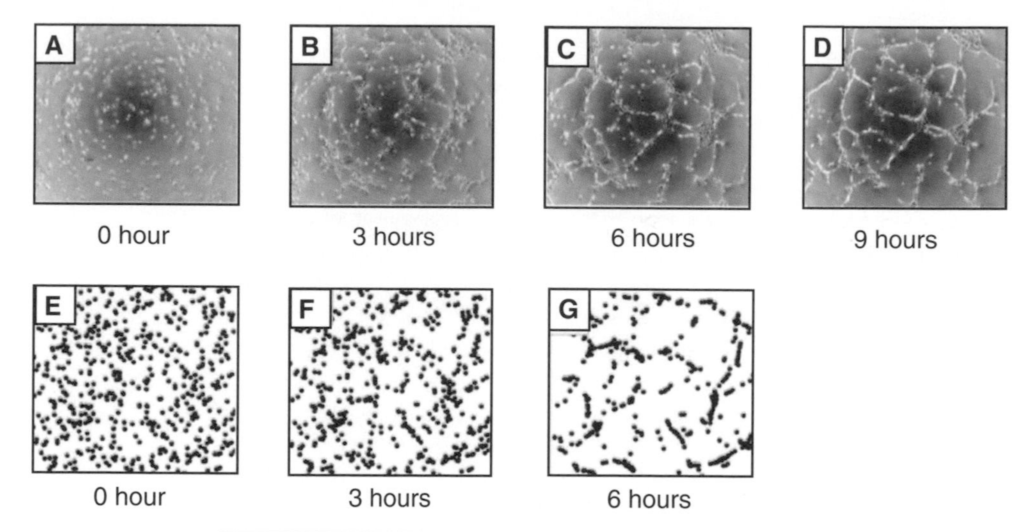

Fig. 8.7 *In vitro* vasculogenesis and, for comparison, the predictions of the model of Bussolino and coworkers. (A–D) Human endothelial cells were plated (at 125 cells/mm^2) on Matrigel, and the time course of network formation was recorded by the time-lapse videomicroscopy of a 4 mm^2 portion of the surface. (E–G) The positions of the cell centroids obtained by the model of Serini *et al.* (2003) (i.e., the two-dimensional analogs of Eqs. 8.3–8.5) using the same number of cells as in panels A–D. In the model, an initially random distribution of cells was prepared by imposing conditions in the form of a set of randomly distributed bell-shaped bumps each having a finite width of the order of the average cell diameter and zero velocity. The dots in panels E–G denote the cell centroids, the centers of these bumps. Nonzero velocities are built up by the chemoattractive term, starting from inhomogeneities in the density distribution. Subsequently, the dynamics amplifies the density variations and forms a capillary-like network. (Reprinted from Serini *et al.*, 2003, with permission.)

the concentration c of VEGF is below its saturation value ($c \approx 20$ nM). The initially randomly distributed cells start to move, interact, and adhere to their neighbors, eventually forming an interconnected multicellular network. The migration of cells is directed toward zones of higher cell concentration. The analogues of the telephone cables in Fig. 6.6 are now the chords formed by the adhering cells. The network has a characteristic mesh size, set by the average chord length. Below the percolation threshold no network forms. If the VEGF concentration is above its saturation limit then the cell receptors become desensitized (Anderson and Chaplain, 1998) and no network forms because cells are unable to follow the chemotactic gradient.

This two-dimensional model of Bussolino and coworkers is based on the following assumptions.

(i) The cell population can be described by a continuous time-dependent density $\rho(x, y; t)$ and velocity vector $\mathbf{v}(x, y; t)$.

(ii) It is assumed that there is no endothelial cell proliferation and so the cell number is conserved during network formation. Although this assumption may not be strictly true, network formation is likely to take place independently of any increase in cell number.

(iii) In the early stages of vascular assembly the cell population can be modeled as a fluid of particles, which interact by way of a chemoattractant (i.e., a form of VEGF) present at concentration $c(x, y; t)$.

(iv) The chemotactic agent is secreted by the cells themselves and thus depends on cell density. It diffuses and degrades with time.

(v) Cells are accelerated in their motion by gradients of the chemoattractant and obey Newton's second law. Note that this assumption implies a linear force–acceleration relation and thus it is not consistent with the linear force–velocity relation that, as argued in Chapter 1, is biologically more realistic for cells moving through viscous media. We comment on this disparity at the end of this section.

Even though the model is two-dimensional, to simplify the mathematics we will analyze the basic equations in one spatial dimension. Let us then cast the above assumptions into mathematical equations.

In order to express the conservation of cell number, point (ii), consider a small volume element ΔV around a given point in space (a small circle in two dimensions, a line segment in one dimension). The number of fluid particles (corresponding to the cells, point (iii) above) crossing the area ΔA enclosing ΔV in unit time is by definition $j\Delta A$, where $j = \rho v$ is the particle current perpendicular to ΔA. The net change per unit time of the number of particles inside ΔV is on the one hand given by $\Delta(\rho \Delta V)/\Delta t$; on the other hand it is equal to the difference between the in-flowing and out-flowing particles (per unit time) through the enclosing area, which can be expressed as $-\Delta(j\Delta A) = -(j_{\rm in} - j_{\rm out})\Delta A$. (Note that by convention a current component in the outward direction is taken as positive.) If particle number is conserved, it can change inside ΔV only as a result of the flow of particles across the enclosing area (i.e., no "sources" or "sinks" of particles are present) and therefore $\Delta(\rho \Delta V)/\Delta t = -\Delta(j \Delta A)$ must hold. If ΔV (and thus ΔA) is fixed, this relationship is equivalent to

$$\frac{\Delta \rho}{\Delta t} = -\frac{\Delta j}{\Delta V / \Delta A}.$$

In one dimension $\Delta V / \Delta A = \Delta x$ and, replacing the finite differences with infinitesimal ones, we finally arrive at the conservation law

$$\frac{\partial \rho}{\partial t} = -\frac{\partial j}{\partial x}. \tag{8.3}$$

Next we consider the dynamics of cellular (i.e. fluid-particle) motion, which obeys Newton's second law, mass times acceleration = $m \Delta_{\mathrm{total}} v / \Delta t$ = net force, where $\Delta_{\mathrm{total}} v$ is the total change in velocity v in time Δt (see below). The net force, according to the model (see point (v)) is provided by the gradient of the chemoattractant and is thus proportional to $\Delta c / \Delta x$. The velocity of a fluid particle depends on both x and t. At a given point in space, v in general changes with time and, at a given instant, v may be different at different locations. The total change $\Delta_{\mathrm{total}} v$ is therefore the sum of changes $\Delta_t v + \Delta_x v$ due to spatial and temporal variations respectively. We can write this as

$$\Delta_{\mathrm{total}} v = \frac{\Delta_t v}{\Delta t} \Delta t + \frac{\Delta_x v}{\Delta x} \Delta x,$$

which in the limit of infinitesimal changes becomes

$$\mathrm{d}v = \frac{\partial v}{\partial t} \mathrm{d}t + \frac{\partial v}{\partial x} \mathrm{d}x.$$

Dividing both sides of this equation by $\mathrm{d}t$, we arrive at

$$\frac{\partial v}{\partial t} + v \frac{\partial v}{\partial x} = \beta \frac{\partial c}{\partial x}. \tag{8.4}$$

Here we have used $\mathrm{d}x/\mathrm{d}t = v$, and the constant β incorporates the reciprocal of the mass and the proportionality constant due to the relationship between force and chemoattractant gradient.

Finally, point (v) can be expressed mathematically as

$$\frac{\partial c}{\partial t} = D \frac{\partial^2 c}{\partial x^2} + \alpha \rho - \frac{c}{\tau}. \tag{8.5}$$

In the above equation, D, α, and τ are respectively the diffusion constant, the rate of release, and the characteristic degradation time of the chemoattractant.

Collectively, the two-dimensional analogues of Eqs. 8.3–8.5 define the dynamics of capillary network formation in the model of Bussolino and coworkers. The authors performed a series of calculations on the time course of vascular network formation based on this model. The similarity between the experimental patterns (Fig. 8.7, upper row) and the model patterns (Fig. 8.7, lower row) is striking. Thus it is reasonable to conclude that the physical mechanism that has been described incorporates the most salient features of network assembly during early capillary formation.

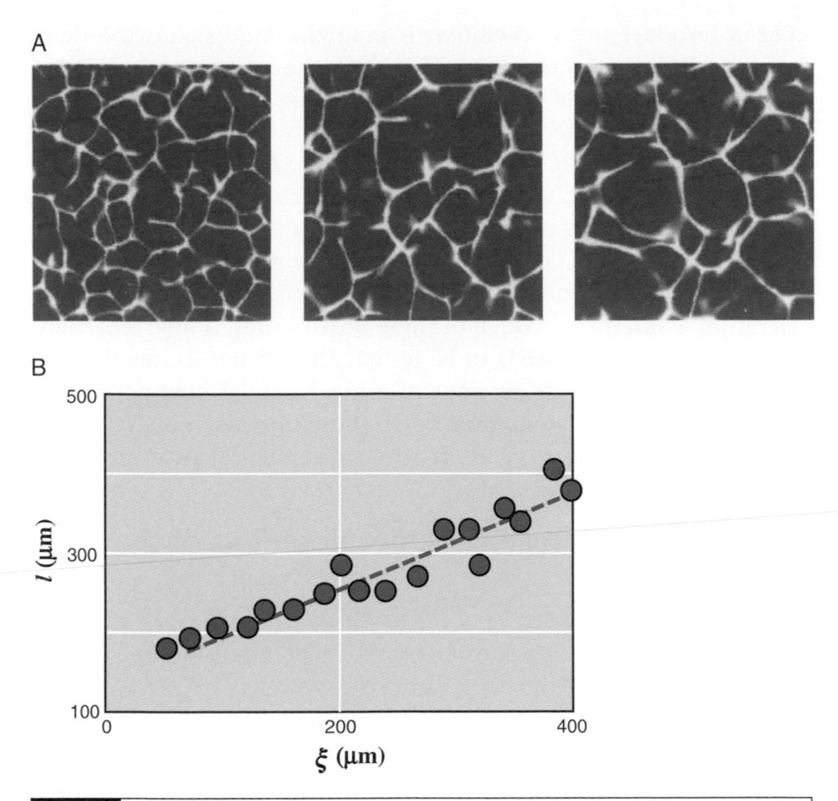

Fig. 8.8 (A) Model networks simulated with different values of the interaction range ξ. (B) Characteristic mesh size l vs. ξ. The values of l were calculated as the average node-to-node distance in the simulation. The mesh size in the model networks of Serini *et al.* (2003) is the analogue of the average chord length in the experiments. (A reprinted from Serini *et al.*, 2003, with permission. B after Serini *et al.*, 2003.)

Equations 8.3–8.5 contain a natural length scale $\xi = \sqrt{D\tau}$, which represents the effective range of interaction between cells mediated by the chemoattractant, and a natural time scale $(D/(\alpha\beta)^{2/3}$ in one dimension), representing the characteristic time for network formation. Simulations of network formation performed with different values of ξ are shown in Fig. 8.8A. The plot in Fig. 8.8B shows that ξ is proportional to the characteristic mesh size l of the assembled model networks, the analogue of the average chord length in the experiments (cf. the upper and the lower rows in Fig. 8.7). Performing additional experiments, Bussolino and coworkers found that, above the percolation threshold, the average chord length is independent of the cell-plating density, in the range $\rho_0 = 100$–200 cells per square millimeter. This is consistent with the finding shown in Fig. 8.8B, according to which the model network's characteristic mesh size depends on the range of cell interaction, set by independent parameters (such as D and τ).

As noted above, the model of Bussolino and coworkers makes the biologically implausible assumption that cells accelerate through the

ECM in response to the chemotactic gradient. Merks and coworkers, using a cellular automata model based on the scheme of Serini *et al.* (2003) found that the effect of dropping the acceleration assumption was that disconnected island-like patterns were produced, rather than realistic networks (Merks *et al.*, 2004). Merks and coworkers also showed, however, that realistic patterns could be obtained when various combinations of cell adhesion, contact-inhibition of motility (resulting from, e.g., crowding), and cell elongation (which in their model could be controlled independently) were substituted for the acceleration assumption. Which of these possibilities, if any, represents the biological reality needs to be tested. This demonstrates that the continuum and discrete versions of a given model may permit the manipulation of different aspects of the simulated developmental process, thereby suggesting alternative experimental strategies.

Branching morphogenesis: development of the salivary gland

Up to now we have largely considered the morphogenetic and pattern-forming capacities of initially uniform epithelial or mesenchymal cell populations. (An exception was the interaction between the notochord and the surface epithelium during neurulation, discussed in Chapter 5). In most cases of organogenesis, however, interaction and cooperation between distinct cell populations is the norm. A common theme is the involvement of both epithelial and mesenchymal components.

Glandular organs, which include the salivary and mammary glands, the pancreas, and anatomically similar structures such as the lung and kidney, form initially from a mass of mesenchymal tissue surrounding a hollow, unbranched, epithelial tube. One or more clefts appear at the tip of the epithelial tube, causing it to split into two or more lobules. Mesenchymal cells deposit ECM and condense (see Chapter 6) around the clefts and stalks of the lobules, while the epithelium continues to proliferate. When the lobules have grown sufficiently large, additional clefting and bifurcations occur, leading to a highly branched structure (Fig. 8.9).

The submandibular salivary glands of rodents have been the subject of many studies devoted to uncovering the mechanisms of branching morphogenesis (reviewed in Hieda and Nakanishi, 1997, and Melnick and Jaskoll, 2000). Treatment with X-rays or chemical inhibitors of DNA synthesis showed that while branching requires the proliferation of epithelial cells, clefting does not (Nakanishi *et al.*, 1987). Salivary gland epithelia separated from their mesenchymes branch normally when recombined with mesenchyme of the same organ type but abnormally or not at all when recombined with mesenchyme from other organs (Spooner and Wessells, 1972; Ball, 1974; Lawson, 1983). These and a number of additional key experimental findings, along with insights from the physical analysis of epithelial and

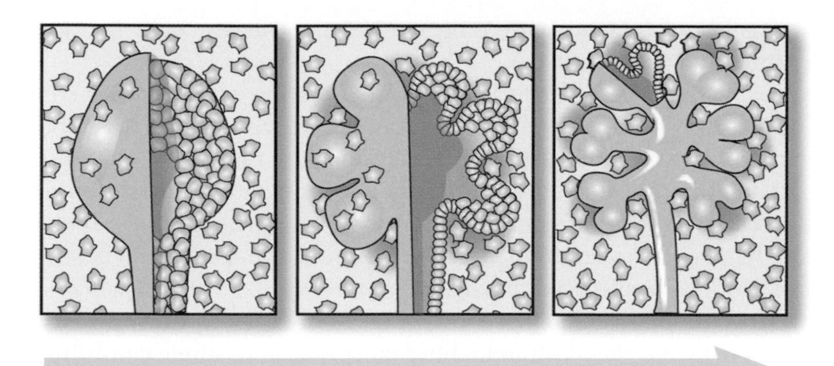

Fig. 8.9 Branching morphogenesis in a salivary gland. On the left, a bud of epithelioid tissue with a simple unbranched shape protrudes into a mass of mesenchyme. The cut-away portion of the bud shows the tightly adhering cuboidal epitheliod cells in its interior. The loose mesenchymal cells exterior to the bud are also seen. Middle, the epithelial bud flattens slightly and splits into two or more lobules by the formation of clefts. The epithelioid cells also begin to rearrange into a single layer surrounding an interior lumen. As a cleft deepens, mesenchyme in and near it condenses and deposits new ECM. When the young lobules have grown sufficiently large, further branching occurs followed by extension and mesenchymal condensation. This process continues until a highly branched structure has formed. (After Lubkin and Li, 2002, and Sakai *et al.*, 2003.)

mesenchymal tissues, provide the basis for a physical model of salivary gland morphogenesis.

Modeling branching morphogenesis: the analysis of Lubkin and Li

Most studies on branching morphogenesis have concentrated on biochemical aspects. However, during this process tissues grow, move, and most importantly change shape; this cannot take place without physical forces. This is particularly evident during cleft formation, a process which is not possible to understand without taking into account biophysical and biomechanical considerations.

On the one hand a number of experiments suggest that ECM and growth-factor components of the mesenchyme by themselves, rather than any mechanical action or support, are sufficient to promote branching morphogenesis of the epithelium (Nogawa and Takahashi, 1991; Takahashi and Nogawa, 1991). These experiments form the basis of the *epithelial theory* of branching morphogenesis (Hardman and Spooner, 1992), which assumes that the forces necessary to drive cleft formation originate from within the epithelium, in particular from the contraction of actin filaments. This theory is supported by experiments in which cultured salivary glands were treated with cytochalasin B (an F-actin disrupting agent), which resulted in the abolishment of clefting (Spooner and Wessells, 1972). Similar results have been obtained with lung epithelium (Nogawa and Ito, 1995; Miura and Shiota, 2000a).

On the other hand it is known that a wide variety of cell types, including fibroblasts, generate traction forces within the extracellular matrix (Stopak and Harris, 1982; Vernon *et al.*, 1992, 1995) that result in deformation and, under certain conditions, pattern formation. The *mesenchymal theory* of branching morphogenesis (Hieda and Nakanishi, 1997) places the origin of cleft formation in the contractile behavior of fibroblasts in the mesenchyme.

Lubkin and Li (2002) proposed that neither the epithelial nor the mesenchymal theory of branching morphogenesis alone can fully account for cleft formation and that branching observed in *in vitro* mesenchyme-free experiments is not mechanically equivalent to cleft formation in mechanically intact rudiments or *in vivo*. They constructed a biomechanical model of cleft formation based on the mechanical properties of both the epithelium and the mesenchyme. Lubkin and Li adopted the earlier proposition by Steinberg that embryonic epithelia in many respects mimic the behavior of liquids (Steinberg and Poole, 1982; Steinberg, 1998), a notion we have already encountered in Chapters 4 and 5. They generalized Steinberg's ideas further to embryonic mesenchymes and treated such tissues also as liquids in their model. We saw in Chapter 6 that certain aspects of mesenchymal morphogenesis indeed can be interpreted by attributing liquid-like behavior to these tissues.

The major assumptions and elements of the Lubkin–Li model are as follows (see also Fig. 8.10).

- **(i)** A branching rudiment is considered as a uniform epithelium inside a uniform mesenchyme. Both tissues are modeled as liquids of uniform density (in both space and time) and characterized by their respective viscosities ($\eta^-$and η^+) and surface tensions. For simplicity only a planar section of the epithelium is modeled.
- **(ii)** Epithelial growth and the formation of lumens (which are generated as the epithelium matures) are ignored; this is consistent with the model's focus on cleft formation rather than the growth of the branching rudiment.
- **(iii)** Shape changes are driven by spatial variations in the interfacial tension γ, between the two tissues, which arise due to point forces, f_i, $f_j, \ldots$, acting at selected sites i, $j, \ldots$ along the interface, denoted by Γ in Fig. 8.10. The origin of these forces, as discussed above, lies in the contractile properties of epithelial cells and fibroblasts and in the biomechanical properties of the ECM and is not modeled explicitly.
- **(iv)** Forces create stresses and deformations in the two tissue fluids, which lead to spatial rearrangements and potentially to cleft formation and branching.

As before, let us cast these assumptions into mathematical language. The physical state of any liquid is described by its density ρ, a velocity field $\mathbf{v}$, and a pressure field p (think of hydrostatic pressure as one of

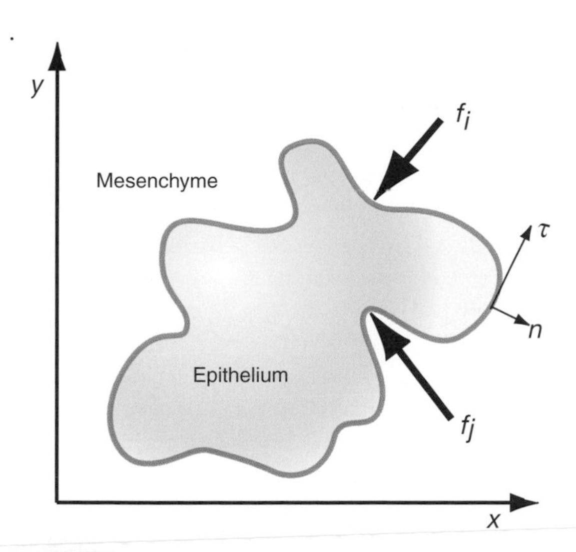

Fig. 8.10 The geometry of the fluid components in the model of Lubkin and Li. The inner fluid, the epithelium, has uniform viscosity η^-. The outer fluid, of uniform viscosity η^+, represents the mesenchyme. The quantities, f_i, f_j are point forces acting at the interface in the directions indicated by the arrows. The normal and tangential directions are denoted by n and τ, respectively. The surrounding region of fluid-mesenchyme is large in relation to the size of the epithelial portion. After Lubkin and Li (2002).

the simplest such fields). We will first set up the "equation of motion of a liquid" in terms of the velocity and pressure field, using Newton's second law. This equation will govern the dynamical behavior of the liquid, which in the present case is the hypothesized mechanism of cleft formation.

The model of Lubkin and Li is two-dimensional (point (i)) Moreover, as discussed in the section on the viscous transport of cells in Chapter 1, inertial forces can typically be neglected in early embryonic processes because of the high viscosity of the materials involved. Therefore the above equation simplifies considerably to ($i = x, y$)

$$f_i - \frac{\partial p}{\partial i} + \eta\left(\frac{\partial^2 v_i}{\partial x^2} + \frac{\partial^2 v_i}{\partial y^2}\right) = 0. \tag{8.6}$$

In the model of Lubkin and Li, Eq. 8.6 applies to both the epithelial and mesenchymal fluids, which are distinguished by their respective viscosities, η^- and η^+. In Eq. 8.6 $\mathbf{f}(x, y)$ is the (inward-directed, see Fig. 8.10) clefting force, and Lubkin and Li assume that this force acts at discrete points, with the same magnitude f_0, along the interface Γ (Fig. 8.10) between the two fluids.

Since the model does not incorporate growth (point (ii)), the fluids' volumes are constant during cleft formation: they are assumed to be incompressible. The condition of incompressibility (ρ is constant in space and time) is easily obtained from the conservation law Eq. 8.3 generalized to two dimensions ($\mathbf{j} = \rho\mathbf{v}$)

$$\frac{\partial v_x}{\partial x} + \frac{\partial v_y}{\partial y} = 0. \tag{8.7}$$

Box 8.2 | Dynamics of viscous liquids: the Navier–Stokes equation

We consider the mass of liquid, $\Delta m = \rho \Delta V$, within a small volume element $\Delta V = \Delta x \Delta y \Delta z$. The liquid inside ΔV experiences stresses (stress = force/area) due to the surrounding liquid. Let the stresses acting on the faces $\Delta y \Delta z$ at $x = 0$ and at $x = \Delta x$ be respectively σ_0 and $\sigma_{\Delta x}$ (Fig. 8.11); σ_0 is seen to have only a normal component, whereas $\sigma_{\Delta x}$ has both normal and tangential components. The magnitudes of the normal components of the stress forces are simply $\sigma_0 \Delta y \Delta z = -p(0)\Delta y \Delta z$ and $\sigma^n_{\Delta x} \Delta y \Delta z = -p(\Delta x)\Delta y \Delta z$, where p is the pressure (the minus sign is due to the fact that the pressure force, by definition, is compressive, thus opposite to the indicated direction of the stresses), whereas the tangential components provide shear (see Chapter 1). Let us first assume that the liquid in ΔV is in an equilibrium state such that its acceleration and velocity are both zero everywhere. In this case no shear forces can be present (remember, shear can be maintained only in flowing liquids, see Chapter 1) and therefore the in-plane components of the stresses must be zero everywhere. If an external force per unit volume **f** acts on the small volume element then the total force acting in the x direction on the small mass Δm, which must be zero, is $f_x \Delta x \Delta y \Delta z - [p(\Delta x) - p(0)] \Delta y \Delta z = 0$ (f_x is the x component of the force **f**). Dividing this equation by ΔV and taking the limit of infinitesimal changes, the equilibrium condition for the liquid is expressed by $f_x - \partial p/\partial x = 0$. Clearly, similar equations are obtained in the y and z directions. Since ΔV is arbitrary, these equations are valid anywhere in the liquid, so long as it is in equilibrium. If, however, the velocity and acceleration are not zero then the shear forces cannot be ignored. The terms that take shear forces into account involve the density, velocity field, and viscosity. Their derivation is beyond the scope of this book and we only quote the result (for details see for example Tritton, 1988). Finally, then, the dynamical behavior of a liquid, as a consequence of Newton's second law, is governed by the equation ($i = x, y, z$)

$$\rho \left(\frac{\partial v_i}{\partial t} + v_x \frac{\partial v_i}{\partial x} + v_y \frac{\partial v_i}{\partial y} + v_z \frac{\partial v_i}{\partial z} \right) = f_i - \frac{\partial p}{\partial i} + \eta \left(\frac{\partial^2 v_i}{\partial x^2} + \frac{\partial^2 v_i}{\partial y^2} + \frac{\partial^2 v_i}{\partial z^2} \right). \tag{8B2.1}$$

Note that here the quantities ρ, v_i, f_i, p in general depend on all three coordinates x, y, z. The above equation is known as the Navier–Stokes equation. The expression in parentheses on the left-hand side is the total acceleration and is the generalization of the result in Eq. 8.4, which, although introduced in a different biological context – the flow of cells in a chemotactic model for angiogenesis – is similarly based on Newton's second law of motion.

Equations 8.6 and 8.7 provide the mathematical basis for the analysis of Lubkin and Li. These equations need to be supplemented by "boundary conditions," physically justified relations between the fundamental quantities at the interface between the two fluids (as well as at the physical boundary of the entire system). Since the two fluids are different, the two dynamical quantities **v** (x, y) and $p(x, y)$ are

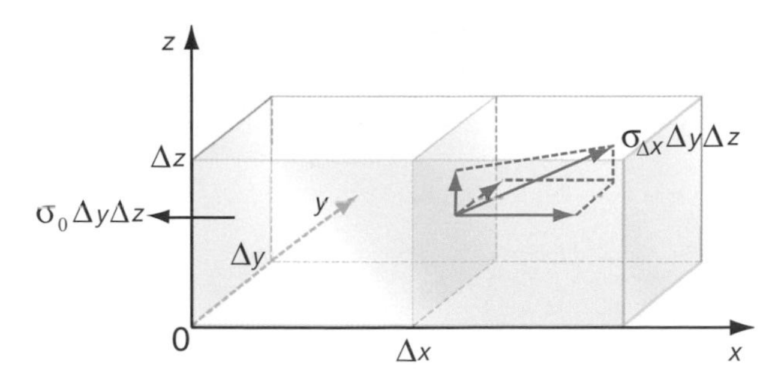

Fig. 8.11 Illustration of the various stresses acting on a small volume element in a viscous liquid, used for clarifying the basis of the Navier–Stokes equation. Forces acting on the volume element are drawn outwards by convention. In general, the force acting on a given face is not perpendicular to the face. The normal and tangential components are related respectively to pressure and shear; thus the force acting at $x = 0$ gives rise only to pressure whereas the one acting at Δx provides both pressure and shear. For more details see Box 8.2.

discontinuous at the interface. These discontinuities can be determined from the condition that no uncompensated forces can arise along the interface (such forces would result in infinite acceleration). For an interface with a complicated shape, such as the one in Fig. 8.10, the mathematical expressions for the boundary conditions are complicated, because the direction of the normal and tangent vectors (along the n and τ directions in Fig. 8.10 respectively) in general do not coincide with that of the main coordinate axis. Therefore we will not dwell on the explicit form of these conditions. Instead, we use simple physical considerations, based on the material of earlier chapters, to illustrate the main points. (For more details see for example Leal, 1992.)

Let us first consider the normal (perpendicular, n-directed) stresses at the interface. If the two liquids are in equilibrium with $\mathbf{v}(x, y) = 0$ everywhere then the shape of the interface is determined by the Laplace equation 2.2: $\Delta p = p_e - p_m = \gamma/R$. Here Δp is the discontinuity (jump) in pressure across the interface, p_e and p_m denote respectively the local pressure (the normal stress) on the epithelial and mesenchymal side of the interface, and R is the radius of curvature at the same location (since the interface is a line, there is only one radius of curvature). When the velocity is not zero it provides an additional stress; therefore an extra term, containing the discontinuity (across the interface) of the viscosity times the velocity gradient in the normal direction, arises in the above Laplace equation. (Stress is proportional to the gradient of the velocity; see Eq. B1.2a in Chapter 1). Next we consider the tangential (in-plane) stresses. We recall (see Eq. 2.6) that an interfacial tension which is spatially varying (along the interface) provides a shear stress proportional to the gradient of the velocity component in the direction of the normal. In conclusion, the boundary conditions along the normal and tangential directions

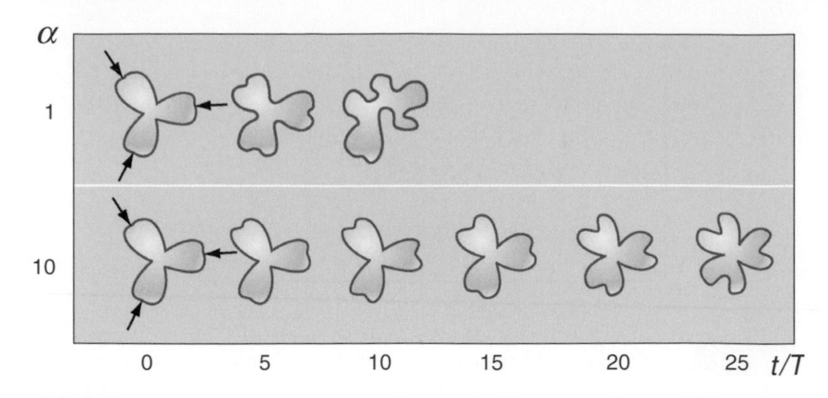

Fig. 8.12 Effect of the viscosity ratio α on the evolution of clefting, as a function of the nondimensional time t/T (where t is the real time and T is the characteristic time), from an initially three-lobed epithelial rudiment, in the model of Lubkin and Li (2002). The directions of the clefting forces are indicated by arrows. Their magnitude $\varphi = 2.5$, whereas the nondimensional surface tension is $\beta = 0.01$. When the epithelium is embedded in a material more viscous than itself, it takes longer to form the same depth of cleft than when it is embedded in a material of the same viscosity.

both involve the interfacial tension and the gradient of the velocity in the normal direction.

The model of Lubkin and Li is thus fully defined by Eqs. 8.6, 8.7, and the equations expressing the boundary conditions. These equations can be rewritten in terms of nondimensional parameters $\alpha = \eta^+/\eta^-$ (the viscosity ratio), $\beta = \gamma_0 T/(\eta^- L)$ (the nondimensional surface tension) and $\varphi = f_0/(\gamma_0 L)$ (the nondimensional clefting force), where L and T are characteristic length and time scales, respectively. The uniform interfacial tension γ_0 characterizes the boundary between the epithelium and the mesenchyme in the absence of any clefting force. The clefting force makes the interfacial tension nonuniform. The magnitudes of L and T reflect the spatial and temporal characteristics of salivary globules and the branching process ($L \approx 100$ μm is the typical diameter of a salivary globule, $T \approx 8$ hr is the typical time between branchings). All the other quantities were estimated on the basis of experimental results (Table 1 in Lubkin and Li, 2002).

Lubkin and Li applied their model to an initially three-lobed rudiment by numerically solving the equations governing cleft formation. They tested the effects of the viscosity ratio, clefting force, and interfacial tension. Figure 8.12 shows the time evolution of cleft formation within the model and the sensitivity of this process to the magnitude of the nondimensional parameter α.

Vertebrate limb development

The vertebrate limb is one of the most intensively studied systems in developmental biology, both experimentally and theoretically.

Because many of the major molecular determinants of limb skeletal pattern formation have been described, and because viable models for this process invoke a variety of physical mechanisms acting in a concerted, integrated, fashion, we will devote the remainder of the chapter to a discussion of this system.

Like the salivary glands, limb morphogenesis involves the co-operation of epithelial and mesenchymal components. In contrast with the glands, however, the pattern of structural elements in the limb is determined with great precision. A salivary gland can perform its function – secretion of its enzyme products into the oral cavity via a hierarchy of tubular ducts – with variable numbers of clusters of secretory cells, lobules, and ducts. Its pattern is not entirely determinate. In contrast, all normal limbs of a given type (e.g., the forelimb of a bird, the leg of a frog) have a fixed and precisely arranged set of structural elements, i.e., bones, muscles, and nerves, and the coordination of these elements is integral to the limb's optimal functioning. Thus, unlike the branching morphogenesis of salivary glands, which may depend on random cues such as are provided by the action of point forces in the mechanism of Lubkin and Li (2002), vertebrate limb development depends on one or more well-regulated mechanisms of spatial pattern formation such as those discussed in Chapter 7.

The limbs take form from mounds of tissue ("limb buds"), which emerge from the body wall, or flank, at four discrete sites – two for the forelimbs and two for the hindlimbs. The mesenchymal tissue of the early limb bud, which gives rise to the skeleton and muscles, forms a paddle-shaped tissue mass referred to as a "mesoblast," surrounded by a layer of simple epithelium, the ectoderm. The skeletons of all vertebrate limbs develop in a *proximodistal* fashion: that is, the structures closest ("proximal") to the body form first, followed, successively, by structures more and more distant ("distal") from the body. For the forelimb of the chicken, for example, this means that the humerus of the upper arm is generated first, followed by the radius and ulna of the mid-arm, the wrist bones, and finally the digits (Fig. 8.13).

The bones of the limb skeleton do not take form directly as bone tissue. The pattern is first laid out as cartilage, which is replaced by bone at a later stage of embryogenesis in most, but not all, vertebrate species. Some salamanders, for example, have limb skeletons composed almost entirely of cartilage.

Before the cartilages of the limb skeleton form, the mesenchymal cells of the mesoblast are dispersed in a hydrated ECM, rich in the glycosaminoglycan hyaluronan. The first morphological evidence that cartilage will differentiate at a particular site in the mesoblast is the emergence of precartilage mesenchymal condensations. The cells at these sites then progress to fully differentiated cartilage elements by switching their transcriptional capabilities. As discussed in detail in Chapter 6, condensation involves the transient aggregation of cells within a mesenchymal tissue. This process is mediated first by the local production and secretion of ECM glycoproteins such as fibronectin (Tomasek *et al.*, 1982; Kosher *et al.*, 1982; Frenz *et al.*, 1989b),

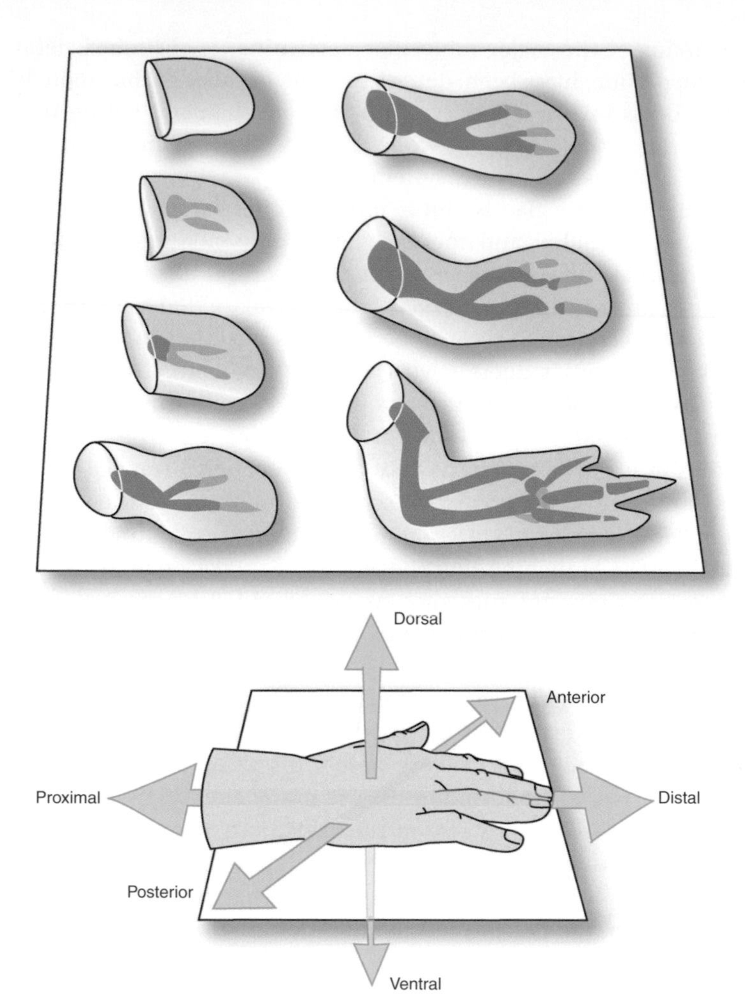

Fig. 8.13 The progress of chondrogenesis in the chick wing bud between four and seven days of development. The limbs are shown as if transparent. The lighter gray regions represent precartilage; the darker-gray regions represent definitive cartilage. (After Newman and Frisch, 1979). The single proximal element that forms first is the humerus (the femur in the leg); the two elements of the mid-wing form next, the radius and the ulna (the tibia and fibula in the leg); the distal-most, last-forming, elements are the digits. The proximodistal, anteroposterior, and dorsoventral axes are indicated on an illustration of a human hand.

which act to alter the movement of cells and trap them in specific places. The aggregates are then consolidated by direct cell–cell adhesion. For this to occur the condensing cells need to express, at least temporarily, adhesion molecules such as cadherins (Oberlender and Tuan, 1994; Simonneau *et al.*, 1995).

Because all the precartilage cells of the limb mesoblast are capable of producing fibronectin and cadherins, but only those at sites destined to form skeletal elements do so, there clearly must be communication among the cells to divide the labor in this respect.

This is mediated in part by secreted diffusible factors of the TGF-β family of growth factors, which promote the production of fibronectin (Leonard *et al.*, 1991) and N-cadherin (Tsonis *et al.*, 1994), though the actual CAM involved may be a different cadherin (Luo *et al.*, 2005). Limb bud mesenchyme also shares with many other connective tissues the significant autoregulatory capability of producing more TGF-β upon stimulation with this factor (Miura and Shiota, 2000b; Van Obberghen-Schilling *et al.*, 1988).

The limb bud ectoderm performs several important functions. First, it is a source of fibroblast growth factors (FGFs) (Martin, 1998). Although the entire limb ectoderm produces FGFs, it is the particular mixture produced by the apical ectodermal ridge (AER), a narrow band of specialized ectodermal cells running in the anteroposterior direction (shown schematically in Fig. 8.14) along the tip of the growing limb bud in birds and mammals, that is essential to limb outgrowth and pattern formation. The AER keeps the precondensed mesenchyme of the "apical zone" (Fig. 8.14) in a labile state (Kosher *et al.*, 1979) and its removal leads to terminal truncations of the skeleton (Saunders, 1948).

The FGFs produced by the ectoderm affect the developing limb tissues through one of three distinct FGF receptors (Fig. 8.14). The apical zone is the only region of the mesoblast containing cells that express FGF receptor 1 (FGFR1) (Peters *et al.*, 1992; Szebenyi *et al.*, 1995). In the developing chicken limb, cells begin to condense at a distance of approximately 0.3 mm from the AER. In this, the morphogenetically "active zone" (Fig. 8.14) FGFR1 is downregulated and cells that express FGFR2 appear at the sites of incipient condensation (Peters *et al.*, 1992; Szebenyi *et al.*, 1995; Moftah *et al.*, 2002). Activation of these FGFR2-expressing cells by FGFs releases a laterally acting inhibitor (that is, it acts in directions peripheral to the condensations) of cartilage differentiation (Moftah *et al.*, 2002). Although the molecular identity of this inhibitor is unknown, its behavior is consistent with that of a diffusible molecule. Finally, differentiated cartilage in the more mature region, proximal to the condensing cells, expresses FGFR3, which is involved in the growth control of this tissue (Ornitz and Marie, 2002). (Since the pattern is fixed in this region it is referred to as the "frozen zone"; Fig. 8.14.) The ectoderm, by virtue of the FGFs it produces, thus regulates the growth and differentiation of the mesenchyme and cartilage.

The limb ectoderm performs one additional function. By itself the mesenchyme, being an isotropic tissue with liquid-like properties, would tend to round up. Ensheathed by the ectoderm, however, it assumes a paddle shape (see the upper part of Fig. 8.13). This is evidently due to a biomechanical influence of the epithelial sheet, with its underlying basal lamina, along with the anteroposteriorly arranged AER, which has a less organized basal lamina (Newman *et al.*, 1981). There is no entirely adequate biomechanical explanation for the control of limb bud shape by the ectoderm (but see Borkhvardt, 2000, for a review and suggestions). The shape of the limb bud will

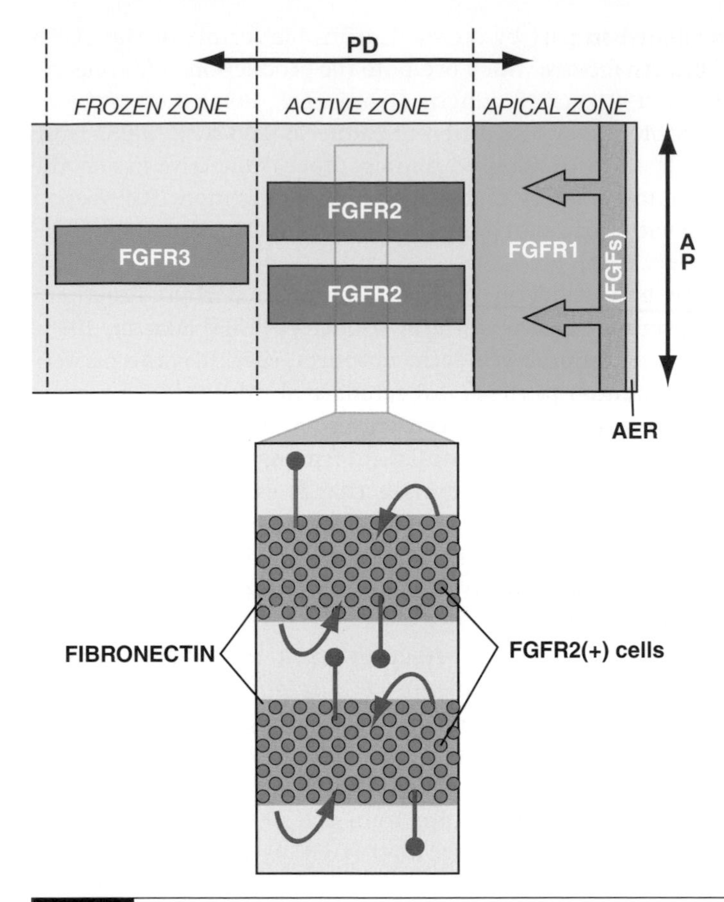

Fig. 8.14 Schematic representation of the biochemical genetic circuitry underlying the pattern-forming instability described in the model of Hentschel *et al.* (2004), superimposed on a two-dimensional representation of the 5-day limb bud shown in Fig. 8.13. The positive autoregulation of TGF-β, the induction of fibronectin by TGF-β, the promotion of precartilage condensation by fibronectin, and the FGF-dependent elicitation of a lateral inhibitor of chondrogenesis from sites of condensation are all supported by experimental evidence. The molecular identity of the inhibitor is unknown, as is the factor or activity it inhibits. The model assumes that the inhibitor acts directly on TGF-β. The colored rectangles represent the distribution of the densities of the indicated cell types, defined by the expression of the various FGF receptors in the different zones. The apical zone contains a high density of cells expressing FGFR1 (green). In this zone, cell rearrangement is suppressed by the FGFs emanating from the AER. The active zone is the site of the spatiotemporal regulation of mesenchymal cell condensation (i.e., pattern formation). Pattern formation begins with the establishment of populations of cells expressing FGF receptor 2 (red). The lower part of the figure gives an enlarged version of part of this zone. The curved arrows show positively autoregulatory activators; the straight lines ending in circles show laterally acting inhibitors. When condensed cells leave the proximal end of the active zone and enter the frozen zone they differentiate into cartilage cells, which express FGFR3 (blue), and their spatiotemporal pattern becomes fixed. At different stages of development the active zone will contain different numbers of elements; eventually the frozen zone will encompass the entire pattern. The length of the dorsoventral axis (normal to the plane of the figure; see Fig. 8.13) is collapsed to zero in this simplified model. PD, proximodistal axis; AP, anteroposterior axis.

thus serve as an assumed boundary condition for the model of skeletal pattern formation that we will now consider.

Skeletal pattern formation: the model of Hentschel *et al.*

The vertebrate limb is clearly the most elaborate of the organs whose development we are modeling in this chapter. It is a truly multidimensional structure within a structure: a complex pattern of bones embedded in a paddle-shaped tissue mass. Not only are the overall external dimensions of this structure changing in time but so also are those of the various interior zones, namely, the noncondensed zone at the distal end, beneath the AER, the subapical zone of condensation, and the more mature proximal zone.

Skeletal patterning in the limb involves cell movement and differentiation: mesenchymal condensation is followed by chondrogenesis (i.e., cartilage differentiation). It is dependent on the interaction of epithelial and mesenchymal tissue types. The spatiotemporal evolution of the skeletal pattern is controlled by several classes of morphogenetic growth factors and their receptors, ECM molecules, and cell adhesion molecules. The above list is far from being complete. It does not take into consideration, for example, the complex but subsidiary questions of the patterning of the limb muscles and nerves (see the references given in Newman, 1988) and the replacement of the cartilage skeleton by bone tissue (processes that we will not explicitly consider here).

A physical model that incorporates all the above ingredients must involve a large number of coupled reaction–diffusion and domain-growth equations, which are nonlinear (owing to complicated feedback mechanisms and differentiation, see Chapter 3) and which must be solved under moving boundary conditions, since the entire limb and its internal domains are constantly changing in size. The model of Hentschel *et al.* (2004) is a synthesis of cell-biological and molecular genetic knowledge in this area and draws on concepts presented in previous chapters as well as earlier attempts to model avian limb development (e.g., Newman and Frisch, 1979; Newman *et al.*, 1988; Miura and Shiota, 2000b; Miura and Maini, 2004). Rather than present this model in elaborate detail, in the spirit of this book we will simply sketch out the approach taken by Hentschel and coworkers in setting up a biologically motivated mathematical formalism for the multifaceted problem of vertebrate limb development and then present some results.

Geometry of the developing avian limb

The developing limb has a smooth, but nonstandard, geometric shape that changes over time. Moreover, different processes take place in different parts of the changing structure. Hentschel and coworkers made the following geometric approximation based on the simplified biological model shown in Fig. 8.14. The limb bud was considered as a parallelepiped of time-dependent proximodistal length, $L(t)$ along the x axis, and fixed length l_y along the anteroposterior (thumb to little finger) direction (the y axis). The dorsoventral (back to front) width (the z axis) was collapsed to zero in this simple model. $L(t)$ was considered to consist of three regions, as described above: an apical zone of size $l_{\text{apical}}(t)$, at the distal tip of the bud and consisting of noncondensing mesenchymal cells, followed by an interior active zone of length $l_x(t)$, which contains differentiating and condensing cell types, and the proximal frozen zone of cartilage cells of length $l_{\text{frozen}}(t)$; $L(t) = l_{\text{apical}}(t) + l_x(t) + l_{\text{frozen}}(t)$. The division of the distal part of the limb into two zones reflects the activity of the AER in suppressing differentiation of adjacent mesenchyme (see above). This is assumed in the model to result from the distribution of FGF, which is highest under the AER and lower at some distance from it. The active zone, therefore, is where differentiating cells, morphogens, and growth factors interact dynamically, giving rise to a time-dependent pattern of condensations. As will be seen below, the length of the active zone, $l_x(t)$, serves as a "control parameter" that determines the character of the condensation pattern. Cells from the proximal end of the apical zone are recruited into the active zone as they divide and move away from the influence of the AER. The active zone loses cells in turn to the proximal frozen zone, the region where cartilage differentiation has occurred and a portion of the definitive pattern has already formed.

Variables: cell types and molecules

Four main types of mesenchymal cell are involved in chick-limb skeletal-pattern formation. These are represented in the model by their spatially varying (as shown in the parallelepiped defined above) and temporally varying densities. The cell types are characterized by their expression of one of the three FGF receptors found in the developing limb. The densities of cells expressing FGFR1 (R_1 cells), FGFR2 (R_2 and R_2' cells) and FGFR3 (R_3 cells) are denoted respectively by c_{R_1}, c_{R_2}, $c_{R_2'}$, and c_{R_3}. The cells of the apical zone are R_1 cells and those of the frozen zone R_3 cells (see Ornitz and Marie, 2002). The active zone contains R_2 cells and the direct products of their differentiation, R_2' cells. These latter cells secrete enhanced levels of fibronectin. The R_1, R_2 and R_2' cells are mobile, while the R_3 (cartilage) cells are immobile. Thus the total density of mobile cells is $c_R = c_{R_1} + c_{R_2} + c_{R_2'}$.

According to the model, the transitions and association between the different cell types are regulated by a number of the gene products that we have encountered previously in early developmental pattern formation: FGFs, TGF-β, a diffusible inhibitor of chondrogenesis, and fibronectin. Their spatially and temporarily varying concentrations are denoted c_F, c_A, c_I, and c_f respectively.

The model of Hentschel and coworkers thus comprises eight variables.

A core mechanism for chondrogenic pattern formation

The major point in formulating a model of a complex biological process such as limb formation is to define clearly which aspects of the process one intends to investigate. The analysis of Hentschel *et al.* concentrates on the mechanism that establishes the gross anatomical structure of the avian limb: the correct spatial arrangement of its various skeletal elements, the humerus, the radius and ulna, and the three digits (see Fig. 8.13). Once the modeling task has been defined, one needs to identify the most relevant factors that govern the process to be modeled. Experimental evidence suggests that the eight variables listed above constitute a "core" set necessary to describe the development of a basic, "bare-bones" skeletal pattern (see discussion above). These variables are intimately interconnected and interacting: diffusing morphogen-type growth factors (i.e., FGFs and TGF-β) and extracellular matrix molecules (i.e., fibronectin) secreted by some cells represent signals for others to move, to differentiate, or to produce other molecules or more of the same molecules. It is the interactions among these variables (Fig. 8.14), typically described in terms of reaction–diffusion equations (see Chapter 7) that finally generate an array of possible patterns. The geometry and other physical constraints of the limb bud select one or another of these patterns, in the biologically relevant case the pattern of the limb skeleton. The model of Hentschel *et al.* involves eight reaction–diffusion equations (see Eq. 7.5),

$$\frac{\partial c_i}{\partial t} = \frac{D_i \partial^2 c_i}{\partial x^2} + F_i(\{c_i\}),$$

where c_i ($i = \mathrm{R}_1, \mathrm{R}_2, \mathrm{R}_2', \mathrm{R}_3, \mathrm{F}, \mathrm{A}, \mathrm{I}, \mathrm{f}$) may denote any of the eight variables identified above, the D_i are the diffusion constants (which in the case of migrating cells may be effective parameters, as discussed in Chapter 1), and the notation $F_i(\{c_i\})$ indicates that the reaction functions may depend on any of or all the variables. (The authors used the term "reactor–diffusion" to emphasize that the active component is a living cell, not just a chemical reaction as in Turing's original formalism (Turing, 1952).)

Although the reaction functions employed in this model incorporate multiple cellular properties and molecular factors, much is left out. Instead of keeping track of each of several FGFs separately, only one parameter, their global density c_F, is used. Furthermore, the extracellular matrix is modeled only by fibronectin, which represents a marked simplification of this hydrated network containing numerous types of macromolecules. Hox transcription factors and nonuniformly distributed morphogens such as Sonic hedgehog and Wnt-7A (reviewed in Tickle, 2003), which modulate cell properties such as differential growth and the production of cell adhesion and ECM proteins, are not considered. These limitations do not affect the ability of this bare-bones model to capture the essential aspect of limb formation, i.e., the proximodistal generation of increasing numbers of parallel skeletal elements.

Mathematical analysis of the limb bud reactor–diffusion system

To find the general solution of a system of eight coupled partial differential equations in a multidimensional space (spanned by all the model parameters, such as diffusion constants and rate constants), is a daunting task. In the present case the difficulty is compounded by the fact that these equations have to describe not only the time-dependent generation of a pattern of chemical concentrations corresponding to the successive positions of prospective skeletal elements but also the process of mesenchymal condensation, involving changes in the local arrangement of mesenchymal cells (as discussed in Chapter 6). Proving the existence of unique symmetry-breaking solutions to a system of this complexity, a highly nontrivial endeavor, has been accomplished by Alber *et al.* (2005). Their result justifies considering the behavior of this model in various approximations.

A number of useful simplifications are possible in formulating a model for this process without significantly departing from the underlying physics and biology. Specifically, we will illustrate how the separation of time scales and linear stability analysis was used to reduce the number of equations in the model of Hentschel and coworkers. We will then indicate how the outlined mathematical formalism leads to the desired pattern.

The equation that describes the spatiotemporal changes in FGF concentration in the model has the form

$$\frac{\partial c_{\mathrm{F}}}{\partial t} = D_{\mathrm{F}}\frac{\partial^2 c_{\mathrm{F}}}{\partial x^2} - k_{\mathrm{F}}c_{\mathrm{F}} + J_{\mathrm{F}}(x, t). \tag{8.8}$$

To simplify the notation, we will indicate the spatial variable in the proximodistal direction only, noting however that the equations must be solved in at least two dimensions. (Variations in the dorsoventral direction, i.e., the z direction, are ignored in this model; this is equivalent to assuming that the cross-sections of the bones are symmetrical along that axis, not a serious departure from biological reality.) In Eq. 8.8, k_{F} and $J_{\mathrm{F}}(x, t)$ are respectively the decay constant and the flux of FGFs. The latter quantity represents the overall (cellular) source of the growth factors. Provided that k_{F} and $J_{\mathrm{F}}(x, t)$ are given and that the boundary conditions on the surface of the developing limb parallelepiped are specified for the function $c_{\mathrm{F}}(x, t)$, the above equation, in principle, uniquely determines the spatial distribution of FGFs everywhere in the growing limb at any instant of time. (In their two-dimensional analysis Hentschel *et al.* assumed zero-flux boundary conditions on the anteroposterior borders of the limb, i.e., that morphogens cannot leak out of those surfaces.) As discussed above, it is the AER that is the major source of FGFs. Therefore, as far as FGF production is concerned, this structure can be represented by a steady boundary flux of these molecules into the apical and active zones. The diffusion of FGFs is assumed to be a "fast process" in comparison to the growth of the limb; that is, the motion and eventual condensation of cells in the active zone is assumed to take place in the presence of an already equilibrated FGF distribution determined by the simplified equation

$$\frac{d^2 c_F}{dx^2} = \frac{k_F}{D_F} c_F. \tag{8.9}$$

This equation, which describes the (time-independent) FGF density profile in the active zone, is supplemented by the boundary condition at the distal tip of the active zone, $D_F \, dc_F/dt = j$, giving the inflowing FGF diffusion current (see Chapter 1); j has a fixed value that is related to the equilibrium value of the flux J_F at the boundary. The approximation used to arrive at Eq. 8.9 from Eq. 8.8 is called the separation of time scales. Since Eq. 8.9 is an ordinary differential equation, its solution is relatively simple. Once $c_F(x)$ is determined from Eq. 8.9, it can be inserted into the other equations in which it appears, thus leading to fewer coupled reaction–diffusion equations.

To illustrate how the dynamics of cells is incorporated into the model of Hentschel *et al.*, we consider the R_2 cells. The dynamics of these cells includes diffusion, haptotaxis, mitosis, and differentiation. It is governed by the following "reactor–diffusion" equation:

$$\frac{\partial c_{R_2}}{\partial t} = D_{R_2} \frac{\partial^2 c_{R_2}}{\partial^2 x} - \chi \left(\frac{\partial c_{R_2}}{\partial x} \frac{\partial c_f}{\partial x} \right) + r c_{R_2}(c_{eq} - c_{R_2}) + k_{12} c_{R_1} - k_{22} c_{R_2}. \tag{8.10}$$

The second term on the right-hand side describes condensation resulting from haptotaxis up the fibronectin concentration gradient and thus depends on $\partial c_f/\partial x$; it also depends on $\partial c_{R_2}/\partial x$, because this factor characterizes the developing pattern of R_2 cell condensation. The factor χ is a constant. Mitosis is modeled by the term $r c_{R_2}(c_{eq} - c_{R_2})$. This specific form implies that an equilibrium density of the R_2 cells is stable in the absence of mesenchymal condensation. Such an equilibrium would come about as a result of the suppression of mitosis at high cell densities, $c_{R_2} > c_{eq}$. The last two terms describe, respectively, the differentiation of R_1 cells into R_2 cells, and the differentiation of R_2 cells into R_2' cells. Since cells differentiate in response to molecular signals, the rate constants k_{12} and k_{22} are functions of FGF and TGF-β (for further details see Hentschel *et al.*, 2004). The equations governing the dynamics of the other molecules (TGF-β, inhibitor, and fibronectin) and cell types are constructed in a similar manner, by taking into account their roles and biological specificity.

The separation of time scales and the use of conveniently scaled density functions finally reduces the number of variables in the model of Hentschel and coworkers to four, the concentrations of the three morphogens FGF, TGF-β, and the inhibitor and the density of the total mobile cell population R. These were first analyzed by using the method of linear stability analysis (for details of this method see the Appendix to Chapter 5 and for a specific application see the model of juxtacrine signaling of Sherratt *et al.* in Chapter 7). As discussed in Chapter 7, linear stability analysis establishes which spatial patterns might develop as possible solutions to the system of equations. These patterns are identified as modes that are exponentially growing in time; they are spatially and temporally varying linear perturbations to the homogeneous solutions. Such modes can be selected by varying the ratio of the length of the active zone and the dimension along the anteroposterior axis, $l(t) = l_x(t)/l_y$. Biology thus dictates how $l(t)$ should change with time. If a skeletal pattern with one skeletal element (i.e., the humerus) is to be followed by two skeletal elements (i.e., the radius and ulna), and then by three (i.e., the digits), then the choice of model parameters should allow $l(t)$ to vary in such a way as to make the modes representing the various skeletal elements grow exponentially.

The function $l(t)$ was constrained, but not uniquely determined, by realistic values for the model parameters. Numerical integration of the three nonlinear reactor–diffusion equations with plausible choices for the model parameters and the time variation of $l(t)$ led to a density profile of mobile cells that varied inversely, and in a "quantal" fashion, with l_x, the proximodistal length of the active zone. Specifically, one, two, and three stripe-like distributions of condensed cells (representing successive modes of the reactor–diffusion system under the chosen parameter and boundary values) formed as the active zone length diminished continuously with time. (Note that the nonlinear dynamics changes the prediction of the linear stability analysis in that the modes saturate in concentration instead of continuing to grow at an exponential rate. For details of the numerical analysis, see Hentschel *et al.*, 2004).

To visualize the actual pattern of cartilage elements produced by the model over the duration of limb outgrowth, simulations of this system were performed with different initial conditions of activator concentration. The simulations were carried out under the biologically justified assumption that the elements in the frozen zone keep growing while successive elements in more distal regions are being generated. The results of three of these two-dimensional simulations are shown in Fig. 8.15A with, for comparison, a diagram of a section of the actual skeletal pattern in the plane spanned by the proximodistal and anteroposterior axes at the completion of chondrogenesis (see Fig. 8.13). Although all the simulations look limb-like, the dependence of the pattern's details on initial conditions point to the incompleteness of the model. The "robustness" (developmental stability; see Chapter 10) of embryonic limb development depends on a multiplicity of biochemical interactions (see Tickle, 2003) that extend beyond the bare-bones mechanism modeled by Hentschel and coworkers.

Simulations of the biologically more realistic three-dimensional version of this model are computationally much more demanding, requiring a "multimodel" computational framework (Chaturvedi *et al.*, 2005; Cickovski *et al.*, 2005). Using a simplified version of the system of Hentschel and coworkers to model the activator and inhibitor chemical fields, Chaturvedi *et al.* (2005) obtained the time series of profiles of the activator (i.e., TGF-β) concentration across successively distal positions in the developing limb shown in Fig. 8.15B.

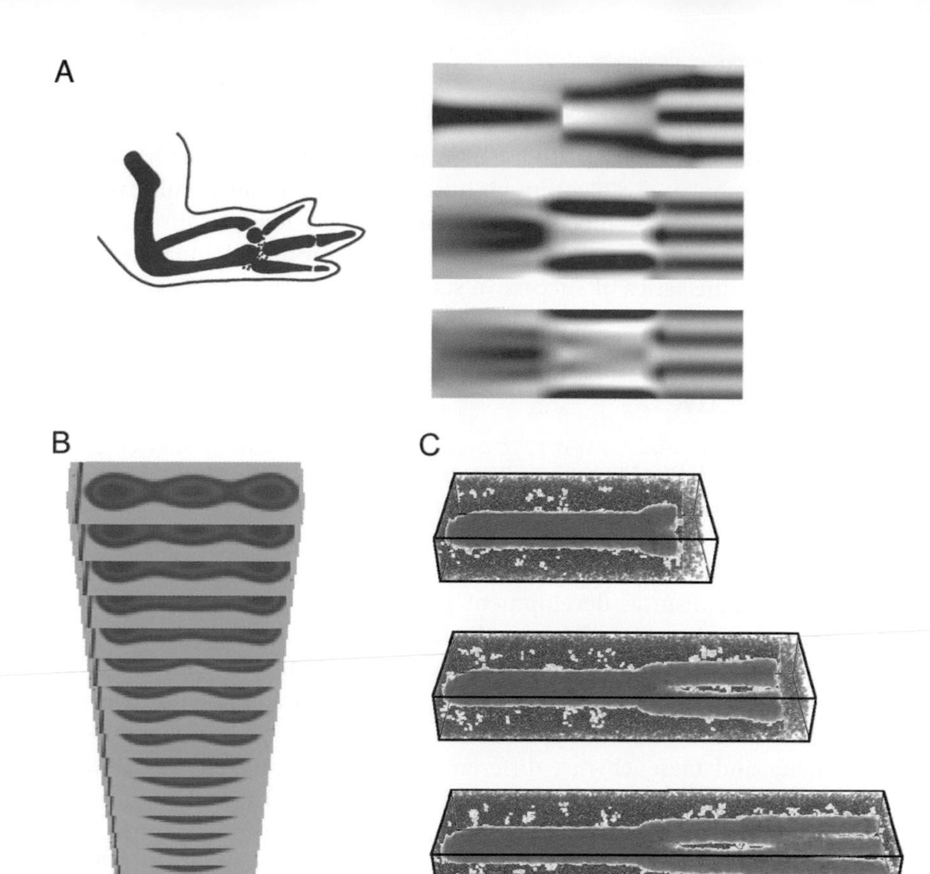

Fig. 8.15 Simulations of limb skeletal development in the model of Hentschel *et al.* (2004). (A) On the left, a longitudinal section of the skeleton of the chicken limb at seven days of development; anterior is at the top of the diagram and posterior is at the bottom. On the right, typical examples of skeletal structures comparable to the seven-day pattern generated by the model, using different initial conditions. The cell density in the active zone was calculated from Eq. 8.10 and then fixed for cells leaving the active zone and entering the frozen zone (see Fig. 8.14). Growth in all zones was assumed to occur at a constant rate. Earlier-formed cartilage elements were thus subject to more growth than later ones. The distribution of cartilage is shown in a continuous grayscale in the simulation panels, black representing the highest cartilage density. The skeletal form in the model is dependent on the parameter values and time-dependent changes in the active zone (which were the same for all three simulations shown) and on the initial conditions (which differed, yielding slightly different patterns). (B) Time series of the concentration of the diffusible morphogen TGF-β displayed in cross-sections of the active zone at successive stages of development, with time increasing in the upward direction, in a three-dimensional model based on the analysis of Hentschel *et al.* (Chaturvedi *et al.*, 2005). (C) A three-dimensional simulation of cell distribution during chick-limb skeletal patterning, based on a modified version of the model of Hentschel *et al.* (2004), using the CompuCell3D multimodel simulation framework (Cickovski *et al.*, 2005). Whereas the original model represented the cell density as a continuous variable, this simulation used the cellular Potts model (see Chapter 6) to represent cell position and motion. Three successive developmental stages are shown, the final stage corresponds to those in panel A. Cells that have undergone condensation are shown here in gray. (The simulations shown on the right in panel A are from Hentschel *et al.*, 2004, by permission. Panel B is from Chaturvedi (2005) by permission. Panel C is given courtesy of T. Cickovski and J. A. Izaguirre, University of Notre Dame).

Cickovski *et al.* (2005), using a slightly different version of the model of Hentschel *et al.* (2004) in conjunction with an energy-minimizing cellular automaton model for cell condensation like that discussed in Chapter 6 (Zeng *et al.*, 2003), obtained the three-dimensional developmental time series of successive stages of limb skeletogenesis shown in Fig 8.15C.

Perspective

During the later stages of embryogenesis the body as a whole becomes more structurally complex and functionally integrated. This means that basic physical mechanisms become correspondingly less applicable to an understanding of the changes in the shape and form of the entire organism as development proceeds. At the same time, newly arising subdomains of the developing organism, the cell clusters that constitute the organ primordia, now become the loci for many of the same physical processes discussed in earlier chapters as determinants of body form. These include the viscoelastic behavior of epithelia and mesenchyme, differential adhesion, cell-state transitions based on multistable transcription-factor networks, and juxtacrine, paracrine, and reaction–diffusion-based pattern-forming systems. The organs serve the body but are also partially independent of it. They are therefore subject to some architectural constraints different from those of body plans. Their functions – transport in the case of the vascular system, secretion for the salivary glands, locomotion and grasping for the limbs, for example – are best accomplished by modular, tubular, or branched structures. Unconstrained by the need to produce an integral body, physical (in concert with genetic) mechanisms of differentiation, morphogenesis, and pattern formation, acting on the epithelial and mesenchymal components characteristic of the advanced, rather than the early, stages of embryogenesis, mobilize genetic processes and products to generate the elaborate organs of complex organisms.

Chapter 9

Fertilization: generating one living dynamical system from two

So far we have followed early development from the first cleavage of the zygote (fertilized egg) to the appearance of fully developed organs. We have reviewed the biology of the most fundamental processes along this path, introduced relevant physical concepts, and used them to build models of the same processes. As the organism matures it eventually arrives at the stage where it is ready to reproduce. At this point the female and the male possess the fully developed sex cells or gametes (i.e., egg and sperm), the fusion of which, termed *fertilization*, sets the developmental process in motion. Fertilization thus can be viewed both as an end and a beginning, the process that simultaneously terminates and initiates the developmental cycle.

The major activities during fertilization that apply generally to any sexually reproducing organism are: (i) contact and recognition between the sperm and the egg; (ii) the regulation of sperm entry into the egg; (iii) fusion of the sperm's and egg's genetic material; and (iv) activation of the zygotic metabolism.

It is clear from this list that fertilization is among the most complicated and spectacular of all developmental processes. It must also be clear that highly sophisticated machinery is needed to carry out the tasks associated with fertilization. The sperm must often travel over great distances relative to its size, it must distinguish between eggs of different related species (in certain marine organisms, sea urchins and abalone for example). Upon encountering the sperm, the egg undergoes a series of changes that alter the electrical (in some species) and mechanical properties of its membrane in order to prevent multiple fertilization ("polyspermy"). Next, the zygote propagates a series of traveling chemical waves followed by a set of mechanical waves – rippling changes in contractility in the egg cortex – that eventually lead to cleavage. How can physics help to unravel these mysteries? Obviously we cannot attempt to construct a physical model of the full process of fertilization. Instead, we will concentrate on those details that clearly require physical mechanisms for their enaction. In particular, we will consider sperm locomotion in a viscous medium, the electrical phenomena associated with egg membrane potential that control sperm entry, the propagation of calcium waves in the egg cytoplasm upon sperm entry, and initiation of the contractile waves in the egg cortex that are the prelude to cleavage. Most of the physics that we will need has already been introduced; the rest will be discussed along the way.

Development of the egg and sperm

The egg and sperm each represent the endpoints of a complex developmental process. In the case of the gametes, or sex cells, this process is referred to as *gametogenesis*. In contrast with the processes of organogenesis described in Chapter 8, in which the functional endpoints were units consisting of many cells, and typically many *types* of cells, gametogenesis gives rise to specialized cells that function individually and at a great distance from their sites of differentiation in the ovaries and testes (collectively, the gonads).

In most sexually reproducing organisms, including insect and vertebrate species, the "germ" cells, which develop into the sperm and eggs, do not even originate in the gonads. They begin their life as a distinct spatially sequestered cell lineage (the "germ line") early in embryogenesis, during blastula formation. The fate of these cells is fixed by the incorporation into their cytoplasm (the "germ plasm") of a set of molecular determinants during cleavage. The germ cells of the developing organism are thus among the earliest cell types to be determined during embryogenesis (De Felici, 2000; Starz-Gaiano and Lehmann, 2001). As with other types of cell differentiation, gametogenesis requires extensive remodeling of chromatin (Kimmins and Sassone-Corsi, 2005).

Germ cells are a migratory population, much like the neural crest cells discussed in Chapter 5. The earliest cells to arise in the germ line, the primordial germ cells, or PGCs, constitute a mesenchymal population that gets conveyed to the gonads by one of several different routes, depending on the species. In frogs and toads the PGCs are first identifiable as a group of cells lining the floor of the blastocoel. They become concentrated in the posterior region of the larval gut and migrate along the dorsal side of the gut, then along the tissue connecting the gut to the inner surface of the abdominal wall (the "dorsal mesentery"), and finally into the gonadal mesoderm (Gilbert, 2003).

In mammals, PGCs arise in a region of the epiblast of the gastrulating embryo that becomes *extra-embryonic* mesoderm i.e., the supportive tissues outside the developing body (Ginsburg *et al.*, 1990); and then have to find their way into the body and ultimately the gonads. They first accumulate in the *allantois* (a sac that develops from the posterior portion of the developing digestive tube), then move into the *yolk sac*, the membranous pouch beneath developing mammalian embryos that is also the source of blood-forming stem cells. They then move through the posterior- or hindgut, up the dorsal mesentery, and into the developing gonads.

Primordial germ cells typically move as cell clusters, like the neural crest (Gomperts *et al.*, 1994), but they may also move as individual cells (Weidinger *et al.*, 2002). *Xenopus* PGCs move by extending a filopodium into which cytoplasm streams, and then they retract this "tail" (Wylie and Heasman, 1993). They appear to be guided by haptotactic cues from the surrounding ECM. However, the PGCs of another frog, *Rana*, (Subtelny and Penkala, 1984) and those of some species of fish (Braat *et al.*, 1999) appear to translocate by a "passive" mechanism that does not utilize individual cell motility, as has also been suggested for neural crest dispersion (see Chapter 6).

Once the organism reaches sexual maturity, the immature germ line cells (termed, at this stage, *oogonia* and *spermatogonia*) resident in the gonads begin differentiating into eggs and sperm. For many species, however, gametogenesis is not completed in the gonads. In mammals, the sperm, even after it is fully formed, must enter the female reproductive tract before it is "capacitated" (physiologically capable of fertilizing the egg), and the *oocyte* (the immature egg) does not even complete meiosis (reduction of its chromosome number to the haploid state, see Chapter 3 and below), until it is fertilized.

Meiosis, like capacitation, is one of the important steps, some of which are in common and some distinctive, that oogonia and spermatogonia ("gonial cells") must take in order to give rise to definitive eggs and sperm. These steps are referred to as *oogenesis* and *spermatogenesis* respectively. Before meiosis occurs, the gonial cells must undergo a series of maturation steps associated with cell division that leads to a population of oocytes or *spermatocytes* (immature sperm). Recall from earlier discussions (Chapters 2 and 3) that the typical cell in a *diploid* organism (sea urchins, fruit flies, amphibian, mammals, and so forth) by definition contains two versions, maternal and paternal, of each chromosome. Like any other cell division, the first meiotic division, meiosis I, is preceded by DNA synthesis, yielding two attached *copies*, the "sister chromatids," of each of the two chromosome *versions*. During the metaphase of meiosis I the corresponding versions of the chromosomes, the "homologs," line up with one another, an event referred to as "synapsis" that only happens in cells undergoing meiosis (Fig. 9.1). When this division occurs, therefore, it is the homologs that are partitioned to the daughter cells, not the sister chromatids as in normal (mitotic) cell division (see Fig. 2.1). Since haploid daughter cells containing only one version of each chromosome are produced from the diploid mother cell, meiosis I (Fig 9.1, middle panel) is referred to as a "reduction division." Meiosis II follows (bottom panel), but this is similar to a mitotic division; the sister chromatids are partitioned to the daughter cell, the end result being a set of haploid cells.

Each initial oocyte or spermatocyte thus gives rise to four progeny cells. One important difference between the egg- and sperm-forming processes in many types of organism, ranging from insects to mammals, is that whereas all four progeny of spermatogenesis are functioning sperm, only one of the four meiotic products of oogenesis is

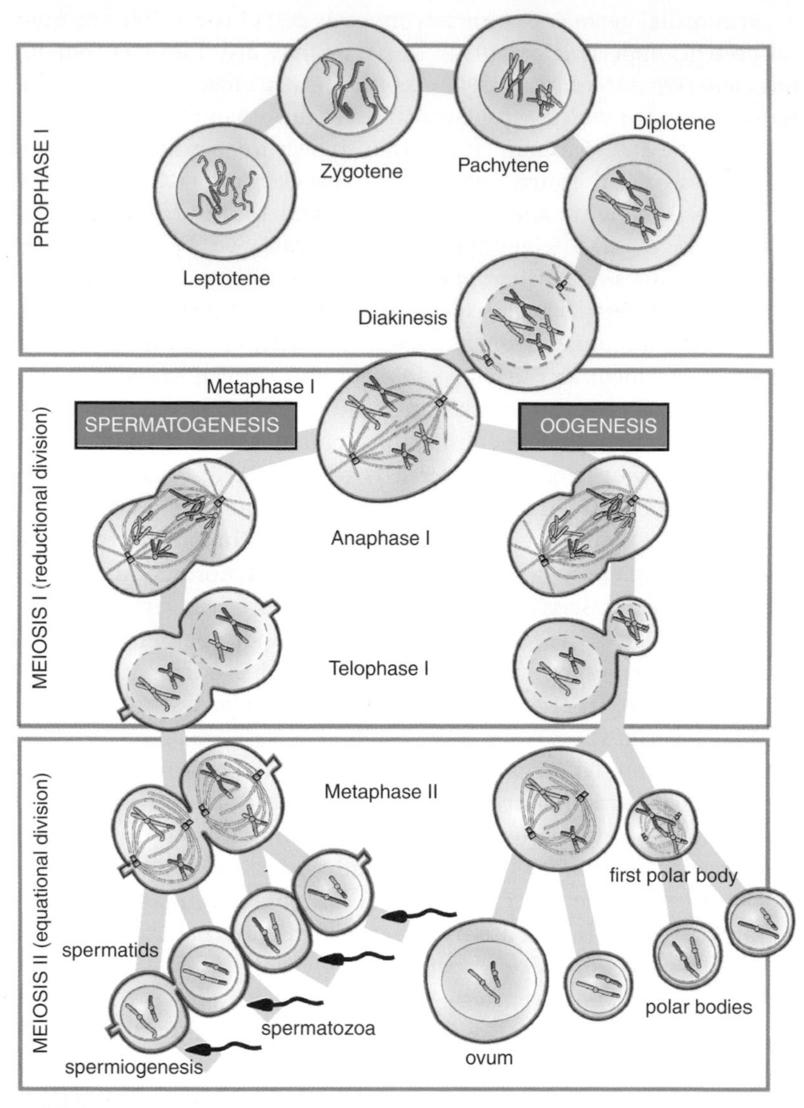

Fig. 9.1 The stages of meiosis. Spermatogonia and oogonia (collectively, "gonial cells"), the diploid cells that are committed to form the male and female gametes, i.e., the sperm and egg, are distinct cell types arising respectively in the testes and ovaries. Their mitotic progeny, the *primary spermatocyte* and *primary oocyte*, undergo a process of reduction division or meiosis (see the main text) in order to give rise to the haploid gametes. The early stages of meiosis are similar in the male and female lineages. Prophase I (top panel) is divided into several steps. In leptotene, the chromosomes assume an extended configuration. (Two homologous pairs of chromosomes are shown; the red pair is descended from chromosomes contributed by the maternal parent and the blue pair from chromosomes contributed by the paternal parent in the fertilization event that gave rise to the individual in which the pictured meiosis is occurring). Each of these chromosomes has already been replicated during the S-phase (not shown) that precedes prophase I. During *zygotene*, the homologs undergo pairing (synapsis), so that the corresponding chromosomes of maternal and paternal origin are physically linked to each other. These linked pairs are called "bivalents." At *pachytene*, the chromosomes become shorter and thicker and the two sister chromatids of each chromosome begin

(*Cont.*)

a functioning egg. The meiotic divisions leading to egg formation are highly asymmetric, reflecting the requirement on the egg to have large stores of cytoplasmic materials – nutrient yolk, protein-encoding, ribosomal, and transfer RNAs, localized morphogenetic factors, cell-cycle regulatory proteins – to support the early stages of development (see Chapter 3). Meiosis I and II thus each produce small nonfunctional cells (the first and second "polar bodies") in addition to the oocyte, which receives almost all the cytoplasm. As noted above, meiosis II does not occur until after fertilization in mammalian eggs.

Meiosis in both egg and sperm is followed by further maturational steps and here, in all species, the steps are quite different for the female and male gametes (McLaren, 1984; McKim *et al.*, 2002; Giansanti *et al.*, 2001). In both cases differentiation is typically aided by accessory cells. In oogenesis, cells of the same oogonial lineage as the oocyte, "nurse cells," and supporting cells of the ovary, "follicle cells," play distinct but equally essential roles (Gosden *et al.*, 1997). In spermatogenesis, all cells of the spermatogonial lineage typically give rise to gametes. As a result of the activities of germ-line and maternal accessory cells the prefertilization oocyte typically expands greatly in size, both in its cytoplasmic compartment and in its nucleus (termed the "germinal vesicle"). While the germinal vesicle contains only a haploid complement of chromosomes, its genome is typically transcriptionally active, producing many different types of RNA molecules (Bachvarova, 1985; Picton *et al.*, 1998). In some cases, such as *Xenopus* and other amphibians, the requirement for ribosomes during the

to separate, remaining attached only by their centromeres (see Chapter 2). Since the two-stranded nature of the individual chromosomes of the bivalents is now apparent, the resulting "double–double" structures are referred to as "tetrads." At this stage "crossing-over" also takes place. Crossing-over refers to the process by which corresponding portions of homologous chromosomes are exchanged with each other, yielding chromosomes distinct from those contributed by either of the individual's parents. At the following stage, *diplotene*, the two chromosomes in each tetrad begin to repel one another. Prophase I ends with *diakinesis*, in which the nuclear envelope breaks down and the mitotic spindle fibers form. Meiosis I (middle panel) is a sequence of steps analogous to mitosis in somatic (non-meiotic) cells, with an important difference. Rather than the sister chromatids being separated during this process, the homologs are separated. This leads to haploid daughter cells, i.e. cells that contain only one version of each chromosome of a homologous pair. In spermatogenesis, the two daughter cells resulting from meiosis I are of equal size and are called *secondary spermatocytes*. In oogenesis, the division is uneven, the larger cell is called the *secondary oocyte* and the smaller one the first *polar body*. Meiosis II (bottom panel) begins as these cells enter prophase (not shown) directly, without undergoing S-phase, the DNA having already been replicated in the S-phase immediately preceding meiosis I. This is followed by anaphase II and telophase II (not shown) and then metaphase II. Meiosis II is similar to somatic cell mitosis: the sister chromatids split apart and are partitioned into daughter cells. In spermatogenesis, these cells are called *spermatids*; the original primary spermatocyte yields four of them. Each spermatid undergoes a series of morphological changes (*spermiogenesis*) to produce a mature gamete, the spermatozoon, or sperm. In oogenesis, meiosis II, like meiosis I, is asymmetric, yielding a single ovum or egg, the single gamete arising from the primary oocyte, and two other polar bodies.

early cleavage stages leads to a several-thousand-fold amplification of the number of copies of the genes specifying ribosomal RNA in the oocyte nucleus during oogenesis (Thiebaud, 1979).

The end result of oogenesis is a large cell (the human and sea urchin eggs are on the order of 100 μm in diameter, that of *Xenopus* is up to 2 mm, and the average length of the two main axes of the ellipsoidal chicken egg is about 5 cm). Yolk consists of energy-rich nutrient molecules, and the amount required for development depends on the species' biology. Mammalian embryos get almost all their nourishment from the maternal blood supply, so the eggs that give rise to them contain very little yolk. Birds' eggs, which develop outside the mother's body, contain huge amounts of yolk. Another typical feature of eggs is an array of secretory vesicles in the cortical region of their cytoplasm, which fuse with the plasma membrane once fertilization occurs. The contents of these "cortical granules" – primarily glycoproteins – provide a barrier to the entry of additional sperms (see below). Finally, the terminally differentiated egg is also surrounded by an extracellular matrix (ECM) layer, consisting of various glycoproteins and other molecules (Wassarman *et al.*, 1999), which is termed, depending on the species, the "jelly coat" (sea urchins, amphibians), the "vitelline membrane" or "vitelline envelope" (fish, birds), and the "zona pellucida" (mammals) (Fig. 9.2).

Spermatogenesis is completed by a final series of steps, termed *spermiogenesis*, which gives rise to the morphologically definitive motile sperm cell, also called the "spermatozoon." During spermiogenesis the sperm loses most of its cytoplasm, develops an extended flagellum, retains its mitochondria, which generate ATP to drive the flagellar motion (although the mitochondrial DNA does not, at least in mammals, get transmitted to the next generation of organism), and develops a single apical secretory vesicle, the *acrosome*, which contains enzymes, released during fertilization in a dramatic exocytotic event (the "acrosome reaction"), that facilitate penetration of the sperm through the egg's ECM (Fig. 9.3).

Physics of sperm locomotion

For the sperm to arrive at the egg it has to overcome the viscous drag it experiences either in sea water (marine vertebrates and

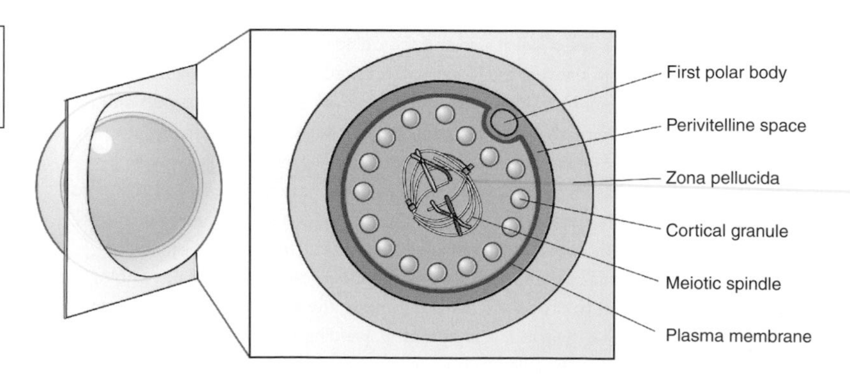

Fig. 9.2 Cross-section of the egg of a mammal; see the main text for details. (After Gilbert, 2003.)

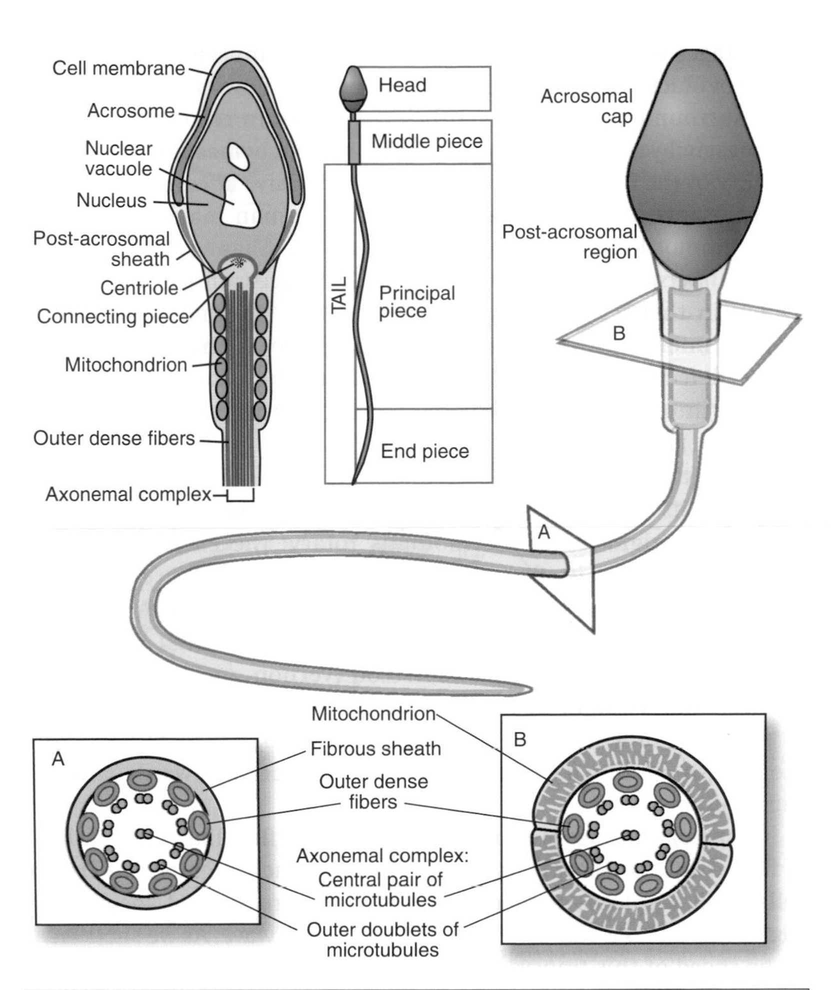

Fig. 9.3 Principal components of a mammalian spermatozoon. The top right panel shows a three-dimensional rendition of the entire spermatozoon; the top center panel indicates the anatomical designations of its three main portions. The top left panel shows a schematic of a cross-section of the sperm head with characteristic subcellular components labeled. Exocytosis of the acrosome (see Fig. 9.5) enables the sperm to penetrate the vitelline membrane (see Fig. 9.2) and brings the lower region of the sperm cell membrane into contact with the egg cell membrane, with which it fuses. The sperm nucleus is incorporated into the zygote; the single, coiled sperm mitochondrion is not. The axonemal complex is the contractile assembly of the sperm tail flagellum. Sections through the middle piece (B) and the tail (A) are shown in detail in the two lower panels, where the relative locations of the mitochondrion and the microtubular components of the axonemal complex are represented. (After Ross *et al.*, 2003.)

invertebrates) or in the female reproductive tract (mammals, insects). The extent to which the sperm's own propulsive activity contributes to its translocation toward the egg differs with different species. In a watery environment external to the female's body (e.g., in sea urchins or fish), the inherent motility of the sperm is all important. In the reproductive tract of a female mammal other factors dominate at

Fig. 9.4B is a viscous drag: $F_{||} = f_{||}v_{||}$ and $F_{\perp} = f_{\perp}v_{\perp}$ (Fig. 9.4D). Here the friction constants $f_{||}$ and $f_{\perp}$ are given in Eq. 1.3. The expressions in Eq. 1.3 are valid for a cylindrical body whose length (l) is much larger than its diameter d. Since the diameter of the flagellar rod is small, the condition $l/d \gg 1$ is fulfilled for the finite small segments in Fig. 9.4B. Moreover, as indicated by Eq. 1.3, for any finite large value of l/d we have $f_{\perp} > f_{||}$, which for a tilt angle $\theta \approx 45^{\circ}$ (when $v_{\perp} \approx v_{||}$) implies that the force needed to move the flagellar segment perpendicular to its axis is larger than that needed to move it parallel to its axis. (In reality the tilt angle is considerably smaller than 45°, which makes $F_{\perp}/F_{||}$ even larger.) It is exactly this disparity between $F_{\perp}$ and $F_{||}$ that makes sperm propulsion possible.

We now decompose $F_{\perp}$ and $F_{||}$ for the up-moving segment in Fig. 9.4B into components along the direction of sperm locomotion (vectors 1 and 4 in Fig. 9.4D) and perpendicular to it (vectors 2 and 3 in Fig. 9.4D). Thus the net force acting on the flagellar segment as it moves to the right is

$$F_{\rightarrow} = F_{\perp} \sin\theta - F_{||} \cos\theta = (f_{\perp} - f_{||}) v \sin\theta \cos\theta. \tag{9.3}$$

Since $f_{\perp} > f_{||}$, the resulting net force produces a forward thrust (opposite to the direction of the wave velocity, see Fig. 9.4A), which in light of Eq. 1.3 is proportional to the viscosity of the liquid. Adding the components perpendicular to the direction of motion we obtain a net downward pointing force that resists the upward motion of the segment,

$$F_{\downarrow} = F_{\perp} \cos\theta + F_{||} \sin\theta = (f_{\perp} \cos^2\theta + f_{||} \sin^2\theta) v. \tag{9.4}$$

Repeating the above exercise for the down-moving segment in Fig. 9.4B, it is easy to see that the net force on it parallel to the direction of motion again points to the right and is equal to $F_{\rightarrow}$ in Eq. 9.3. (Since both v and θ now change sign in Eq. 9.3 the direction of the net thrust does not change). The net force perpendicular to the direction of motion does change sign and is $F_{\uparrow} = -F_{\downarrow}$, the latter being given in Eq. 9.4.

A similar analysis can be carried out for each segment along the flagellar rod. The net outcome is that, because of the shape of the flagellum, hydrodynamic forces alone are capable of propelling the sperm along a steady direction towards its target, the egg. (Note that if there is an integral number of wavelengths along the flagellum then, according to the above analysis, the net force acting on it perpendicular to the direction of motion is zero.) In some marine species, in which the sperm navigates using a species-specific chemotactic gradient produced by the egg (Yoshida *et al.*, 1993), and perhaps in mammals as well (Spehr *et al.*, 2003), the gradient sets the direction of the sperm's velocity vector, for which the sperm itself, using the mechanism described, supplies the vector's magnitude.

Interaction of the egg and sperm

Once the sperm arrives at the egg it has to penetrate its ECM (vitelline membrane, jelly coat, zona pellucida) and fuse with its plasma membrane. The initial interaction, in species as varied as fruit flies, clams, sea urchins, and humans, is the acrosome reaction. This sudden exocytotic event, in which the acrosomal vesicle fuses with the sperm plasma membrane, spilling out its enzyme contents (Fig. 9.5), is triggered by the binding of proteins and/or glycoproteins in the apical membrane of the sperm to glycoproteins of the egg ECM (Wassarman, 1999). The sperm then moves through the ECM until its new apical membrane, previously the "floor" of the acrosomal vesicle, meets up with the egg's plasma membrane and fuses with it. This event, called "syngamy," permits the sperm nucleus, also called the male "pronucleus," to enter the egg, where it meets up with the female pronucleus. (In mammals, the second meiotic division must take place before the female pronucleus can form.) Once these

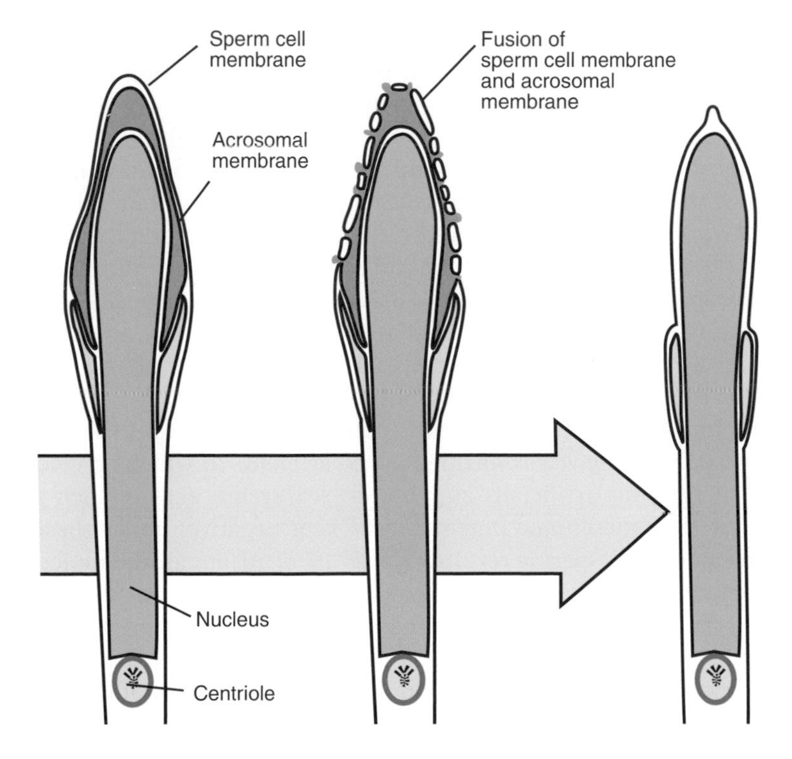

Fig. 9.5 Schematic illustration of the acrosome reaction in a mammalian sperm. The bounding membrane of the acrosome (a large intracellular vesicle) and the apical region of the sperm-plasma membranes fuse, forming discontinuous vesicles and thereby releasing the fluid contents of the acrosome (orange). This fluid contains enzymes that locally break down the zona pellucida (the vitelline envelope) of the egg, permitting the sperm-plasma membrane, which remains intact throughout the acrosome reaction, to approach and eventually fuse with the egg-plasma membrane (see also Fig. 9.7). (After Yanagimachi and Noda, 1970.)

haploid pronuclei join together, the fertilized egg contains the genome of a new diploid individual. It is now termed a "zygote," and is ready to begin the series of cleavages described in Chapter 2.

Typically, thousands of sperms reach the egg more or less simultaneously. If more than one male pronucleus is deposited into the egg's cytoplasm (a situation known as "polyspermy"), the amount of genetic material and number of chromosomes would be uncharacteristic of the species and the gene expression levels (which often differ for maternally and paternally contributed versions of a gene) would be unbalanced. Polyspermic zygotes are not viable, and two major protective mechanisms have evolved to prevent this from occurring: a "fast block" to polyspermy, triggered by sperm–egg surface contact and involving a transient electrical change in the egg's plasma membrane (Glahn and Nuccitelli, 2003); and a "slow block" to polyspermy, triggered by a self-organized traveling chemical wave, leading to exocytosis of the egg's cortical granules and the secretion of a long-lived ECM barrier (Sun, 2003).

Membrane potential and the fast block to polyspermy

A few seconds after the first sperm binds, the egg's resting electric membrane potential, which is typically negative (about -70 mV for the sea urchin), shifts to a positive value of about $+20$ mV. This is accomplished by the opening of sodium ion channels (for the composition of the plasma membrane see Chapter 4), which allows the influx of Na^+. (Note that by convention the electric potential difference across the plasma membrane is measured relative to the extracellular milieu). Sperms cannot fuse with a plasma membrane having a positive resting potential. The sudden change in the membrane potential provides a fast block to polyspermy, but it is transient: after a few minutes its original value is reestablished.

The importance of the electrical potential difference in the fast block to polyspermy was confirmed by experiments in which an electrical current was artificially supplied to sea urchin eggs in such a way that their membrane potential was kept negative; under these circumstances polyspermy resulted (Jaffe, 1976). Although the molecular details of how the sperm triggers electrical changes at the egg surface are not well understood, the physics of all such membrane phenomena are governed by a fundamental mechanism described by the Nernst–Planck equation, which determines the value of the membrane's resting potential.

Ultimately, it is the ions in the close vicinity of the membrane, held there by their electrical attraction to the oppositely charged counter-ions, that give rise to its potential (Fig. 9.6). Ions can move across the membrane, through specific ion channels, for two distinct reasons: there must be an imbalance either in the concentration of particular ions (i.e., a concentration gradient) or in the net charge (i.e., an electric field gradient) between the two sides of the membrane. It is easy to see that these two gradients lead to opposite effects. Diffusion along a concentration gradient will move molecules from the

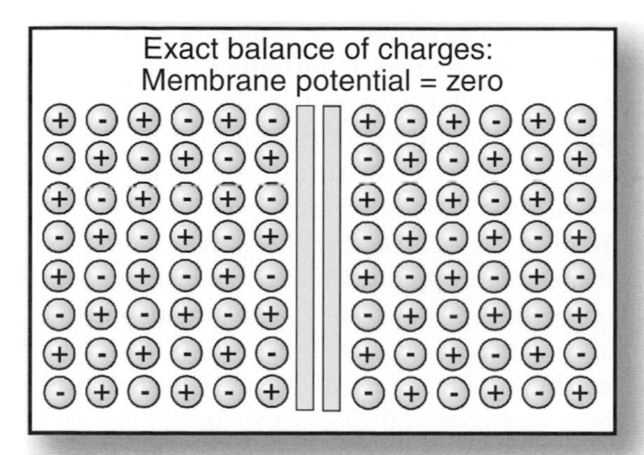

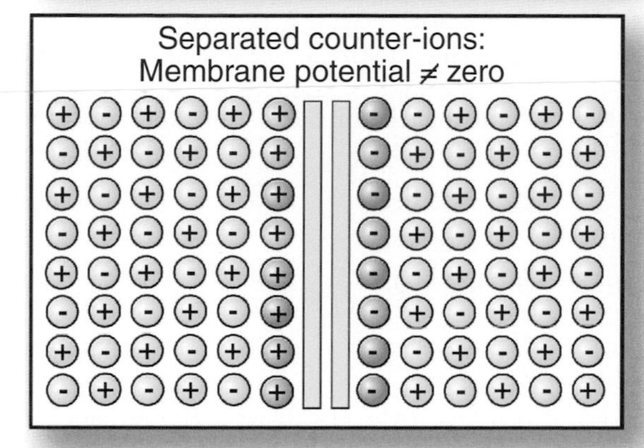

Fig. 9.6 Origin of the membrane potential. A nonzero potential across the membrane arises when the balance of charges on its intracellular and extracellular sides (upper panel) is disrupted by the migration of a few positive (or negative) ions through the membrane, leaving negative (or positive) counter-ions behind (lower panel).

higher concentration side to the lower one. If the molecules carry charge, as in the case of ions, this motion will lead to a charge imbalance and thus to an increase in the electric field gradient, which, in turn, will resist the motion due to the concentration gradient. The flow of ions is thus driven by the combination of the two gradients, namely, the electrochemical gradient introduced in Chapter 1. When the electrochemical gradient is zero there is no net flow of ions. The electric potential difference at which this equilibrium is established is the resting, or equilibrium, membrane potential.

A nonzero electrochemical gradient implies a particle current (see Chapter 1)

$$j(x) = -D\frac{\Delta c(x)}{\Delta x} + v_{\mathrm{d}}c(x). \tag{9.5}$$

(Note that for simplicity we have assumed a steady, time-independent, concentration.) The first term on the right-hand side of Eq. 9.5 is the diffusion current, the second is the current due to external forces (see Chapter 1). In the case of the movement of ions with charge q due to electric forces, the drift velocity (also introduced in Chapter 1) is $v_d = F/f = qE/f = \mu E$, where E is the electric field established by the ion distribution and μ is the ion mobility. Equation 9.5 implies the presence of a single ionic species, with diffusion coefficient D and charge q. If there are several types of ions, each will be represented by a separate term in Eq. 9.5. For simplicity we assume that there are two kinds of ions with opposite charge $q_+ = -q_- = q$ and corresponding diffusion coefficients D_+ and D_-. We also assume that the concentration gradients of these two ions are equal. Thus, using the Einstein–Smoluchowski relation, $D = k_B T/f$ and the relationship between the electric field and potential $E = -\Delta V/\Delta x$, Eq. 9.5 takes the form

$$j(x) = -qc(x)(\mu_+ - \mu_-)\left[\frac{\Delta V}{\Delta x} + \left(\frac{D_+ - D_-}{\mu_+ - \mu_-}\right)\frac{\Delta \ln c(x)}{\Delta x}\right]. \tag{9.6}$$

Here "ln" stands for the natural logarithmic function and we have used the relation

$$\frac{\Delta c}{c\Delta x} = \frac{\Delta \ln c}{\Delta x}.$$

Finally, at equilibrium, for which $j = 0$,

$$\Delta V = V_i - V_o = -\frac{D_+ - D_-}{\mu_+ - \mu_-}\ln\frac{c_i}{c_o}, \tag{9.7}$$

where the subscripts i and o refer to the inside and the outside of the cell respectively. Equation 9.7 is called the Nernst–Planck equation and the ratio of the concentrations, c_i/c_o, is the Donnan ratio. In the hypothetical case when the membrane potential is due to a single positively charged ionic species, Eq. 9.7 can be written as

$$V_i - V_o = \frac{-k_B T}{q}\ln\frac{c_i}{c_o},$$

Since the logarithmic function varies slowly with its argument and for realistic concentrations gives a number of the order unity, the resting potential (for ions at 37 °C having unit positive charge) is determined by the factor $k_B T/q = 26.7$ mV. The resting potential for most membranes other than the fertilized egg is indeed around this value and varies between −20 and −200 mV.

Propagation of calcium waves: spatiotemporal encoding of postfertilization events

The electrical fast block to polyspermy, which has been quantitatively studied in sea urchins (Jaffe, 1976) and frogs (Cross and Elinson, 1980), probably does not occur in mammalian eggs (Jaffe and Cross, 1986). Even in species in which it occurs, it is transient: the membrane potential of the fertilized sea urchin egg remains positive for only about a minute (Jaffe and Cross, 1986; Glahn and Nuccitelli, 2003).

However, within a minute after sperm–egg fusion another sequence of ion-dependent events takes place (this time involving calcium ions, Gilkey *et al.*, 1978), leading first to the concerted release of the contents of the cortical granules – the "cortical reaction" (Fig. 9.7). The exocytosed material that thereby comes to lie in the perivitelline

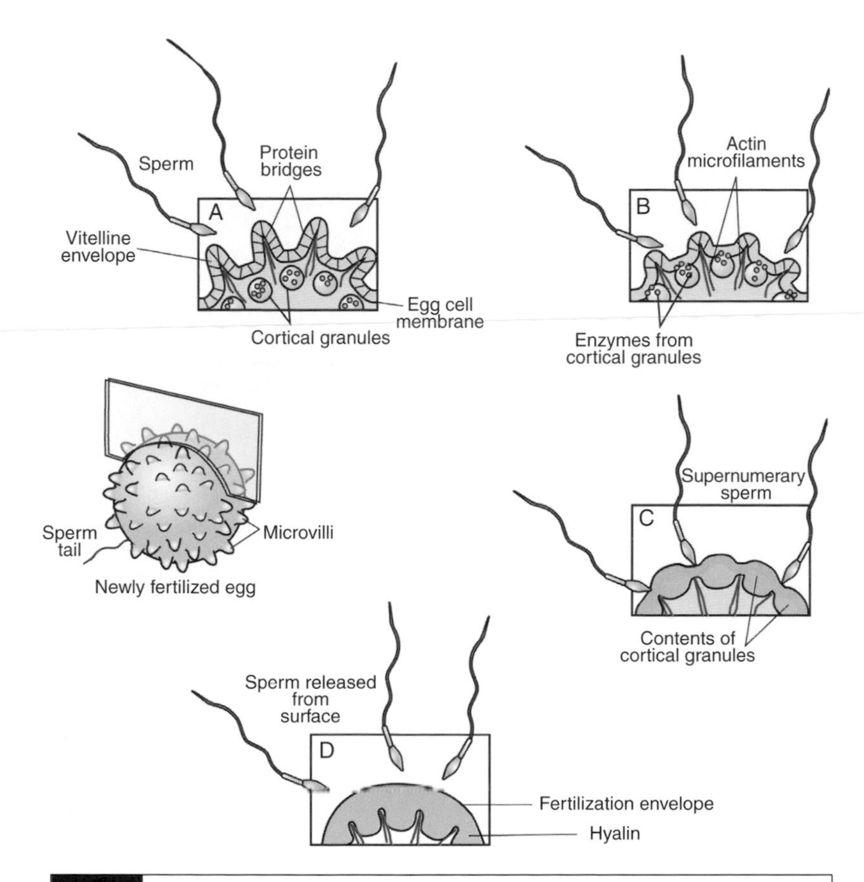

Fig. 9.7 Schematic illustration of the steps involved in the slow block to polyspermy. The leftmost panel gives a three-dimensional representation of a fertilized egg (the microvilli are shown at a larger than true scale) with the tail of the fertilizing spermatozoan (or sperm) protruding from it. Panels A–D show the cortical reaction at successive times following fertilization (viewing the interior of the egg through the cut indicated). Prior to the stage shown in panel A the fertilizing sperm has penetrated the vitelline envelope by means of the acrosome reaction (see Fig. 9.5), and its plasma membrane has fused with the egg plasma membrane (see the main text for a description). This membrane fusion initiates the cortical reaction. (A) Three supernumerary (non-fertilizing) sperms approach the vitelline envelope. Beneath the egg-plasma membrane cortical granules move along radial actin microfilaments (magenta) toward the inner surface. (B, C) These granules fuse with the plasma membrane and release their contents, enzymes that cleave the protein bridges connecting the vitelline envelope to the egg-plasma membrane, as well as the protein hyalin, osmotically active proteoglycans that cause water to enter the perivitelline space and swell the vitelline envelope, and additional enzymes that cause the newly deposited ECM to harden. (D) The resulting matrix, the "fertilization envelope," prevents further penetration of the supernumerary sperms, which are consequently released from the egg surface. (After Gilbert, 2003.)

space (see Fig. 9.2) between the egg's plasma membrane and the egg's ECM (e.g., the zona pellucida) provides a barrier, termed the *fertilization envelope*, to additional sperm entry. This is the slow block to polyspermy, and it happens in virtually all species studied. Then, depending on the species, successive waves of calcium ion concentration go on to trigger other events of early development. In mammals these include the completion of meiosis, the initiation of mitosis, the initiation of protein synthesis from stored maternal mRNAs (Runft *et al.*, 2002) and the initiation of surface waves of cortical contractility (Deguchi *et al.*, 2000) (Fig. 9.8).

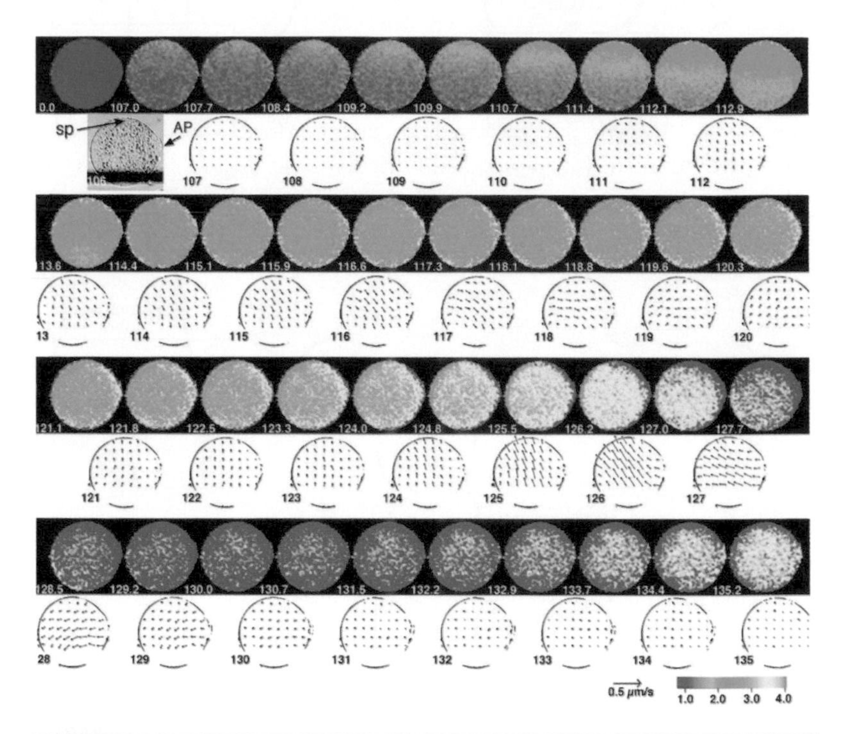

Fig. 9.8 A traveling wave of Ca^{2+} concentration (colors) and cytoplasmic movement (black-and-white) during the rising phase of the initial Ca^{2+} wavefront in the mouse egg, following fertilization. The position of the sperm-fusion site and that of the animal pole are indicated by arrows labeled "sp" and "AP" respectively in the bright-field image (second row, at the left). The egg was loaded with the Ca^{2+}-sensitive dye calcium green-1 dextran and subjected to fluorescence microscopy. The (pseudo-)colors represent the intensity of the emitted fluorescence, relative to a basal value (see the color bar at the bottom right), obtained from images just before the rise in Ca^{2+} concentration. The number shown at the bottom left of each image is the time (in seconds) of acquisition. The zero of time is defined by the time when the image for the basal value of fluorescence was taken, before the initial rise in intensity. The arrows in the black-and-white images represent the direction and magnitude of the local cytoplasmic velocity, as detected from the analysis of the time series of bright-field images taken at 0.5 s intervals (for reference see the arrow at the bottom right). After the reference image at 106 s only the contour of the egg and selected lattice points, approximately 11 micrometers apart, are shown. (Reprinted from Deguchi *et al.*, 2000, with permission from Elsevier.)

Unlike the ion transients that occur during the fast block to polyspermy, the source of which is in the external medium, the source of ions for the later calcium transients is in intracellular "stores" (Bugrim *et al.*, 2003), membranous compartments that are under the control of intracellular signals, most notably inositol trisphosphate (IP_3) (Berridge *et al.*, 2000). Once the release of these stores has been initiated in the fertilized egg, a periodic series of self-sustaining waves of elevated Ca^{2+} concentration travel through the egg's cortical cytoplasm (Kubota *et al.*, 1987; Miyazaki *et al.*, 1993; Eidne *et al.*, 1994; Jones, 1998; Deguchi *et al.*, 2000; Dumollard *et al.*, 2002). Several of the major postfertilization events mentioned above are initiated by different numbers of Ca^{2+} waves, while their completion requires a greater number of Ca^{2+} waves than their initiation (Ducibella *et al.*, 2002). This suggests that there is informational content in the spontaneous Ca^{2+} waves that follow fertilization. Moreover, it seems that a single dynamically organized signaling system can regulate the different cellular events associated with early development that must occur in a distinct temporal sequence.

Modeling Ca^{2+} oscillations in the egg

Concomitant with the transient depolarization of the egg's plasma membrane brought about by the fusion of the egg and sperm is an approximately ten-fold increase in the intracellular calcium ion concentration of the egg. This increase occurs (depending on the species) in the form of one or more traveling waves of elevated Ca^{2+}, which start at the point of sperm entry. The initial effect of the elevated calcium ion concentration is the triggering of the cortical granule reaction, in the course of which these organelles beneath the plasma membrane fuse with it and discharge their protein contents. This process establishes the slow block to polyspermy (Jaffe *et al.*, 2001), as discussed above. The initial and subsequent waves, in species where they occur, cause major intracellular restructuring and eventually bring the postfertilization machinery into motion.

Transmitting information in the form of calcium oscillations is a ubiquitous means of both intercellular and intracellular signaling. Experimental evidence indicates that both the temporal and the spatial behavior of these oscillations has information content. Strikingly, the magnitude of their frequency (typically in the range 10^{-3}–1 Hz) and amplitude controls the specificity of gene expression (Dolmetsch *et al.*, 1998), which may, in turn, activate the initial developmental events.

The fertilization wave, as observed in sea urchins or *Xenopus*, is a continuous wave of well-defined amplitude, which sweeps through the egg in a short time (around 30 seconds in the sea urchin (Hafner *et al.*, 1988) and 2.5 minutes in *Xenopus* (Fontanilla and Nuccitelli, 1998)). The generation and propagation of these waves is due, as noted above, to the release of calcium ions from the endoplasmic reticulum, which is triggered when the cytosolic concentration of the ions reaches a threshold value, a phenomenon known as Ca^{2+}-induced

Ca^{2+} release. At the threshold, specialized channels (i.e., IP_3 receptors) are activated. These channels are inactivated as the local Ca^{2+} concentration rises further and subsequently remain closed during a refractory period. Finally, cytosolic Ca^{2+} is resequestered into the endoplasmic reticulum via specialized pumps.

It is not entirely clear what initiates the fertilization calcium wave (Dumollard *et al.*, 2002). According to one hypothesis, the transient increase in calcium concentration at the point of sperm entry is due to the release of Ca^{2+} by the sperm itself (Stricker, 1999). Another proposal is that an as yet unidentified "sperm-factor" activates local calcium release upon fusion of the sperm's and egg's plasma membranes (Oda *et al.*, 1999; Sardet *et al.*, 1998; Stricker, 1999; Jaffe *et al.*, 2001). It is also possible that the interaction of the sperm and the egg activates a special class of receptors on the egg surface and that this results in the production of IP_3, which opens IP_3-dependent calcium channels in the egg's endoplasmic reticulum near the fertilization site (Miyazaki *et al.*, 1993; Stricker, 1999; McDougall *et al.*, 2000). Experimental results seem to favor this last hypothesis (Bugrim *et al.*, 2003).

The activation of Ca^{2+}-releasing channels by a threshold concentration of cytosolic Ca^{2+} is another manifestation of the nature of the cell as an excitable medium (Lechleiter *et al.*, 1991, 1998). Numerous models have been constructed to describe the formation and propagation of calcium oscillations (for a review see Schuster *et al.*, 2002). As we have seen in Chapter 5, to describe excitability mathematically, nonlinear differential equations are needed. A relatively simple model of this sort, the "fire–diffuse–fire" (FDF) model of Dawson *et al.* (1999), gives an account of the formation and propagation of continuous postfertilization calcium waves.

In the FDF model the release of calcium from the intracellular stores is assumed to take place through an array of release sites represented by point sources corresponding to actual storage vesicles with regulated channels. (A similar model was discussed by Bugrim *et al.*, 1997). These sites are spaced at a distance d from one another and are embedded in a continuum (the cytosol) in which calcium ions are assumed to diffuse. The release of calcium takes place while a channel is open, which defines the chemical time scale τ. Another time scale (the intersite diffusion time) is defined by d^2/D (cf. Eq. 1.1), where D is the Ca^{2+} diffusion coefficient. Whenever the cytosolic Ca^{2+} concentration in the vicinity of a release site reaches a threshold value $[Ca^{2+}]_T$ above the basal concentration $[Ca^{2+}]_B$, the site starts releasing calcium ions at a rate σ/τ. Here σ is the total number of ions released in time τ by a single site. (Some of the released calcium ions are buffered by binding to proteins and thus do not participate in wave generation and propagation. This is incorporated into the model by using a number smaller than σ and a buffered diffusion coefficient, details we will ignore here.)

Ca^{2+} release and diffusion is described by a single nonlinear equation, which is a reaction–diffusion equation similar to those encountered in Chapters 7 and 8 (details can be found in Dawson *et al.*, 1999, and in Pearson and Ponce-Dawson, 1998). Mathematical analysis shows that the dynamics depends on only two dimensionless parameters (instead of the six parameters d, τ, D, σ, $[Ca^{2+}]_T$, $[Ca^{2+}]_B$ introduced above), defined by the following expressions:

$$\Gamma = \frac{\sigma/d^3}{[Ca^{2+}]_T - [Ca^{2+}]_B}, \tag{9.8}$$

$$\beta = \frac{D\tau}{d^2}. \tag{9.9}$$

The meaning of these parameters is easy to understand. Since σ/d^3 is the release concentration (the amount of Ca^{2+} released per site divided by the volume per site), Γ is the ratio of the release concentration to the difference between the threshold and basal concentrations. Intuitively it is clear that no traveling wave can be sustained if $\Gamma < 1$. The parameter β is the ratio of the time for which a site is open, τ, to the intersite diffusion time d^2/D. This parameter controls the shape of the propagating wave. If the chemical reaction involved in release is the rate-limiting process (the "reaction-limited" case, $\tau \gg d^2/D$) then $\beta \gg 1$ and the wave is continuous. If $\beta \ll 1$, the "diffusion-limited" case, the wave is saltatory (i.e., it has an abruptly changing shape). In the former case the wave travels without observable change in its shape, whereas in the second case its shape changes in time.

Using experimental results for τ, d, D, σ, $[Ca^{2+}]_T$, and $[Ca^{2+}]_B$, the FDF model predicts that the fertilization wave in *Xenopus* is a continuous wave, in accord with the experimental findings (Fontanilla and Nuccitelli, 1998) and has the shape shown in Fig. 9.9. (Saltatory waves have been observed in *Xenopus* oocytes prior to fertilization; Callamaras *et al.*, 1998).

In summary, the FDF model emulates the active process of calcium wave propagation by a combination of the passive diffusion of Ca^{2+} and its active release from point-like sources. This is a reasonable representation of the biological reality. As discussed in Chapter 1, in the crowded intracellular environment simple diffusion can be a viable transport mechanism only on a limited spatial scale, even for small molecules such as Ca^{2+}. In the FDF model the cytosol is assumed to be homogeneous apart from point-like release sites and thus intersite diffusion is assumed to be unhindered. The close spacing of the sites, however, ensures that free diffusion takes place only between the release sites and therefore on a small spatial scale. On the scale of the entire egg the reinforcement of the wave by Ca^{2+}-induced Ca^{2+} release is an active process. The important message from the work of Dawson *et al.* (1999) is that a relatively simple model based on biologically plausible assumptions can explain the excitable character of the zygote (with respect to Ca^{2+} wave generation and propagation) and provide testable predictions of the shape of postfertilization calcium waves.

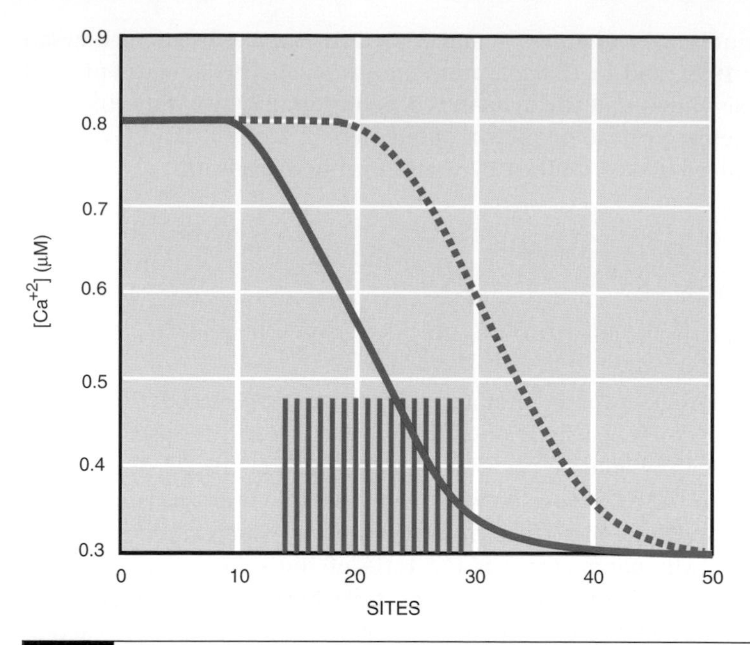

Fig. 9.9 Two "snapshots" of a continuous Ca^{2+} wave in the FDF model of Dawson *et al.* (1999). The concentration profile is shown along a streak of release sites; 0 marks the initial site. The green vertical lines indicate the sites that are simultaneously firing at the time of the first snapshot (solid red line). The broken line corresponds to a later time. The wave is traveling with a velocity of 5.2 μm/s (corresponding to parameter values $\tau = 9$ s, $d = 3$ μm, $D = 50$ μm^2/s, $\sigma = 1.35 \times 10^{-14}$ μmol, $[Ca^{2+}]_T = 0.4$ μM, and $[Ca^{2+}]_B = 0.3$ μM, taken from the literature on the fertilization wave in *Xenopus* eggs (Fontanilla and Nuccitelli, 1998)). Note that the waveform travels without observable deformation.

Surface contraction waves and the initiation of development

We have seen that fertilization begins by the mechanical activity of the sperm (its flagellar-driven motility) followed by a chemically induced sperm exocytotic event (the acrosome reaction). This is followed by a sequence of chemical excitations in the egg cytoplasm; the most dramatic are traveling waves of Ca^{2+} concentration of varying frequency and amplitude. These waves trigger subsequent events of fertilization and zygote formation, prominent among which is the egg's own exocytotic event, the release of the cortical granule contents resulting in the slow block to polyspermy. The appearance of rippling waves on the egg surface (see Fig. 9.8), which immediately precedes the cleavage process, takes us back to the point at which the zygote began its developmental excursion.

In sea urchins, for example, successive waves of microvilli elongation and stiffening (Cline *et al.*, 1983) propagate with the same directionality and speed as the initial calcium wave (Suzuki *et al.*,

1995). At about nine minutes after fertilization, actin filaments detach from the cortex and translocate away from the surface into the deeper regions of the cytoplasm (Terasaki, 1996). Within 15 minutes, the egg cortex is transformed from a fairly flat, soft, layer studded with short micropapillae (broad microvilli) into a stiff layer containing numerous surface microvilli and a new set of cortical vesicles, replacing the exocytosed cortical granules (Sardet *et al.*, 2002). In *Xenopus* eggs, within minutes after fertilization, dynamic actin "comet tails" accumulate around intracellular vesicles, which then begin to move through the cytoplasm (Taunton *et al.*, 2000), possibly representing the same phenomenon seen in the sea urchin.

In essentially all species, the postfertilization cortical microfilament cytoskeleton (comprising actin) is reorganized and contracts in a wave-like manner starting from the site of sperm entry. This mechanical motion of the egg cortex is based on its ability to undergo cycles of calcium-dependent contraction and relaxation (Sardet *et al.*, 1998; Roegiers *et al.*, 1995, 1999; Benink *et al.*, 2000). The cortical reorganizations and cytoplasmic flows that occur between fertilization and first cleavage – primarily the translocation of cortical and subcortical materials parallel to the plane of the plasma membrane (Eidne *et al.*, 1994) – appear to be driven by interactions between the microfilament and microtubule cortical cytoskeletons (Canman and Bement, 1997; Benink *et al.*, 2000).

The cortex of the fertilized egg possesses altered mechanical and viscoelastic properties. For one thing, it is generally thicker than the unfertilized cortex, since the cores of the newly arising microvilli have microfilament bundles that extend and intermingle with the microfilament meshwork underlying the membrane (Wong *et al.*, 1997). The cortical microfilament network contracts and relaxes during specific phases of the meiotic and mitotic cell cycles, a process that appears to be regulated by the presence of microtubules (Mandato *et al.*, 2000).

Recall that the nature of the "astral signal," proposed by White and Borisy (1983) as triggering rearrangement of the cortical cytoplasm and eventually initiating cleavage, is unknown (see Chapter 2). It would be satisfying if one or more of the postfertilization Ca^{2+} waves that sweep across the zygote served this function. That things might be more complicated was suggested by the studies of Wong and coworkers (1997), who used the drug cytochalasin D in sea urchins to block dynamical changes in cortical actin organization downstream of the Ca^{2+} waves and showed that cytokinesis for the first cleavage occurred nonetheless. The establishment of the contractile apparatus for cytokinesis of the first cleavage division (the function of the putative astral signal) thus appears to be independent of the Ca^{2+}-induced waves of surface contraction, though possibly not of other effects of the calcium waves.

In *Xenopus*, these early surface contraction waves are apparently required for the cytoplasmic rearrangement leading to localization of the germ plasm (Quaas and Wylie, 2002), the maternally synthesized determinant of the primordial germ cells discussed at the beginning of this chapter. It is possible that cortical waves are involved in the localization of other cytoplasmic determinants of future pattern as well.

The mechanochemical coupling of Ca^{2+} traveling waves and shape changes in the cortical cytogel occurs across all classes of animals. Biologically, it represents an anticipatory link from the stage immediately after the gametes join in fertilization to the stage at which the gametes that will give rise to the next generation begin to form. Because it also represents a physical phenomenon not encountered until now in our presentation, we will conclude this chapter with a model for this process.

Mechanochemical model for cortical activity in fertilized eggs

The generation and propagation of chemical waves (as described for example by the FDF model) can be formulated in terms of reaction–diffusion equations like the ones discussed in Chapter 7. In order to produce morphogenetic modifications of the embryo, chemical waves must stimulate the mechanical rearrangement of cell or tissue components. As we have seen, calcium waves in the zygote in fact trigger mechanical waves of cortical expansion and contraction that precede the first cleavage (Takeichi and Kubota, 1984; Deguchi *et al.*, 2000; Ducibella *et al.*, 2002).

Living cells, such as the zygote, represent excitable media, which respond actively to external perturbations. The typical outcome is the formation and propagation of chemical or mechanical waves. An example of mechanical excitability was discussed in Chapter 5 in connection with gastrulation, and examples of chemical excitability were discussed in Chapter 7 in relation to pattern formation and earlier in this chapter in connection with the formation and propagation of calcium waves. Here we encounter an example of cell excitability that combines mechanical and chemical effects: cortical expansion and contraction and eventually the cleavage of the zygote. To model the Ca^{2+}-wave-generated shape changes of the embryo, the coupling of chemical and mechanical waves is needed; this goes beyond the usual reaction-diffusion formalism discussed in Chapter 7. Oster and coworkers (Cheer *et al.*, 1987) proposed a mechanochemical model for cortical activity in the postfertilized embryo. Later, Ballaro and Reas (2000) extended the model to incorporate newer experimental findings.

Since cellular shape changes are driven by dynamical rearrangement of the cytoskeleton, e.g., the polymerization–depolymerization of the actomyosin network, the mechanochemical model must incorporate the concentrations of Ca^{2+} and of the molecules that control its release, sequestration, and resequestration (e.g., IP_3, cAMP), as well as the physical state of the cytoskeletal filaments with their cross-linker proteins and of the motor proteins that can move these filaments (e.g., myosin). Calcium influences the cytoskeleton through its interaction with the cross-linkers. At low Ca^{2+} levels actin remains in a gel state but, as the Ca^{2+} level increases and cross-linkers preferentially bind with calcium, actin filaments are broken down and the gel changes into a sol. It is this Ca^{2+}-driven sol–gel transition that is responsible for the expansion–contraction waves in the cortex. The modeling of such complex phenomena, involving numerous molecular species, is not straightforward. Here (following Cheer *et al.*, 1987) we only outline how this program can be formulated. Mathematical details can be found in the cited references.

As discussed in Chapter 7 (see Eqs. 7.5–7.7), the general equations for the concentrations $c_i, i = 1, 2, \ldots, N$, of N molecular species whose spatial distributions are governed by both chemical reaction and diffusion are

$$\frac{\partial c_i}{\partial t} = D_i \frac{\partial^2 c_i}{\partial x^2} + F_i(c_1, c_2, \ldots, c_N). \tag{9.10}$$

For the modeling of mechanochemical wave propagation in the cortex of the embryo, depending on the complexity of the model the c_i may denote the concentrations of Ca^{2+}, IP_3, cAMP, the actomyosin network, and the cross-linking, capping, and severing proteins (the last three types of molecule are collectively called solation factors because their concentration determines whether the cytoskeletal network is in the gel or sol phase). The rate constants of the various molecular interactions determine the functions F_i; for specific examples of such expressions, see Cheer *et al.* (1987) or Ballaro and Reas (2000). The reaction–diffusion equations 9.10 now have to be coupled to the mechanical properties of the cortex.

Following Cheer *et al.*, let us consider a small volume element ΔV of the cortical actomyosin gel, which is essentially a network of cross-linked polymer fibers (i.e., filamentous F-actin). As the gel is placed in solvent it swells owing to the difference in the osmotic pressure, P_O, between the solvent and the gel. Owing to the cross-linking of filaments, the gel has elastic properties and thus resists swelling. The total swelling pressure in the gel, P_S, is the sum of the osmotic pressure, which tends to expand it and the elastic stress, P_E, which restrains it: $P_S = P_O + P_E$. (Note that in this expression P_E should be considered as a negative quantity since it acts in opposition to P_O.) The addition of calcium to ΔV diminishes the concentration of solation proteins. Thus the gel weakens and the magnitude of its elastic pressure decreases. This in turn leads to an increase in P_S. As the amount of calcium falls, the gel strengthens and P_S decreases. A wavelike periodic variation in calcium concentration, as described earlier, thus modulates the sol–gel transition in the actin network, leading to expansion–contraction cycles (a solation wave) and the movement of the egg cortex. As the solation wave passes through a volume element ΔV, it displaces it. We denote by $\varepsilon(\theta, t)$ the tangential displacement of the volume element ΔV that is initially at latitude θ on the surface of the spherically symmetric embryo (Fig. 9.10A). Its variation in time is given by the following equation:

$$\eta \frac{\partial \varepsilon}{\partial t} = P_O + P_E + P_A. \tag{9.11}$$

Chapter 10

Evolution of developmental mechanisms

The view of embryonic development presented in the preceding nine chapters is rather different from accounts to be found in other modern developmental biology textbooks. We have focused on the phenomena of transitions between cell types, changes in the shape of tissues, and the generation of new arrangements of cells and have approached them as problems in physics. In contrast, when these subjects are dealt with in most contemporary treatments of development it is primarily as problems in regulated differential gene expression. While we have by no means ignored the roles of gene products and gene regulatory systems in our account of development, and while the notion of the embryo as a physical system is not entirely absent from the more standard accounts, the different emphases of the two perspectives could not be clearer.

Developmental biology has advanced to its current high level of sophistication with little explicit analysis of the physical dimension of the questions it treats. This is changing, however. The DNA sequencing initiatives of the last decade of the twentieth century confirmed that genes number in the range 10 000–50 000 in all the species traditionally studied by developmental biologists. Analysis with microarrays (computer-interfaced devices for quantifying the abundance of mRNAs in tissue samples) has shown that many genes are simultaneously expressed during any significant developmental event (Bard, 1999; Montalta-He and Reichert, 2003). The sheer number of gene–gene interactions that are now known to occur during embryogenesis has generated a need for new methods to handle such complexity as well as new concepts that could help investigators "see the forest for the trees."

One response to this has been an increase in computational and complex systems-based approaches to the analysis of gene expression data sets (Bard, 1999; Davidson *et al.*, 2002; Oliveri and Davidson; 2004). But while concepts of dynamical systems of the sort discussed in Chapter 3 sometimes enter into these analyses, such approaches continue to be largely gene-centered. That is, they are concerned primarily with the control of gene activities by other gene activities. The physics of soft condensed materials – diffusion, viscoelasticity, network formation, phase separation – are the concepts we have used to explain epithelial and mesenchymal morphogenesis, pattern formation, and organogenesis in earlier chapters. This physics has been slower to enter into mainstream developmental biology. It is one of the purposes of this chapter to explore what it is about the organization of development that has permitted the genetic aspects to stand in, until now, for the whole. It has taken a reconsideration of the connection between the development of biological form and its evolution (Gould, 1977; Raff, 1996; Salthe, 1993), a subject that was out of fashion for most of the twentieth century, to help bring the physics of materials back into the analysis of development, where it once occupied a more prominent place. (See Newman, 2003a, for a brief history of "systems" approaches to development.) In what follows, we address the connections between development, physics, and evolution.

The physical origins of developmental systems

Since the "physical" and "gene-centered" approaches deal with identical subject matter – the developing embryo – it makes sense to try to understand the relationship between them. Specifically, what is it about the construction of organisms and embryos that has permitted a relatively coherent (but, we would suggest, mechanistically incomplete) body of knowledge to be generated that focuses on causal chains acting at the level of gene interactions?

On the one hand, viscoelastic materials and chemically and mechanically excitable systems – the objects of the major explanatory framework used in this book – existed prior to the world of organisms. The "generic" physical forces and mechanisms used to account for developmental phenomena in the preceding chapters are capable of acting on living tissues and nonliving materials alike to produce similar shapes, patterns, and dynamical activities (a view closely associated with the biologist D'Arcy W. Thompson (Thompson, 1917)). In principle, a rough sort of "development," that is, a sequence of form-changes in an aggregate of cells, could be driven solely by physics. On the other hand, it is also the case that developmental steps in any real embryo are typically associated with the appearance of one or more specific gene products at a particular time and place.

Many key gene products that participate in and regulate multicellular development emerged over many millions of years of evolution in a world containing only single-celled organisms. These gene products (organized with other molecules into cells and their ECMs) constitute the material acted upon by physical mechanisms during embryonic development. Physical mechanisms acting on materials by vritue of their inherent physical properties, not the genes or their products on their own, are what produces biological forms. But as an embryo develops, the operation of physical mechanisms is constrained and focused by hierarchical systems of coordinated gene activities – so-called "developmental programs." This implies (since developmental programs evolved gradually, after the origination of multicellularity) changes in form and cell type in ancient cell aggregates were not subject to the precise control seen in present-day multicellular organisms.

For these reasons, the relationship between the physical and genetic aspects of development can be understood only in an evolutionary context. In the remainder of this chapter we will explore scenarios for the origin and evolution of developmental systems, with emphasis on the physical aspects.

The first metazoa

Multicellular organisms first arose more than 1.5 billion years ago (Knoll, 2003). The earliest of these were filamentous algae, plants whose rigid cell walls would have prevented them from being susceptible to many of the viscoelasticity-based morphogenetic mechanisms of animal development discussed in this book. (However, see Nagata *et al.*, 2003, for application of a number of related processes to the development of plants.) The first "metazoa" (multicellular animals) appear in the fossil record earlier than 700 million years ago (Ma). The most extensively studied of these fossils, dating from 580–543 Ma, have been described as relatively simple, flat, quilt-like creatures, probably without body cavities (Seilacher, 1992; Brasier and Antcliffe, 2004). Modular body subdivisions often exhibited a fractal branching pattern instead of the segmentation seen in modern organisms (Narbonne, 2004; see Chapters 6 and 8 for a discussion of fractals). By approximately 540 Ma the "Cambrian explosion" had occurred, a term denoting the fact that virtually all the general categories of body organization seen in modern organisms had burst into existence in the preceding 25–30 million years, a mere blink of the eye geologically speaking (Conway Morris, 2003).

Metazoan bodies are characterized variously by axial symmetries and asymmetries, multiple tissue layers, interior cavities, segmentation, and various combinations of these properties. Each species can be assigned to one of approximately 35 body plans (Raff, 1996; Arthur, 1997) or organizational categories (Minelli, 2003). These are essentially the same as the "phyla" of the standard taxonomic system (Valentine, 2004). The organs of an animal are constructed using morphological motifs similar to the body plans (see Chapter 8). While the early world contained many unoccupied ecological settings ("niches") within which new organismal forms could flourish, this alone cannot account for the rapid profusion of body plans once multicellularity was established, nor for the particular forms assumed by bodies and organs.

Nonliving materials, such as liquids, clays, polymer melts, soap bubbles, and taut strings, by virtue of their inherent physical properties can take on only limited, characteristic, sets of shapes and configurations – sessile drops, vortices, standing and traveling waves, and so forth. The most ancient cell aggregates, lacking the highly integrated genetic programs of modern-day organisms, would have been molded in a similar fashion, by physical forces and determinants inherent to their material properties and scale. Can we reconstruct the forms likely to have been assumed by these ancient cell aggregates?

Multicellular aggregates differ from single cells in their linear dimensions (i.e., size) and in their susceptibility to physical determinants. As we saw in Chapter 2, for example, liquid surface tension *per se* is not an important determinant of individual cell shape, and free diffusion (Chapter 1) has only limited scope as a mode of intracellular transport. And while individual cells may assume distinct states based on the dynamics of their metabolic or transcriptional circuitry (Chapter 3), such states obviously cannot cooperate with one another or assume different tasks in an organism that consists of only one cell.

The cell aggregates that eventually evolved into the metazoa were a different story. Their cells, like those in modern embryos, were likely to have been free to rearrange with respect to one another. Primitive aggregates, therefore, would have been likely to exhibit liquid-like behaviors such as rounding-up and spreading (Chapter 4). These cell clusters would also have provided a medium for the diffusion of secreted proteins and thus for the generation of global (e.g., aggregate-wide) diffusion gradients (Chapter 7). The single-cell organisms that existed before the emergence of multicellularity must have been metabolically active, thermodynamically open, systems, much like present-day cells (the evolution of eukaryotic cells had been under way for more than a billion years before metazoans appeared; Knoll, 2003). Since the premetazoan aggregates would thus have been composed of chemically excitable cells (Chapter 3), they were not only viscoelastic and chemically heterogeneous, they were also excitable media. This means they would have had the potential to elaborate self-organized spatial patterns of cells with different biochemical states (Chapter 7). And in cases where these cell types exhibited different amounts of adhesive molecules, adhesion-based sorting would have been inevitable. The evolution of polarized cells and planar basal laminae would have led, as a physical side-effect, to aggregates with lumens or to elastic cellular sheets (Chapter 5).

In our earlier discussions (Chapters 1–8), which dealt entirely with individual cells and multicellular organisms of the present-day type, not their ancient counterparts, we saw that viscosity, elasticity, dynamical-state transitions, differential adhesion, and reaction-diffusion coupling all continue to be factors in embryonic development, notwithstanding the constraints of evolved architecture and multisystem integration. In the earliest arising multicellular forms these physical determinants must have acted in an even less constrained fashion, potentially allowing new forms to arise with little or no genetic change (Newman and Müller, 2000; Newman, 2003b).

In the following section we discuss how some of these physical processes might have first become relevant to the world of multicellular organisms. We note that, unlike most of the descriptions in previous chapters, in which physical concepts were linked, wherever possible, to experimentally determined properties of cells and known molecules, we are not on such solid ground in attempting to characterize the origination of multicellular forms. While the discussion that follows is based on plausible biological and physical arguments, it is explicitly in a hypothetical mode, as is frequently the case with issues of causation in evolutionary biology.

Physical mechanisms in the origin of body plans

Let us imagine an ancient aggregate of cells before the emergence of true metazoa. If a group of cells in one region of this cluster released a protein product at a higher rate than their neighbors – either randomly or because some external cue stimulated them to do so – the aggregate would thereby have become chemically nonuniform from one end to the other. Let us further assume that in some of these cases the cells happen to react differently to different concentrations of the molecule in question (we will call it a "proto-morphogen") and thus assume different states in a fashion that depends on the concentration of the proto-morphogen. Under these circumstances, a spatially non-homogeneous distribution of a chemical (e.g., a gradient) would have given rise fortuitously to a nonhomogenous distribution of cell states (Fig. 10.1). We can thus see how "generic" physics (i.e., diffusion) acting on a scaled-up biological system (i.e., a multicellular aggregate) can give rise to an incipient developmental process.

But how could such a haphazardly determined effect be perpetuated from one generation to the next? In present-day embryos the position of the embryonic "organizer" (i.e., a cell or group of cells that is a unique source of a diffusible morphogen) is often determined by maternally deposited cues, or some other genetically influenced process, in conjunction with external cues, such as the sperm entry point. In such cases, hereditary transmission of the relevant genes or gene variants creates reproducible conditions for the recapitulation of the event from one generation to the next. In our hypothetical ancient form, lacking such a genetic program, recurrence of the developmental event could have been perpetuated by less formal means. If, for example, the cells in the primitive aggregate had a 1% chance of randomly producing and secreting the proto-morphogen, then half of all 50-cell aggregates would have a (proto-)organizer cell. These variants would "develop," that is to say, they would self-organize into a nonuniform distribution of cell states.

In this scenario, if there were a selective advantage to having a phenotype containing nonuniformly distributed cell types then cell clusters whose genotype inclined them to produce proto-organizer cells at a higher frequency would become more prevalent. This tendency would be balanced by the fact that if all cells became organizers there would be no gradient. And this, in turn, would put a premium on genetically variant clusters in which proto-organizer cells limit the appearance of other proto-organizer cells, that is, produce a lateral inhibitory factor simultaneously with the proto-morphogen.

In order for such biochemical circuits to generate stable spatial patterns they must be part of a reaction–diffusion or (taking cells

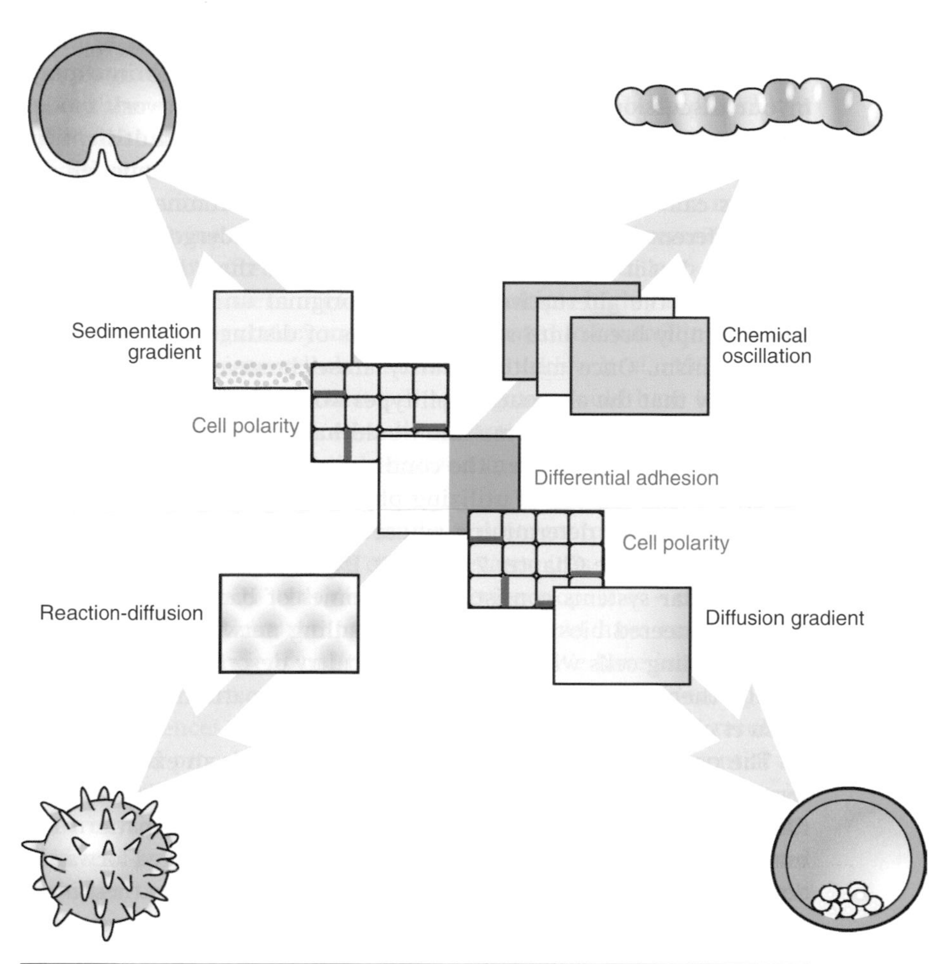

Fig. 10.1 Schematic representation of the hypothesized origination of body plans via the morphogenetic consequences of linking the regulation of cell–cell adhesion to various physical and chemical pattern-forming mechanisms. Red lettering, cell properties; black lettering, pattern-forming mechanisms. The central box denotes the effects of differential adhesion in causing the formation of boundaries within a tissue mass, across which cells will not mix (see Chapter 4 and Fig. 4.5). The polarized expression of adhesion molecules leads to cavities and other lumenal structures (see Fig. 4.2). All the peripherally arranged boxes denote pattern-forming mechanisms which, when deployed in conjunction with differential adhesion, can lead to standard body-plan organizational motifs. The sedimentation of a dense cytoplasmic component or the diffusion of a morphogen (see Fig. 7.7) are ways in which an egg or early blastula can become spatially nonuniform; if these nonuniformities lead to the generation of the expression of differentially adhesive cell populations, gastrula-like structures will form (see Fig. 5.6). Similarly, chemical oscillations (see Fig. 7.3) or pattern-generating reaction–diffusion processes (see Fig. 8.15), if they come to regulate adhesive differentials between cells, can lead to segmentation or other periodic structures. (After Newman, 1994.)

So far, no cell-autonomous oscillations have been identified during the segmentation of invertebrates, in contrast with the situation in vertebrates. However, the sequential appearance of gene expression stripes from the posterior proliferative zone of short-germ-band insects and other arthropods such as spiders and horseshoe crabs, has led to the suggestion that these patterns arise from a segmentation clock like that found to control vertebrate somitogenesis (Stollewerk *et al.*, 2003; see also the discussion in Chapter 7). This clock would regulate, directly or indirectly, downstream genes such as *engrailed* (*en*), a segment polarity gene that specifies a transcription factor, in cells as they leave the proliferative zone (Fig. 10.2).

On both theoretical grounds (Boissonade *et al.*, 1994; Muratov, 1997) and experimental grounds (Lengyel and Epstein, 1992) it has long been recognized that the kinetic properties that give rise to a chemical oscillation (the Hopf bifurcation; see Chapter 3) can, when one or more of the components is diffusible, also give rise to standing or traveling spatial periodicities of chemical concentration (the Turing bifurcation, see Chapter 7). If the *Drosophila* embryo were patterned by a reaction–diffusion system, it would be tempting to hypothesize that the *Drosophila* mode of segmentation arose by an evolutionary change (i.e., delayed cellularization of the blastoderm) that placed the ancestral oscillator in a context in which diffusion was possible.

Things are not this simple, however. In reality, *Drosophila* segmentation is controlled by a hierarchical system of genetic interactions that has little resemblance to the self-organizing pattern-forming systems associated with reaction–diffusion coupling. Only by examining the *Drosophila* segmentation hierarchy in comparison with the means used to achieve the same morphological outcome in short-germ-band insects is it possible to formulate a plausible hypothesis for the mechanistic and evolutionary relationships between the two systems.

Molecular mechanisms of segmentation in insects

The formation of overt segments in *Drosophila* (see St Johnson and Nusslein-Volhard, 1992, and Lawrence, 1992, for reviews) requires the prior expression of a stripe of Engrailed in the cells of the posterior border of each of 14 presumptive segments (Karr *et al.*, 1989). By means of a gene network (the *segment polarity module*), cells that express *en* cause cells that do not acquire a distinct adhesive affinity, leading to a boundary of cell immiscibility (Rodriguez and Basler, 1997).

The positions of the engrailed stripes are largely determined by the activity of the pair-rule genes *even-skipped* (*eve*) and *fushi tarazu* (*ftz*), which exhibit alternating, complementary, seven-stripe patterns prior to the formation of the blastoderm (Frasch and Levine, 1987; Howard and Ingham, 1986). *Even-skipped*, for example, as its name suggests, is expressed in only odd-numbered "parasegments." (The parasegments are developmental modules consisting of the posterior half of each segment and the anterior half of the next segment posterior to it; see Lawrence, 1992, and Fig. 10.3). The stripe patterns of the pair-rule genes are generated by a complex set of interactions among

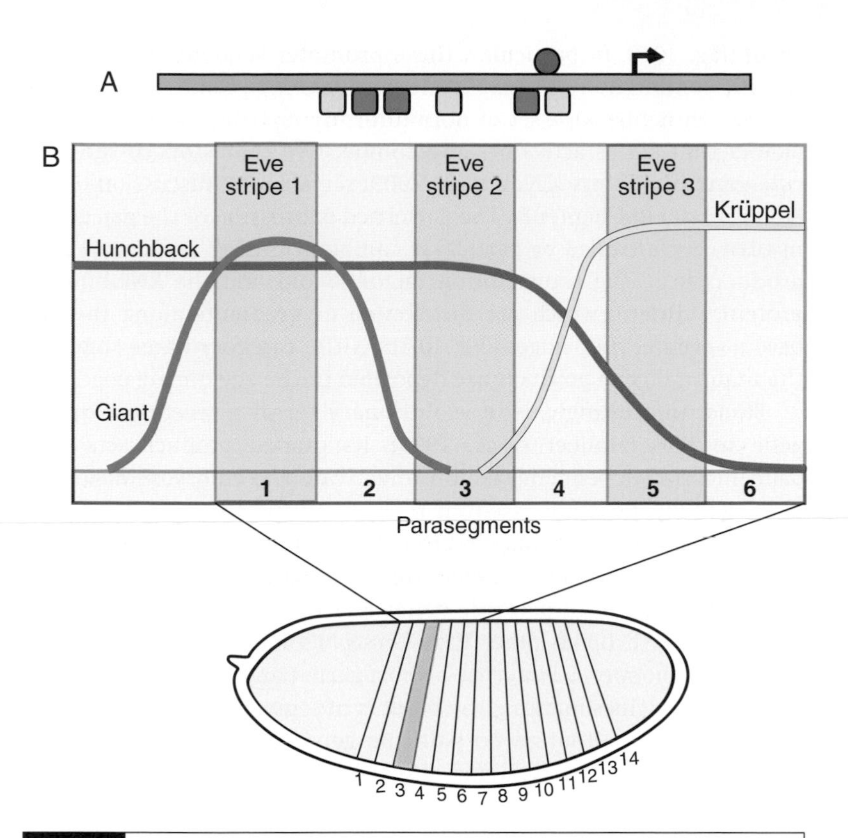

Fig. 10.3 (A) Schematic representation of a portion of the *even-skipped* gene, including the promoter and transcription start site (cf. Fig. 3.9). Contained within the promoter is a subregion (the stripe-2 enhancer) responsible for the expression of the second eve stripe (see Fig. 10.2). This sequence contains binding sites for gap-gene-class transcription factors that positively (Hunchback, red) or negatively (Giant, blue; Krüppel, yellow) regulate *eve* expression. (B) Lower part: the *Drosophila* syncytial blastula. The upper part shows, schematically the distribution of Giant, Hunchback, and Krüppel proteins in the region containing the first six prospective parasegments. At the position of prospective parasegment 3 (which comprises the posterior half of the last head segment plus the anterior half of the first thoracic – upper body – segment) the levels of the activator Hunchback is high and those of the inhibitors Giant and Krüppel are low. This induces *eve* stripe 2. To either side of this position Giant and Krüppel are high, restricting the expression of *eve* to a narrow band. (Panel A is based on Small *et al.*, 1992; panel B is after Wolpert, 2002.)

transcription factors in the syncytium that encompasses the entire embryo. A simple reaction–diffusion model would suggest that all the stripes (which are quite similar-looking, Fig. 10.2) are produced by chemically identical conditions emerging at evenly spaced locations. Instead, individual stripes are generated by stripe-specific, rather than generic, mechanisms (Akam, 1989). The formation of *eve* stripe number 2, for example, requires the existence of sequences in the *eve* promotor that switch on the *eve* gene in response to a set of spatially distributed morphogens that under normal circumstances have the requisite values only at the stripe 2 position (Small *et al.*, 1991, 1992,

1996) (Fig. 10.3). In particular, these promoter sequences respond to specific combinations of "gap" gene products (e.g., Giant, Krüppel, embryonic Hunchback), a set of nonuniformly distributed transcription factors that act as activators and competitive repressors of the pair-rule gene promoters (Clyde *et al.*, 2003; see also the discussion of the Keller model in Chapter 3). The patterned expression of the gap genes, in turn, is controlled by particular combinations of "maternal" gene products (e.g., the transcription factor Bicoid and the RNA-binding protein Staufen), which are distributed as gradients along the embryo at even earlier stages (Fig. 10.2). As the category name suggests, the maternal gene products are deposited in the egg during oogeneis.

Bicoid, interestingly, is in evolutionary terms a recently acquired gene in flies (Stauber *et al.*, 1999). Its graded product acts in a concentration-dependent fashion to activate the embryos's *hunchback* gene. But Bicoid is not essential for performing this function. In its absence, an ancient regulatory circuit involving maternal Hunchback protein can properly activate embryonic *hunchback* (Wimmer *et al.*, 2000).

The expression of pair-rule genes and *engrailed* during the development of arthropods other than *Drosophila* has also been explored. In the grasshopper, *Schistocerca*, a short-germ-band insect, Engrailed is localized in stripes marking the borders of segments (Patel *et al.*, 1989, 1992; Fig. 10.2), although no pair-rule genes have been found to be expressed in stripes (Dawes *et al.*, 1994). In *Tribolium*, an "intermediate-germ-band" beetle (having both a growth zone and a syncytium in different regions of the embryo), the pair-rule genes *eve* and *ftz* are expressed in a striped pattern in both syncytial and growth zone regions and this is followed by the expression of *en* along the posterior margin of each segment (Brown *et al.*, 1994a, b; see also Patel *et al.*, 1994). While the modes of segmentation may have changed over the course of evolution, the expression patterns of the segment-polarity gene *engrailed*, and to a lesser extent, the pair-rule genes, appear to be conserved. Based on the conservation of these gene expression patterns throughout the insects, and their similarities with those found in other groups, it seems reasonable to assume that the last common ancestor of *Drosophila* and modern intermediate-germ-band insects was itself intermediate germ-band and probably had a pattern of pair-rule and *en* expression in the syncytial region of its embryo similar to that found in present-day *Drosophila*.

Physical mechanisms and the evolution of insect segmentation

As noted above, the kinetic properties that give rise to a limit-cycle chemical oscillation can also give rise to standing or travelling spatial periodicities of chemical concentration, when one or more of the components is diffusible. Whether a system of this sort exhibits purely temporal, or spatial, or spatiotemporal periodicity depends on particular ratios of reaction and diffusion coefficients. A simple dynamical system that exhibits temporal oscillation or standing waves, depending on whether diffusion is permitted, is shown in Fig. 10.4.

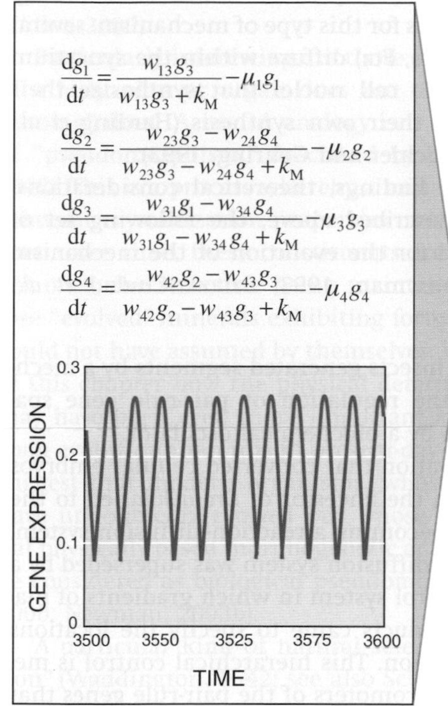

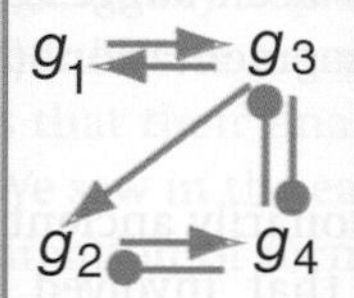

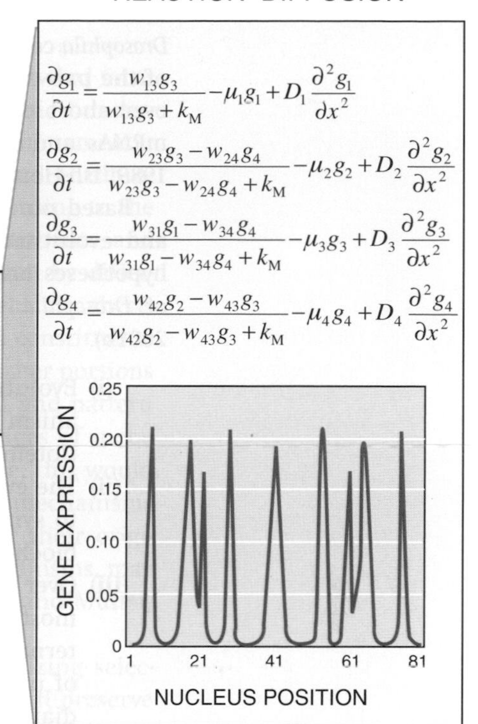

Fig. 10.4 Example of a network (central box) that can produce, for the same parameter values, sequential stripes when acting as an intracellular biochemical clock in a one-dimensional cellularized blastoderm with a posterior proliferative zone, or simultaneously-forming stripes when acting in a one-dimensional diffusion-permissive syncytium. The arrows in the central box indicate positive regulation and the lines terminating in circles indicate negative regulation. In the upper parts of the blue boxes the equations governing each of the two behaviors are shown. The four genes involved in the central network diagram, as well as their levels of expression in the equations, are denoted by g_1, g_2, g_3, and g_4. In the reaction–diffusion case, g_1 and g_2 can diffuse between nuclei (note that the two sets of equations differ only in the presence of diffusion terms for the products of genes 1 and 2). The lower boxes indicate the levels of expression of gene 2 for the two systems. On the left, for the intracellular clock, the horizontal axis represents time t whereas on the right, in the reaction–diffusion system, this axis represents the space variable x. The patterns produced by the two different behaviors are not exactly equivalent because the reaction–diffusion system pattern has a small dependency on the initial conditions. In the pattern shown, the initial condition consisted of all gene product levels being set to zero except gene 1 in the central of 81 nuclei, which was assigned a small value (the exact quantity did not affect the pattern). The patterns shown were found when the following parameter values were used: $k_M = 0.01$; $W_{13} = 0.179$; $W_{23} = 0.716$; $W_{24} = -0.704$; $W_{31} = 0.551$; $W_{34} = -0.466$; $W_{42} = 0.831$; $W_{43} = -0.281$; $\mu_1 = 1.339$; $\mu_2 = 2.258$; $\mu_3 = 2.941$; $\mu_4 = 2.248$. For the reaction–diffusion case, the same parameter values were set but in addition the values $D_1 = 0.656$ and $D_2 = 0.718$ were taken. (Redrawn from Salazar-Ciudad *et al.*, 2001b.)

(e.g., maternal and gap-gene products), seen in *Drosophila* is therefore not inconsistent with the origination of this pattern as a reaction–diffusion process.

Evolution of developmental gene networks: the model of Salazar-Ciudad *et al.*

We asked above why modern-day *Drosophila* does not use a reaction–diffusion mechanism to produce its segments. In the light of the discussion in the previous paragraphs, we can consider the following tentative answer: pattern-forming systems based on reaction–diffusion are inherently unstable to environmental and random genetic changes and would therefore, under the pressure of natural selection, have been replaced, or at least reinforced, by more hierarchically organized genetic-control systems. A corollary of this hypothesis is that the patterns produced by modern, highly evolved, pattern-forming systems would be highly robust in the face of further genetic change.

The first part of this evolutionary hypothesis has been examined computationally in a simple physical model by Salazar-Ciudad and coworkers (2001a). The model consists of a fixed number of communicating cells in a line or of nuclei arranged in a row within a syncytium. Each nucleus has the same genome (i.e., the same set of genes) and the same *epigenetic system* (i.e., the same activating and inhibitory relationships among these genes). The genes in these networks specify either receptors or transcription factors that act within the cells or syncytial nuclei that produce them or paracrine factors that diffuse between cells or syncytial nuclei (Fig. 10.5). These genes interact with each other according to a set of simple rules that embody unidirectional interactions in which an upstream gene activates a downstream one, like those represented in Figs. 10.2 and 10.3, as well as reciprocal interactions in which genes feed back (via their products) on each other's activities (Fig. 10.5). This formalism was based on a similar one devised by Reinitz *et al.* (1995), who considered the specific problem of segmentation in the *Drosophila* embryo.

The effect of gene i on gene j ($i, j = 1, 2, \dots, N$; N is the total number of genes in the model) is characterized by the parameter W_{ij} (the gene–gene coupling constant). These constants are formal representations of cases in which a gene product directly affects another's activity by binding to or chemically modifying it or a gene changes the amount of another gene's product by acting as a transcription factor (as in the Keller model described in Chapter 3). In general the W_{ij} are not symmetrical, $W_{ij} \neq W_{ji}$: the effect of gene i on gene j may be different from the effect of gene j on gene i and could be zero (representing no effect). Some of the genes specify diffusible factors and these factors are assigned diffusion constants.

One may ask whether a system of this sort, with particular values of the gene–gene coupling and the diffusion coefficients, can form a spatial pattern of differentiated cells. Salazar-Ciudad and coworkers performed simulations on systems containing 25 nuclei and a

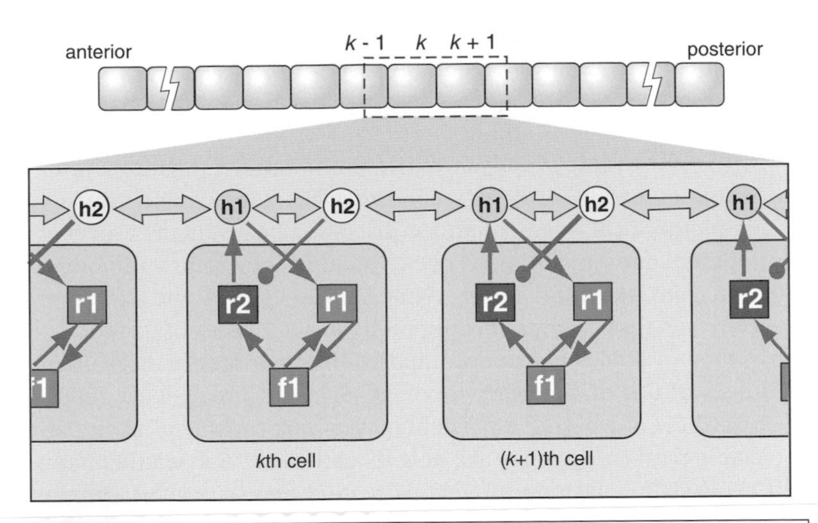

Fig. 10.5 Schematic illustration of the gene–gene interactions in the model of Salazar-Ciudad *et al.* (2001a). A line of cells is represented at the top. (The model equvralently represents a row of nuclei in a syncytium.) Below, types of genes are illustrated. Genes whose products act solely on or within the cells that produce them (r, receptors; f, transcription factors) are represented by squares; diffusible factors (h, paracrine factors or hormones) that pass between cells and enable genes to affect one another's activities, are represented by circles. Activating and inhibitory interactions are denoted by small arrows and by lines terminating in circles respectively. The double-headed green arrows denote diffusion. (After Salazar-Ciudad *et al.*, 2001a.)

variety of arbitrarily chosen but fixed parameters W. A pattern was considered to arise if the system's dynamics lead to a stable state in which different nuclei stably expressed one or more of the genes at different levels. The systems were isolated from external influences (zero-flux boundary conditions were used, that is, the boundaries of the domain were impermeable to diffusible morphogens) and initial conditions were set such that at $t = 0$ the levels of all gene products had zero value except for that of an arbitrarily chosen gene, which for the nucleus at the middle position had a nonzero value.

Isolated single nuclei, or isolated patches of contiguous nuclei expressing a given gene, are the one-dimensional analogue of isolated stripes of gene expression in a two-dimensional sheet of nuclei, such as those in the *Drosophila* embryo prior to cellularization (Salazar-Ciudad *et al.*, 2001a). It had earlier been determined that the core mechanisms responsible for all stable patterns fell into two non-overlapping topological categories (Salazar-Ciudad *et al.*, 2000). These are the hierarchical and emergent categories referred to earlier in this chapter. Hierarchical mechanisms form patterns by virtue of the unidirectional influence of one gene on the next in an ordered succession, as in the "maternal gene product induces gap gene product induces pair-rule gene" scheme described above for the early stages of *Drosophila* segmentation. In contrast, in emergent mechanisms reciprocal positive and negative feedback interactions give rise to the pattern. As noted above, emergent systems are equivalent to self-organizing dynamical systems, as seen for the transcription-factor networks discussed in Chapter 3 and the reaction–diffusion systems discussed in Chapters 7 and 8.

This is not to imply that an entire embryonic patterning event – the formation of a pair-rule stripe in *Drosophila*, for example – must be wholly hierarchical or emergent. As discussed below, many aspects of development are "modular" – decomposable into semi-autonomous functional or structural units. What Salazar-Ciudad and coworkers observed was that each patterning event in the classes of genetic networks they studied could be decomposed into modules that are unambiguously of one or the other topology (Salazar-Ciudad *et al.*, 2000).

Regardless of whether a particular system is capable of giving rise to a stable pattern, one may ask how its ability to do so would change if it evolved in a fashion analogous to real developmental systems. The evolution of organisms is reflected in permanent changes (not just changes in expression) of their genes. In terms of the model of Salazar-Ciudad and coworkers, evolution of a developmental system occurs if its gene–gene coupling constants W_{ij} change. Since gene change in biological systems is typically undirected, it is of interest to determine whether these model systems gain or lose pattern-forming capacity over successive "generations," where a new generation consists of replicas of a set of developmental systems with random changes introduced in the coupling constants W_{ij}. Salazar-Ciudad and coworkers therefore used such systems to perform a large number of computational "evolutionary experiments." Here we will describe a few of these that are relevant to the questions raised by the evolution of segmentation in insects, discussed above.

In their computational studies of the evolution of developmental mechanisms the investigators found that an arbitrary emergent network was much more likely than an arbitrary hierarchical network to generate complex patterns, i.e., patterns with three or more (one-dimensional) "stripes". This was taken to suggest that the evolutionary origination of complex forms, such as segmented body plans, would have been more readily achieved in a world of organisms in which self-organiging (e.g., multistable, reaction–diffusion, oscillatory) mechanisms were also present (Salazar-Ciudad *et al.*, 2001a) and not just mechanisms based on genetic hierarchies.

Following up on these observations, Salazar-Ciudad and coworkers performed a set of computational studies on the evolution of developmental mechanisms after a pattern had arisen. First they identified emergent and hierarchical networks that produced a particular pattern, e.g., three "stripes" (Salazar-Ciudad *et al.*, 2001a). (Note that patterns themselves are neither emergent nor hierarchical – these terms apply to the mechanisms that generate them.) They next asked whether given networks would "breed true" phenotypically, despite changes to their underlying circuitry. That is, would their genetically altered "progeny" exhibit the same pattern as the unaltered version? Genetic alterations in these model systems consisted of point mutations (i.e., changes in the value of a gene–gene coupling constant), duplications, recombinations (i.e., the interchange of coupling-constant values between pairs of genes), and the acquisition of new interactions (i.e., a coupling constant that was initially equal to zero was randomly assigned a small positive or negative value).

To evaluate the consequence of such alterations it was necessary to define a metric of "distance" between different patterns. This was done by specifying the state of each nucleus in a model syncytium in terms of the value of the gene product forming a pattern. Two patterns were considered to be equivalent if all nuclei in corresponding positions were in the same state. The distance between two patterns was one unit if they differed in the state of one of their similarly positioned nuclei, and so forth. The degree of divergence between patterns could then be quantified (Salazar-Ciudad *et al.*, 2001a).

It was found that hierarchical networks were much less likely to diverge from the original pattern (after undergoing simulated evolution as described) than emergent networks (Salazar-Ciudad *et al.*, 2001a). That is to say, a given pattern would be more robust (and thus evolutionarily more stable) under genetic mutation if it were generated by a hierarchical, rather than an emergent network. Occasionally it was observed that networks that started out as emergent were converted into hierarchical networks with the same number of stripes. The results of Salazar-Ciudad and coworkers on how network topology influences evolutionary stability imply that these "converted" networks would continue to produce the original pattern in the face of further genetic evolution. Recall that this is precisely the scenario that was hypothesized above to have occurred during the evolution of *Drosophila* segmentation (see also Salazar-Ciudad *et al.*, 2001b).

Subject to caveats about what is obviously a highly schematic analysis, the possible implications of these computational experiments for the evolution of segmentation in long-germ-band insects are the following: (i) if the ancestral embryo indeed generated its seven-stripe pair-rule protein patterns by a reaction–diffusion mechanism and (ii) if this pattern were sufficiently well adapted to provide a premium on breeding true then (iii) organisms resulting from genetic changes that preserved the pattern but converted the underlying network from an emergent one to a hierarchic one (as seen in present-day *Drosophila*) would have come to represent an increasing proportion of the population.

We can now return to the question raised at the beginning of this chapter, what is it about the construction of organisms and embryos that permits a useful and relatively coherent body of knowledge to be generated that primarily considers causal chains acting at the level of gene interactions? An answer can be suggested based on the preceding discussion. Let us take pattern formation as an example, since we have just been considering it in detail. (We could equally well apply these ideas to cell differentiation or to the shaping of epithelial or mesenchymal tissues). A given pattern-forming mechanism may have originated as a physical process – the differential-adhesion-based sorting of cells, the diffusion-based generation of nonuniformity, reaction–diffusion coupling. But the more hierarchical genetic

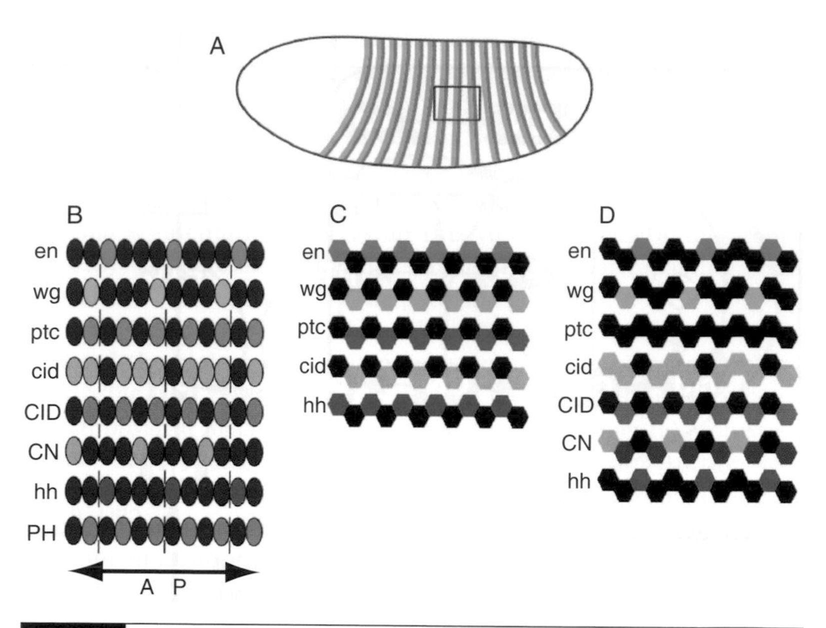

Fig. 10.7 Gene expression patterns of the segment-polarity genes in the *Drosophila* embryo and simulations in the model of von Dassow and coworkers. (A) The spatial expression pattern of *wingless* (*wg*) and *engrailed* (*en*) superimposed on an outline of the embryo. The parasegmental boundaries divide columns of *wg*-expressing cells (green) from columns of *en*-expressing cells (blue). (B) A schematic representation of the experimentally observed embryonic pattern of expression of the genes used in the model (abbreviations as in Fig. 10.6). For simplicity a single strip of twelve cells, corresponding to the boxed region in A, is shown. In each row, only cells expressing the indicated gene are shown in color; the others are shown in black. The full segment-polarity gene expression pattern is obtained by superimposition of the eight rows. (C) The best simulation pattern that was achieved with the interactions indicated by the solid lines in Fig. 10.6. (D) The pattern achieved by addition of the interactions indicated by the broken lines in Fig. 10.6. The simulated patterns achieved with this set of interactions were, in general, truer to the embryonic system (A), though in this particular simulation the expression of *patched* was suppressed. The experimental and model patterns are distinguished by the use of different symbols (ovals and hexagons respectively). (After von Dassow *et al.*, 2000.)

segment-polarity module it was not feasible to include all the numerous gene interactions that characterize this module. The initial decision to choose a hierarchy of unidirectional interactions was motivated by the existence of such a hierarchy for the generation of the pair-rule expression patterns (see above) directly preceding the operation of the segment-polarity module.

Significantly, when von Dassow and coworkers used only the hierarchical interactions represented by the solid lines in Fig. 10.6, despite extensive calculations they found no suitable parameter sets to arrive at the target patterns of gene products (i.e., the true patterns of key gene products generated by the *Drosophila* segment-polarity module). The majority of randomly chosen parameter sets caused either the model components to oscillate strongly or some components to be expressed ubiquitously while others were repressed everywhere. The closest they could come to the target pattern (i.e., obtaining it once in about 3000 randomly chosen parameter sets) is the pattern shown in Fig. 10.7C. No parameter sets produced stable asymmetric patterns. When, however, the investigators added two additional interactions, a positive feedback loop by which *wg* activates its own production and a negative feedback loop by which CN inhibits the expression of *en*, they were able to achieve remarkable success in finding parameter sets that reproduced the experimental patterns. An example of one such pattern is shown in Fig. 10.7D.

Equally remarkable was the tolerance to parameter variation exhibited by this modified model. Because the actual values of the gene product synthesis and the decay rates and other parameters were not known, simulations were performed over the range of biologically plausible values, i.e., several orders of magnitude. Nonetheless, for any of the 48 model parameters a random choice of its value was compatible with the desired behavior roughly 90 percent of the time. For some parameters, reasonable matching to the target pattern occurred despite variation of 100–1000 in their values (von Dassow *et al.*, 2000). The authors concluded that the network's ability to produce a realistic segment-polarity pattern is intrinsic to its topology (i.e., the links and feedback relationships in the pattern of gene–gene interactions) rather than to specific quantitative tuning.

Despite containing many hierarchical interactions, the core pattern-forming mechanism embodied in the gene network proposed by von Dassow and coworkers (Fig. 10.6) is "emergent" in the terminology of Salazar-Ciudad *et al.* (2000, 2001a, b). This distinguishes it from the largely hierarchical gene network that produces the earlier forming pair-rule pattern in *Drosophila* (Figs. 10.2 and 10.3). The feedback loops present in such emergent networks (for example, the broken lines in Fig. 10.6) predispose these networks to exhibit bistable behavior (Ingolia, 2004), i.e., small displacements of a system in its state space can cause it to switch abruptly between alternative stable nodes or orbits (see Chapter 3). Thus the cells in the model of von Dassow *et al.* (2000) can each produce either Engrailed or Wingless but not both at the same time in a way that is consistent with the true expression pattern. Ingolia (2004) has suggested that it is dynamical bistability (as a consequence of the network topology), rather than any other specific aspect of the von Dassow model, that is responsible for its robustness and that a large proportion of realistic models that contain this feature will be similarly resistant to perturbations.

Assuming this class of emergent models is a valid representation of the segment-polarity module, evolution has a wide latitude in altering key parameter values by genetic mutation, so long as the bistable dynamics is preserved. For any parameter choices within the acceptable range, the module will be buffered against developmental "noise," i.e., random biochemical fluctuations that inevitably occur during the course of embryogenesis and other biological processes (Barkai and Leibler, 2000; Vilar *et al.*, 2002).

As described above, in certain cases evolution may have transformed an originally self-organizing developmental mechanism into a hierarchical system of gene interactions like the one that generates the *Drosophila* pair-rule pattern or specifies the mesoderm and endoderm during gastrulation in the sea urchin embryo (Oliveri and Davidson, 2004). In exchange for the molecular precision afforded by such hierarchical mechanisms, systems evolved in this fashion must necessarily relinquish the kind of robustness inherent to emergent mechanisms. In such cases other mechanisms of robustness may evolve, such as gene duplication (Wilkins, 1997) and other mechanisms of functional redundancy (Tautz, 1992; Cooke *et al.*, 1997), in which distinct processes act in parallel to serve the same developmental ends. Since complex forms (e.g., the original segmented arthropod, see above) plausibly originated by dynamical mechanisms acting in concert with genetic mechanisms (Newman and Müller, 2000; Salazar-Ciudad et al., 2001a, b), the replacement of emergent by hierarchical mechanisms would also have had the result that organisms became less "evolvable" (Riedl, 1977; Wagner and Altenberg, 1996; West-Eberhard, 2003). As the example of the segment-polarity module shows, however, by retaining aspects of their physical character, living systems can remain self-organizing and responsive to the external world. This sets limits on the tendency of living processes to become fully assimilated into genetic hierarchies and, in turn, justifies formulating a biological physics of even modern-day, highly evolved organisms.

Perspective

Organismal forms have not always been generated by the highly integrated genetic programs characteristic of modern multicellular species. The ancient progenitors of modern metazoan animals were clusters of nucleated cells that, despite being in aggregates, retained their ability to locomote and change their relative positions. Once these emerged more than 700 Ma ago, physical processes irrelevant or marginal to single-cell life came into play. These included liquid-like tissue behavior, the capacity to sustain diffusion over distances on the scale of the whole organism, reaction–diffusion coupling, differential adhesion, and the cyclical regulation of intercellular adhesion. These mechanisms, in turn, made developmental gradients, tissue multilayering, lumen formation, and segmentation inevitable, thus providing the basic ingredients of modern-day developmental systems. The morphological templates that arose by means of basic physical mechanisms in the early period of metazoan evolution were consolidated and reinforced by later genetic change, with hierarchically organized genetic mechanisms often coming to substitute for conditional physical mechanisms. This may have led to some developmental modules, such as the early stages in *Drosophila* segmentation, taking on the character of "hard-wired" genetic programs. No contemporary embryonic system, however, has evolved away from dependence on the physical, and therefore conditional and dynamical, properties of living cells and tissues for generation of its patterns and forms.

Glossary

Absolute temperature The temperature measured in kelvins, K (0 K ≈ −273 °C).

Acrosome A large vesicle in the sperm head that releases its contents by rapid exocytosis upon encountering the vitelline membrane. The acrosomal contents, a set of enzymes, locally digest the vitelline membrane enabling the sperm to reach the egg cell membrane.

Actin A globular protein (G-actin) that can associate into polymers (F-actin) to form cytoskeletal microfilaments involved in contractile and motile functions of cells.

Activation energy The minimum amount of energy needed to bring potentially reacting chemicals to a state in which they can react.

Activator A molecular component in a biochemical reaction network or circuit that enhances the production of itself or another component.

Active transport The movement of a molecular component across a cell membrane by a process that requires expenditure of free energy.

Adherens junction A macromolecular adhesion complex between apposing sites on two adjacent cells, consisting of cadherins and accessory proteins linking the complex to the actin cytoskeleton.

Allantois A sac that develops from the posterior portion of the developing digestive tube in reptiles, birds, and mammals; in mammals it participates in the formation of the umbilical cord.

Amino acid A class of organic molecules in which a carbon atom (alpha-carbon) is attached to a hydrogen atom, an amino group ($-NH_3$), a carboxy group (-COOH), and a side chain. The genetic code specifies 20 amino acids with distinct side chains, which are used in the synthesis of proteins.

Amino terminus (N terminus) One of two ends of a protein: the end containing a free (uncombined) amino group.

Amoeboid locomotion A type of cell movement that depends on changes in state and flow of cytoplasm.

Amphibian A type of vertebrate organism (e.g., frogs, toads, salamanders) adapted to life in aquatic and terrestrial environments at different stages of their life-histories.

Amphipathic A molecule (e.g., a membrane phospholipid) that contains both hydrophilic and hydrophobic regions.

Anaphase A stage of cell division in eukaryotic cells, immediately after the chromosomes are aligned at the metaphase plate, during which the two sister chromatids of each replicated chromosome are pulled apart by the mitotic apparatus toward the opposite poles of the spindle.

Angioblast A mesenchymal cell that gives rise to blood vessels.

Animal pole The point on the surface of a telolecithal egg (i.e., one with nonuniform distribution of yolk) or early embryo at the center of the hemisphere containing the nucleus (in the case of an egg) or greater number of cells (in the case of an embryo); opposite the yolky vegetal pole.

Anteroposterior axis The direction in a bilaterally symmetric organism or embryo defined by an arrow pointing from the head to the tail.

Anterior Referring to the region of an embryo containing the head; opposite to posterior.

Antibody A protein produced by the immune system of vertebrate organisms that combines specifically with another, often foreign, molecule (antigen). Labeled antibodies are commonly used as tools to detect the presence and distribution of proteins in cells and embryos.

Antigen A molecule that can evoke the production of antibodies in an immune-competent organism.

Apical surface The portion of a polarized epithelial cell that contacts neither adjacent cells nor the basal lamina.

Archenteron The cavity enclosed by the invaginating endoderm during gastrulation in organisms such as amphibians and sea urchins that later becomes the gut lumen.

Association constant The ratio for a bimolecular reaction, at chemical equilibrium, of the concentration of the reversibly formed complex of two molecules and the product of the concentrations of the non-complexed molecules; also called the affinity constant.

Autocatalysis The promotion of the production of a molecular species by the same molecular species. In a strict sense, autocatalysis only applies to the situation in which the molecule acts as a catalyst of the reaction that produces it, but the term is often used in lieu of the more general "positive autoregulation."

Autoregulation Influence (positive or negative) of a molecular species on its own production.

Bacterium A category of cell that lacks a nucleus. Bacteria constitute one of the three major organismal domains. The other two are eukaryotes, organisms with nucleated cells (e.g., protozoa, slime molds, yeast, animals and plants), and archaea, which also lack nuclei.

Basal surface The region of an epithelial cell opposite the apical surface that attaches to the basal lamina.

Basal lamina A sheet-like complex of extracellular matrix macromolecules produced by, and interposed between, an epithelial cell layer and a second epithelium or a connective tissue.

Basement membrane A thick extracellular sheet at the interface of epithelial and connective tissues, comprising a basal lamina and an additional fibrous layer produced by the connective tissue.

Basin of attraction A region in the vector field of a dynamical system in which all trajectories of the system converge to a stable fixed point or orbit or confined chaotic behavior.

Bifurcation An abrupt change in the number or type of attractors (e.g., fixed-point, limit-cycle) of a dynamical system as a function of a characteristic parameter (such as a rate constant if the system is composed of reacting chemical species).

Blastocoel A fluid-filled cavity in a hollow blastula-stage embryo.

Blastoderm The layer of cells, resulting from cleavage in mammalian and a variety of other species. The blastoderm lies on the surface of the yolk of avian or reptilian eggs; in long-germ-band insects like *Drosophila*, it first takes the form of a single layer of nuclei beneath the egg cell membrane (acellular blastoderm) and only later a single layer of cells (cellular blastoderm).

Blastomere Any of the cells produced by cleavage of a fertilized or activated egg, that make up a blastula.

Blastopore An indentation on the surface of a blastula-stage embryo that represents the site of initiation of gastrulation.

Blastula An early-stage embryo during or at the end of cleavage, before gastrulation begins.

Bond energy The free energy released upon the formation of a chemical bond; equivalently, the energy it takes to break the bond.

Brownian motion The non-oriented motion of a single molecule fueled by its thermal energy and influenced by the collisions with surrounding molecules (such as the motion of a solute particle in a solvent).

Cadherin The collective name of a class of specialized cell adhesion molecules, forming mostly, but not exclusively, homophilic bonds (i.e., bonds between pairs of identical molecules).

Carboxyl terminus (C-terminus) One of two ends of a protein: the end containing a free (uncombined) carboxyl group.

Cartilage A skeletal connective tissue. Cartilage is an elastic solid. Lacking blood vessels of its own, it is nourished by diffusion from surrounding tissues. In animals with bony skeletons, much of the skeleton, including that of the limbs and vertebral column, develops first as cartilage and is later replaced by bone.

Caudal Toward the tail. In embryology, usually synonymous with posterior.

Cdc genes (cell-division-cycle genes) Genes that specify one of a set of proteins that regulate one or more steps in the eukaryotic cell cycle.

Cdk (cyclin-dependent protein kinase) A protein kinase (product of a cdc gene) that is active when complexed with a cyclin.

Cell adhesion molecule (CAM) Any member of several classes of integral membrane proteins that bind to other such molecules on adjacent cells to mediate cell attachment.

Cell cortex A dense layer of cytoplasm, rich in actin microfilaments and actin-binding proteins, that lies directly beneath a cell's plasma membrane.

Cell cycle The sequence of molecular and structural events involved in the replication of DNA and cell division.

Cell division The process by which a cell produces two daughter cells.

Cell fusion The joining of two or more separate cells into a multinucleated cell with a single plasma membrane.

Cell junction One of several membrane-associated macromolecular complexes that permit cells to attach to one another or to a substratum.

Cell line A cell population derived from a single progenitor cell that has accommodated to culture conditions and can grow and replicate indefinitely outside the organism of its origin.

Cell motility The movement of cells by amoeboid locomotion.

Cellular automaton A computer program to study the behavior of a complex system by representing it in terms of abstract "cells" programmed to change state or interact according to specific rules.

Centromere The region of a mitotic chromosome where the sister chromatids are attached to each other and the kinetochore is assembled.

Centrosome An organelle of animal cells, located near the nucleus, that organizes the microtubules during mitosis.

Chemotaxis Type of cellular motion driven by a chemical gradient and powered by the motile apparatus of cells (as opposed to the analogous process of diffusion, fueled by thermal energy).

Chordates The group of organisms that contains vertebrates and other animals with a notochord.

Chromatid One copy of a duplicated chromosome after DNA replication when it is still joined to the other copy (its sister chromatid) at the centromere.

Chromatin A complex of DNA, histones, and nonhistone proteins in eukaryotic cells; present in interphase nuclei and mitotic chromosomes.

Chromosome An individual structure composed of chromatin in a eukaryotic cell. The chromosomes are long and tangled during interphase, but become condensed and individually distinguishable during mitosis.

Cilium (plural **cilia**) A hair-like extension of a cell containing a central bundle of microtubules and capable of performing repeated beating movements.

Cleavage A type of cell division in early development in which the embryo is subdivided with no overall change in mass.

Codon A contiguous group of three nucleotides in a DNA or RNA molecule that specifies a particular amino acid according to the genetic code.

Collagen A family of extracellular proteins, each having significant stretches of amino acid sequence of the form Gly-X-Y, where Gly is glycine, X often proline, and Y often hydroxyproline.

Community effect An inductive interaction in embryonic development in which the ability of a cell to respond to an inductive signal is enhanced by, or even dependent upon, other neighboring cells differentiating in the same way at the same time.

Compartment A population of cells arising during development exclusively from a small number of founder cells, none of whose progeny move across boundaries with adjacent compartments.

Competence The capability of a cell or tissue to perform a specific developmental role, often related to their capacity to respond to a specific set of signals.

Constitutive Referring to a molecule, cellular structure, or activity that is not subject to a given type of regulatory mechanism, i.e., that is produced in all cell types, or at all phases of the cell cycle.

Cortical tension The apparent tension of the plasma membrane, due primarily to the contractile microfilaments of the cell cortex.

Convergent extension The rearrangement of a tissue mass characterized by intercalation of individual cells in one direction resulting in the elongation of the mass in an orthogonal direction.

Cortical rotation A phenomenon in newly fertilized amphibian eggs in which the cortical cytoplasm becomes displaced by a characteristic angle (30° in *Xenopus*) relative to the cytosolic core.

Covalent bond Strong chemical connection between atoms in a molecule formed by shared electrons.

Cyclins A class of proteins that regulate the eukaryotic cell cycle by binding to specific protein kinases and whose concentrations vary in a periodic fashion.

Cytokinesis The division of the bulk of a cell (cytoplasm and plasma membrane) into two, as distinct from the division of the nuclear contents, which is mitosis.

Cytoplasm The interior fluid compartment of the cell, within the plasma membrane but outside the organelles.

Cytosol The fluid layer of a cell's cytoplasm, in distinction to the cell cortex.

Dalton Unit of molecular mass, approximately equal to the mass of a hydrogen atom, 1.66×10^{-24} g.

Derivative The rate of change of a function in terms of its variable (for example, the rate of change of concentration in terms of position).

Desmosome A specific type of junctional complex by which adjacent epithelial cells attach to one another. Desmosomes are connected to intermediate filaments within the attached cells.

Determination The state of a developing cell or tissue in which a reversible or irreversible decision has been made to follow one of several possible developmental fates, without an overt change in the cell's or tissue's phenotype.

Diffusion The net drift of molecules in the direction of lower concentration owing to thermal movement.

Differential adhesion hypothesis (DAH) An interrelated set of assertions about cell mixtures and tissues in which cells move randomly and readily past one another. One component of the DAH is that cells take up consistent relative positions in an aggregate solely on the basis of their relative adhesive strengths; a second component is that such cell mixtures and tissues behave like immiscible viscoelastic liquids.

Differential equation A mathematical relationship between a function and its derivatives.

Differentiation The process by which a cell acquires a specialized type (e.g., muscle, neuron).

Dimensional analysis An approach based exclusively on the units of physical parameters used to establish a quantitative relationship between them.

Diploblast Category of animal generated from an embryo with two germ layers, ectoderm and endoderm.

Diploid The state of a cell in which it contains two sets of homologous chromosomes, one from each parent, and thus two copies of each gene.

Dissipative system A system whose energy is not conserved over time.

DNA Deoxyribonucleic acid; the genetic material of eukaryotic cells.

Dominant negative Referring to a gene mutation or a transgene, where the specified protein blocks the function of its normal counterpart.

Dorsal mesentery A strip of connective tissue that connects the dorsal surface of the digestive tube to the inner surface of the dorsal body wall.

Dorsoventral axis The axis of an embryo or body defined by an arrow pointing from its upper or back (dorsal) surface to its under (ventral) surface.

Drift velocity The average terminal velocity with which diffusing molecules move under the influence of a constant force.

Dupré equation An equation that relates the interfacial energy to the work of adhesion and the work of cohesion.

Dynein A microtubule-associated motor protein that operates in chromosome movement during mitosis, organelle transport, and the motion of cilia and flagella, using energy released from the hydrolysis of ATP.

Dynamical system A system of interacting components (e.g., chemical substances) evolving, typically in a complex way, in time.

Ectoderm The outermost germ layer of an embryo, which gives rise to the skin and, depending on the species, the nervous system.

Einstein–Smoluchowski equation The equation that relates the diffusion constant of a molecule to the viscosity of the liquid in which it diffuses.

Elasticity Characteristic physical property of elastic materials, which is manifest in their ability to recover fully or partially from a deformation such as elongation or compression.

Electrochemical gradient The driving force that causes an ion to move across a cell membrane. It is due to the combined effect of a difference in concentration and in electrical potential across the membrane.

Embryogenesis The formation of an embryo from a zygote, and the subsequent processes of morphogenesis and differentiation, leading to the formation of a fully developed organism.

Endocytosis Incorporation of material into a cell by its inclusion in a membrane-bound vesicle arising from invagination of the plasma membrane.

Endoderm Innermost layer of a diploblastic (two-layered) or triploblastic (three-layered) embryo, giving rise to the lining of the digestive tube and, depending on the species, the accessory organs of the digestive system.

Endoplasmic reticulum (ER) A set of internal membranes in eukaryotic cells that are sites of lipid synthesis (smooth ER) and the ribosomal attachment and synthesis of integral membrane and secreted proteins (rough ER).

Endothelium The characteristic cell layer lining blood vessels and the heart.

Enzyme A protein that acts as a chemical catalyst.

Epiblast The upper layer of the blastoderm in birds and mammals (distinct from the hypoblast), which gives rise to all three germ layers during gastrulation.

Epiboly A morphogenetic movement in some forms of gastrulation in which a layer of cells moves over and encloses an interior ball of cells.

Epithelioid A type of tissue in which cells make direct contact with their neighbors via membrane-bound adhesion proteins.

Epithelium A type of epithelioid tissue in which there is a free surface not attached to other cells.

Equilibrium constant The ratio of the product of the concentrations of the resulting chemical species to that of the reactants in a reversible chemical reaction when equilibrium has been reached, i.e., when the rates of the forward and reverse portions of the reaction are equal.

Exocytosis The process by which most materials are secreted from a eukaryotic cell; it involves packaging of the material into a membrane-bound vesicle that then fuses with the plasma membrane.

Extracellular matrix (ECM) A complex network of proteins (such as collagen and fibronectin) and polysaccharides (such as glycosaminoglycans); the major structural element of connective tissues and the basement membrane.

Facilitated diffusion Selective diffusion based on the shape or hydrophilicity of the diffusing molecule.

Fate map A map of an embryo showing cells or areas that are destined to develop into specific later-stage structures or adult tissues and organs.

Fertilization Fusion of the egg and the sperm and subsequently of the maternal and paternal genetic material.

Fertilization envelope The layer of extracellular matrix formed upon fertilization when the contents of the egg's cortical granules are exocytosed between the plasma and vitelline membranes; constitutes a barrier to polyspermy.

Fibroblast A type of cell found in connective tissues that produces an extracellular matrix rich in type I collagen.

Filopodium A long, thin extension of the cell surface, including membrane and cytoplasm.

Fixed point A special solution of systems of first-order differential equations (containing only first derivatives) at which the derivative of each quantity simultaneously vanishes.

Flagellum (plural **flagella**) Long, cilium-like protrusion whose undulations drive a cell through a fluid medium.

Fluid The collective name for phases of matter in which atoms and molecules do not (as in solids) bind tightly to each other, such as liquids and gases.

Follicle cell Accessory cells, not of the oogenic (i.e. egg) lineage, closely associated with the developing ovum.

Fractal A special geometric object, with the property of being self-similar under arbitrary magnification.

Free energy The energy that can be extracted from a system to drive reactions under specific experimental conditions, which thus takes into account changes both in energy, entropy, and other characteristic properties (volume, enthalpy, number of molecules) specified by the experimental conditions.

Frictional force A force acting between two surfaces moving past one another, which is typically proportional to the velocity (as opposed to the acceleration) of the relative motion.

Gametes The cells that participate in fertilization and jointly produce the founding cell (zygote) of a new organism. In animals the gametes are the ova (eggs) and the spermatozoa.

Gap genes A class of nonuniformly expressed genes (e.g., *giant*, *Krüppel*) involved in segmentation in long-germ-band insects such as *Drosophila*. Gap genes are generally controlled by maternal gene products; their products, in turn, are involved in the control of pair-rule genes.

Gap junction A tubular structure consisting of several connexin protein subunits that forms a transmembrane communication channel between adjacent cells.

Gastrula A two-layered (in diploblasts) or three-layered (in triploblasts) stage of embryonic development, which arises from the blastula by a sequence of cellular and morphogenetic changes.

Gastrulation The series of developmental steps leading to the formation of a gastrula.

Gel point The instant when a solution containing polymerizing monomers develops a spanning network of interconnected polymers with apparent elastic properties; at the gel point a special percolation transition (i.e., the sol–gel transition) takes place.

Gene One or more segments of DNA that together specify the sequence of a specific RNA or protein.

Genetic code The relationship, employed in protein synthesis, of the 64 nucleic acid triplet codons to the 20 primary amino acids.

Genome The totality of the genetic material in the chromosomes of a particular organism.

Genotype The specific genetic makeup of a cell or organism, as contrasted with its phenotype.

Germ cells The cells that give rise to eggs and sperm.

Germ layers The tissues of the early-stage animal embryo (in diploblasts the ectoderm and endoderm and in triploblasts the ectoderm, mesoderm, and endoderm) that will give rise to the animal's definitive tissues and organs.

Germ line The cell lineage in animal embryos that includes only the germ cells and gametes.

Glycocalyx The extracellular carbohydrate-rich coat of the cell surface, composed of a network of glycosaminoglycans, proteoglycans, and glycoproteins.

Glycoprotein A protein covalently linked to one or more oligosaccharides. The oligosaccharides of transmembrane glycoproteins are located on the extracellular domain.

Glycosaminoglycan One of several nitrogen-containing polysaccharides (polymers of sugars) found in the extracellular matrices of animal tissues. Examples include hyaluronan, heparan sulfate, and chondroitin sulfate.

Golgi apparatus Cell organelle, consisting of membranous sacs, involved in the modification and packaging of secreted materials and membrane proteins.

Gonads The reproductive organs of animals (female, ovaries; male, testes).

Gradient Continuously varying distribution of a substance as a function of another variable, for example the concentration of a morphogen as a function of position.

Gray crescent The region on the surface of the early frog embryo, generated by rotation of the cortical cytoplasm, where gastrulation is initiated.

Growth factor A secreted protein that influences cell proliferation or differentiation.

Haploid The state of a cell, derived from the diploid state by meiosis, that contains only one set of chromosomes and thus only one copy of each gene. The haploid state contains half the number of chromosomes of the diploid state.

Haptotaxis The movement of cells in response to an insoluble extracellular molecular gradient.

Hemidesmosome A cell junction that attaches an epithelial cell to an underlying basal lamina. The portion adjoining the cell is morphologically similar to one face of a desmosome.

Hensen's node The regional thickening of cells at the anterior end of the primitive groove on the dorsal surface of the embryos of birds, reptiles, and mammals, through which gastrulating cells migrate anteriorily to form tissues in the future head and neck; the functional equivalent of the Spemann–Mangold organizer of amphibian embryos.

Heterodimer A complex of two molecules (e.g., proteins) in which the monomers are nonidentical.

Hindgut The posterior region of the digestive tube.

Histone One of a group of five distinct low-molecular-weight, basic (i.e., positively charged) proteins that organize DNA into chromatin subunits (nucleosomes) in all eukaryotic nuclei.

Hooke's law The physical law that expresses the property of elastic materials, according to which a deforming force (such as tension) is proportional to the deformation it causes (such as elongation).

Homeobox genes Genes that specify a set of transcription factors whose DNA-binding region is an evolutionarily conserved motif called the *homeodomain*. A homeodomain contains about 60 amino acid residues and the corresponding region of a homeobox gene, the *homeobox*, is therefore (see *genetic code*) about 180 base pairs in length.

Homologous genes Genes that are evolutionarily related to one another through common ancestry.

Homologous chromosomes Chromosomes containing the same sets of genes. A homologous pair of each chromosome (other than the sex chromosomes) is present in diploid cells. With respect to sex chromosomes, in mammals a homologous pair of X chromosomes is present in the cells of females; one X and one nonhomologous Y chromosome are present in the cells of males.

Hopf bifurcation The bifurcation taking a fixed point into a limit cycle in a dynamical system.

Housekeeping genes Genes that specify the proteins that perform general cell functions; housekeeping genes are potentially transcriptionally active in every cell.

Hox genes A subset of homeobox genes that is involved in specifying regional identity along the body axis of vertebrates, arthropods, and other metazoans, as well as in the limbs and other organs of these groups.

Hydrolysis Cleavage of a molecule by the splitting of a bond with the addition of the hydrogen and hydroxide ions of a water molecule to the respective newly exposed sites.

Hydrophilic A molecule or portion of a molecule that has strong affinity for water.

Hydrophobic A molecule or portion of a molecule that lacks affinity for water.

Hypoblast The sheet of cells in early-stage avian and mammalian embryos that lies beneath the epiblast in the blastoderm. The hypoblast gives rise to extra-embryonic structures.

Imaginal discs Flattened epithelial sacs, on the surface of larvae of certain insects, that give rise to adult appendages during metamorphosis.

In vitro Literally, "in glass;" in developmental biology, referring to studies of isolated tissues or cells.

In vivo Literally, "in life;" in developmental biology, referring to studies of tissues or cells in the intact embryo.

Induction Alteration of the composition or shape of a tissue by a second tissue or its products.

Inertial force A force equal to the mass of an object times its acceleration.

Ingression A cell rearrangement in which a population of cells moves from the surface of a tissue mass into an interior lumen, as in some forms of gastrulation.

Inner cell mass Population of cells in a mammalian blastula that gives rise to all the structures of the organism's body.

Inositol A sugar molecule that plays several different cellular roles. In phosphorylated forms (e.g., IP_3) and phospholipid-linked forms (inositol phospholipids) it is a component of signaling molecules, which regulate intracellular calcium ion deployment. Its phospholipid-linked forms are also components of the glycosylphosphatidylinositol (GPI) membrane anchors of certain cell surface proteins.

Integral membrane protein A protein that in its functional form contains one or more domains inserted into a membrane.

Integrin Collective name for a class of special adhesion molecules that connect the cell to the extracellular matrix.

Intermediate filaments A family of cytoskeletal filaments of diameter smaller than microtubules and larger than microfilaments. Examples include keratin (epithelial cells), vimentin (fibroblasts), desmin (muscle cells), neurofilaments (neurons).

Interfacial energy The amount of energy required to increase the contact area between two distinct substances by one unit.

Interfacial tension The same as interfacial energy in the specific case when one or both of the contacting substances is a liquid. The liquid–vacuum (in practice, the liquid–air) interfacial tension is called surface tension.

Interphase Phase of the eukaryotic cell cycle in which an intact nucleus (as opposed to mitotic chromosomes) is present.

Involution Tissue rearrangement in which a cell layer undergoes folding at its edges.

Ion channel Configuration of proteins in the plasma membrane or other cell membranes that permits the passage of specific ions.

IP_3 (inositol trisphosphate) An intracellular signaling molecule involved in calcium ion deployment. (*See also* inositol.)

Isologous diversification Behavior of interacting copies of initially identical dynamical systems, leading to stable diversification of the different copies with respect to their dynamical state.

Joule Unit of energy and work equal to 1 meter $\times$ 1 newton.

Juxtacrine Type of local communication between cells involving direct cell–cell contact.

Kinase An enzyme that links a phosphate group to another molecule. A protein kinase links a phosphate to a protein.

Kinesin A motor protein that moves cargoes along microtubules using energy released from the hydrolysis of ATP.

Kinetochore Portion of the chromosome centromere that serves as an attachment site for spindle microtubules.

Knock-out The biological outcome of inactivating or deleting a specific gene; also used to refer to the organism in which this has been accomplished.

Lamin Any of a specific class of intermediate filament-type proteins that form the fibrous nuclear lamina beneath the nuclear envelope during interphase.

Lateral inhibition Suppression of the propagation of an activity in a spatially extended dynamical system, either by another activity emanating from the same locus or by the local depletion of resources required for the initial activity.

Ligand A molecule that binds to a different molecule, its receptor.

Laplace equation The equation that expresses the equilibrium condition of a liquid surface, a consequence of Newton's second law.

Limit cycle A special set of solutions of a system of first-order differential equations that is located along a closed orbit (and thus periodically traversed) in the state space of the functional variables comprising the system.

Linear stability analysis A mathematical method applied to a system of equations to classify the spatial patterns they describe as functions of the various parameters they contain (e.g., the diffusion coefficients and rate constants in reaction–diffusion equations).

Lipid A molecule of biological origin that is insoluble in water.

Lipid bilayer Planar configuration of lipid-containing amphipathic molecules.

Loss angle (δ) A quantity constructed from the loss and storage moduli, defined by $\tan\delta = G''/G'$; the absolute value of δ (between 0 and $\pi/2$) gives a measure of how liquid (sol-like) or elastic (gel-like) is the viscoelastic material.

Loss modulus (G'') One of the physical parameters characterizing a viscoelastic material, namely, its ability to dissipate energy.

Lumen An open space within a tissue.

Lysis The breakdown of subcellular or supracellular molecular structure.

M phase The phase of the eukaryotic cell cycle during which mitosis occurs.

M-phase-promoting factor A regulatory molecule in the eukaryotic cell cycle consisting of a dimeric complex of the proteins cdc2 and cyclin.

Macromere A large cell that arises as a result of unequal early cleavages.

Marginal zone The middle region of an amphibian embryo, including the equator defined in relation to the animal and vegetal poles.

Maternal gene A gene that specifies a protein or RNA assembled into the egg during oogenesis (e.g., *bicoid*, *hunchback*), often in a spatially nonuniform fashion.

Matrix-driven translocation A phenomenon identified in model extracellular matrices constructed *in vitro*, in which, following the establishment of a sharp boundary between two matrices of differing composition (e.g., containing or lacking particles), one matrix partially engulfs the other.

Mediolateral The axis in a bilaterally symmetric embryo or organism defined by the direction from the central axis (e.g., notochord, vertebral column) to the sides of the body. *See also* convergent extension.

Mediolateral intercalation Movement of cells between one another along the embryo's mediolateral axis during convergent extension so as to form a longer, narrower cell mass.

Meiosis The process of two consecutive cell divisions in the diploid germ cells. Meiosis results in four haploid progeny of the original progenitor cell, among which are the male or female gametes.

Membrane A sheet-like structure. Cells have external (plasma) and internal molecular-scale membranes consisting of phospholipid bilayers. Embryos are surrounded by macroscopic tissue membranes.

Membrane channel One of a variety of protein complexes in phospholipid membranes that allows the specific transport of small molecules.

Membrane potential The electric potential difference across a membrane separating two otherwise insulated compartments.

Mesenchymal condensation Rearrangement of mesenchymal cells that brings them into direct contact with one another, usually transiently, at locations that serve as primordia for structures such as cartilage elements, feathers, or hairs.

Mesenchyme A type of tissue, common to the early embryo and early stages of organogenesis, in which cells are loosely suspended in a dilute extracellular matrix.

Mesoderm The middle germ layer in tripoblasts, which gives rise to muscle, blood, and skeletal and other connective tissues.

Metabolism The totality of the biochemical processes that sustain a living cell or organism.

Metaphase The stage of mitosis or meiosis at which the chromosomes are aligned at the cell center prior to separation.

Metazoa The class of multicellular animals.

Micromere Small cell that arises as a result of unequal early cleavages.

Microtubule organizing center Characteristic region within the cells (such as the centrosome), from which microtubules emanate, typically in a radial pattern.

Midblastula transition A set of early embryonic changes characterized by the iniation of transcription from the zygotic genome and pregastrulation cell movements.

Microfilament A cytoskeletal filament composed of a polymer of the protein actin.

Microtubule The polymerized form of the protein tubulin; a constituent of the cytoskeleton that plays central role in vesicular transport and cell division.

Microvillus (plural **microvilli**) A small, nonmotile extension of the cell surface involved in absorption or other surface-dependent functions.

Mitochondrion Membrane-bound organelle of eukaryotic cells, in which free energy is extracted from cellular fuels and stored, by oxidative phosphorylation, in the form of ATP.

Mitogen A biological molecule that stimulates the division of one or more cell types.

Mitosis Stage of the cell division cycle (other than in germ cells), in which the separation of chromosomes and the eventual separation of the daughter cells take place.

Mitotic spindle A special cellular structure made of microtubules that forms in the course of mitosis and meiosis and provides the architectural basis for the separation of chromosomes.

Mole A quantity of a substance corresponding to its molecular weight in grams.

Monomer A subunit of a larger structure (e.g., dimer, polymer).

Monte Carlo method A computational approach for studying the behavior of a complex system by subjecting a model representation of the system to small random perturbations and simulating its evolution through states with successively lower energies.

Morphogen A secreted biological molecule that influences tissue pattern or form.

Morphogenesis The shaping of an organism's body or organs.

Mutant A gene that is altered in its internal sequence relative to a reference gene.

Navier–Stokes equation The equation of motion of a viscous fluid, derived by the application of Newton's second law.

Nernst equation The equation that relates the ionic concentration difference to the electrical potential difference across a membrane such as the cell plasma membrane.

Neural crest The portion of the neural plate that detaches and disperses through the embryo, giving rise to the peripheral nervous system, portions of the cranial skeleton and the adrenal gland, pigment cells of the skin, and several other cell types.

Neural folds The pair of ridges at the periphery of the neural plate that eventually fuse to form the neural tube.

Neural plate The thickened region of ectoderm on the dorsal surface of an embryo where neurulation is initiated.

Neuron A nerve cell; characterized by electrical excitability and extended processes (axons and dendrites).

Neurulation The set of cell differentiation and morphogenetic events that lead to the formation of the neural tube, and eventually the spinal cord, in chordates.

Nieuwkoop center A population of cells that forms opposite the sperm entry point in the frog embryo and serves as an inducer of the Spemann–Mangold organizer.

Notochord A stiff rod-shaped tissue in the embryos of chordates that forms from the central mesodermal layer and acts as an inducer of neurulation in the overlying ectoderm.

Nuclear envelope A structure that consists of two concentric phospholipid bilayer membranes surrounding the nucleus which, in conjunction with the nuclear pores that traverse it, provides a selective barrier between the interior of the nucleus and the cytoplasm.

Nuclear pore One of several thousand eight-fold symmetric multiprotein complexes that traverse the nuclear envelope and serve as selective conduits for the trafficking of molecules between the nucleus and cytoplasm.

Nucleic acid A polymer constructed of nucleotides, e.g., DNA, RNA.

Nucleotide An organic molecule consisting of a sugar, a phosphate group, and a heterocyclic nitrogenous base that serves as the subunit of nucleic acids such as DNA and RNA.

Nucleus The cell organelle that contains the major portion of genetic material and is the site of the transcription and processing of RNA.

Ontogeny The generation, over the course of development, of an individual organism, to be contrasted with phylogeny.

Oocyte A cell in the lineage giving rise to the ovum that precedes the first (primary oocyte) or second (secondary oocyte) meiotic divisions.

Oogenesis The sequence of biosynthetic and macromolecular assembly processes that leads to the development of an ovum.

Oogonia Early cells of the female germline that undergo meioisis and ultimately give rise to ova (eggs).

Open system A material system that exchanges chemicals and energy with its environment (the opposite of a closed system, in which interactions exist only between the constituents of the system itself).

Ordinary differential equation A differential equation that relates a function of one variable to its first-order derivatives (in a first-order differential equation) or to its higher-order derivatives (in a higher-order differential equation).

Organizer A set of cells in an embryo and in some fully formed invertebrate organisms, such as hydra, that influences the movement or differentiation of adjacent tissues.

Organogenesis A set of cell differentiation, pattern-formation, and morphogenetic events that lead to the generation of an organ during embryonic development.

Osmosis The net movement of solvent across a semipermeable membrane (i.e., one that is permeable to solvent, e.g., water, but not solute), driven by a difference in solute concentration on either side.

Osmotic pressure The pressure that must be exerted on the side of a semipermeable membrane containing the higher concentration of solute to prevent the flow of solvent (e.g., water) across the membrane due to osmosis.

Ovum The egg, or female gamete.

Pair-rule genes A class of nonuniformly expressed genes (e.g., *even-skipped, fushi tarazu*) involved in segmentation in insects. The products of pair-rule genes are generally involved in the regulation of segment polarity genes.

Paracrine A type of local communication between cells that involves released factors.

Phenotype The outward expression of an organism's or cell's characteristics, to be contrasted with its genotype.

Planar polarity The structural asymmetry of cells in a tissue that constrains the direction of their elongation, movement, or growth to a preferred plane or stack of planes.

Partial differential equation A differential equation that relates a function of more than one variable (e.g., time and distance) to its derivatives.

Percolation theory/transition The mathematical representation of a (second-order) phase transition in the connectivity properties of systems composed of randomly or irregularly arranged subunits (e.g., filaments, monomers, metallic islands in an insulating matrix, etc.). Below the transition the system contains isolated clusters of the subunits, whereas above the transition, arrived at by increasing the number of subunits, an interconnected network of the subunits forms that spans the entire system.

Phagocytosis The engulfment of a particle (e.g., a bacterium or cell fragment) by a cell.

Phase transition A major alteration in material characteristics (e.g., from liquid to gas, from one crystal structure to another, from insulator to conductor), typically triggered by a change in an external physical parameter (e.g., temperature, pressure, chemical concentration).

Phosphorylation Attachment of a phosphate group to another molecule.

Phylogeny The generation, over the course of evolution, of a given type of organism, to be contrasted with ontogeny.

Plasma (or cell) membrane The lipid bilayer that constitutes the boundary between the interior of a cell and its microenvironment.

Pluripotent cell A cell that is capable of giving rise to a variety of terminally differentiated cell types.

Piconewton (pN) Unit of force equal to 10^{-12} N.

Polar body The products of meiosis I and meiosis II other than the secondary oocyte or ovum.

Posterior Referring to the region of an embryo containing the tail; opposite to anterior.

Polyspermy An abnormal condition whereby an egg is fertilized by more than one spermatozoon.

Pressure The force acting on unit area in the normal direction (the outward normal in the case of a closed surface).

Prokaryote A class of organisms whose DNA is not organized into nuclei.

Promoter Sequence of DNA within a gene that binds transcription factors and related proteins, thereby mediating control of the gene's activity.

Prophase The initial stage of mitosis when chromatin condenses into discrete chromosomes, the nuclear envelope breaks down, and spindle fibers form at opposite poles of the cell.

Protein A polymer of amino acid residues linked by peptide bonds. The amino acid sequences (primary structure) of different proteins are specified by corresponding DNA sequences contained in genes.

Protein kinase An enzyme that transfers a phosphate group to a protein.

Protein phosphatase An enzyme that removes a phosphate group from a protein.

Proteoglycan A macromolecular component of the cell surface or extracellular matrix, consisting of a (core) protein with one or more glycosaminoglycan chains attached to it.

Proximodistal axis The axis of an appendage such as the vertebrate limb, defined by an arrow pointing from the site of its attachment on the body to its tip.

Protozoa Class of free-living single-celled eukaryotes.

Radial intercalation Movement of cells from different layers between one another along the embryo's radius during convergent extension so as to form a thinned cell mass.

Reaction–diffusion mechanism A specific means of pattern formation in which chemical components react with one another as well as diffuse at different rates.

Receptor A protein or protein complex that mediates a biological activity upon binding to another molecule, its ligand.

Replication The process by which a cell's DNA is duplicated.

Rostral Toward the nose. More generally, toward the head or anterior region of the embryo.

Rostrocaudal axis The axis defined by an arrow pointing from the head (rostral) region to the tail (caudal) region of an animal or embryo.

S phase The phase of the cell cycle during which chromosomal DNA is synthesized.

Scale-free network Spatial arrangement of objects (e.g., molecules) with no dominating characteristic distance between the objects (in contrast with, e.g., a regular crystalline lattice, which has a well-defined lattice constant).

Scaling The characteristic power-law dependence of a physical quantity (e.g., mass, conductivity, length of shore line, etc.) on the size of the units (i.e., the scale) of a quantity with which it varies (e.g., concentration, temperature, length, etc.).

Segmentation The transformation of a uniform tissue mass into tandemly arranged tissue blocks.

Segment polarity gene One of several genes in the *Drosophila* body segmentation process (e.g., *engrailed*, *wingless*) whose expression distinguishes the anterior and posterior portions of the developing segment.

Separatrix A geometric structure (a line in two dimensions, a plane in three dimensions) dividing distinct basins of attractions in a vector field.

Signal transduction The transmission of biochemical or mechanical changes within a cell in response to an external stimulus, typically leading to changed cell behavior.

Sister chromatids The two joined chromatids of a single replicated chromosome.

Somatic cell A cell of the body or its organs, in contrast to a cell of the germ line.

Somatopleure The outer of the two sheets into which the paraxial tissue (that lying on either side of the notochord) of the three-layered vertebrate embryo splits subsequent to gastrulation. The somatopleure contains the somatic mesoderm and the ectoderm and eventually forms the body wall (cf. *splanchnopleure*).

Somite A block of tissue, present in a series of pairs, arrayed along the central axis of vertebrate embryos. The somites give rise to the vertebral bones, associated connective tissues, intervertebral discs, and body wall and limb muscles.

Sorting The propensity of distinct living tissues composed of adhering and motile cells to separate and form boundaries of "immiscibility" across which no significant motion of cells takes place.

Spanning cluster An interconnected cluster of subunits that forms at a percolation transition and extends over the entire system.

Spemann–Mangold organizer A signaling center located at the anterior or rostral blastopore lip of the frog embryo that initiates gastrulation movements.

Spermatogonia Early cells of male germ line that undergo meiosis and eventually give rise to spermatozoa.

Spermatozoon (pl. **spermatozoa**) The male gamete.

Splanchnopleure The inner of the two sheets into which the paraxial tissue (that lying on either side of the notochord) of the three-layered vertebrate embryo splits subsequent to gastrulation. The splanchnopleure contains the splanchnic mesoderm and the endoderm and eventually forms the digestive tube (cf. *somatopleure*).

Stable fixed point A fixed point of a dynamical system that attracts trajectories in the vector field.

State space Coordinate system spanned by axes measuring the values of the functional variables (e.g., time-dependent concentrations) characterizing a physical system.

Stationary state (or **steady state**) A state of a dynamical system in which the time derivatives of all the independent variables vanish. It is equivalent to a fixed point of the system.

Stem cell A cell in an embryo or adult tissue that is capable of dividing asymmetrically so as to produce both another stem cell and a cell committed to differentiating into one or more specialized cell types.

Stokes' formula The mathematical expression for the force acting on a spherical particle moving with a given velocity in a viscous fluid. It is derived from the Navier–Stokes equation.

Storage modulus (G') One of the physical parameters characterizing a viscoelastic material, namely, its capacity to store elastic energy.

Strain A relative deformation caused by stress (such as the relative elongation of a wire under tension).

Strange attractor An attracting region in the multidimensional state space of a dynamical system. Once the system arrives in this region it wanders within it in a random fashion without ever leaving it (cf. *limit cycle*).

Stress force A force acting on a unit area of a material in a particular direction (e.g., a shear force acting parallel to the surface).

Surface tension The interfacial tension between a liquid and a vacuum or air.

Syncytium A cell in which many nuclei are contained in a common cytoplasm.

Telophase The terminal stage of mitosis or meiosis in which chromosomes decondense, the spindle breaks down, and nuclear membranes form around the daughter nuclei.

Thermal energy The energy possessed by an object as a consequence of its being at nonzero absolute temperature.

Thermal velocity The randomly oriented velocity of a molecule associated with its thermal energy.

Tight junction A junctional complex between two epithelial cells, located near their apical surfaces, that both plays a barrier role and is a locus of intercellular signaling.

Totipotent cell A cell, e.g., the zygote or a blastomere, capable of giving rise to all an organism's cell types, including those of extraembryonic tissues.

Transcription The process by which a sequence of RNA is synthesized from a DNA template.

Transcription factor A member of any one of several classes of proteins that bind to promoter regions of specific DNA sequences (transcriptional units) and participate in the control of their rates of transcription.

Transgenic organism An organism that has incorporated into its somatic and germline cells a DNA sequence characteristic of a different species.

Tripoblast A category of animal generated from an embryo with three germ layers, ectoderm, mesoderm, and endoderm.

Turing mechanism A dynamical system involving reaction and diffusion in which there is a bifurcation from a spatially uniform to a spatially nonuniform stationary solution.

Tyrosine kinase A protein kinase that attaches a phosphate group to a tyrosine side chain of its target protein.

Unstable fixed point A fixed point of a dynamical system that repels trajectories in the vector field.

Yolk Nutritive material stored in an ovum, generally consisting of various phosphoproteins, glycoproteins, lipoproteins, and polysaccharides.

Yolk sac An extra-embryonic membrane in birds and mammals. In avian embryos it surrounds the yolk.

Young's equation The equation that expresses the equilibrium condition of contacting fluids and their substrata, a consequence of Newton's second law.

Young's modulus A material property that characterizes the response of an elastic material to tension.

Vector field A map that assigns a vector function (with magnitude and direction) to each point in the state space of a dynamical system and thus defines all possible trajectories of the system through its state space.

Vegetal pole The point on the surface of a telolecithal (i.e., having nonuniform distribution of yolk) egg or early embryo at the center of the hemisphere containing the yolk (in the case of an egg) or yolky cells (in the case of an embryo); it lies opposite to the animal pole.

Ventral Referring to the underside, or belly, of an organism or embryo.

Viscosity A characteristic physical property of fluids arising from molecular friction (due to the adhesive interaction between the molecules), which gives rise to their resistance to flow.

Viscoelastic material A substance that manifests simultaneously elastic and viscous liquid behavior (such as a gel).

Vitelline membrane The extracellular matrix of an unfertilized egg; also called the zona pellucida and the shell membrane.

Wild-type A reference sequence for a particular gene, usually the one most commonly found outside the laboratory.

Work of adhesion The effort required to separate distinct bounding substances across a unit area of their contact.

Work of cohesion The effort required to separate the molecules of a substance so as to produce two free surfaces (i.e., interfaces with the embedding medium) of unit area.

Zygote The fertilized egg or ovum; the earliest embryonic stage of a diploid organism.

Zona pellucida The vitelline membrane of mammalian eggs.

References

Adams, C. L., Nelson, W. J., and Smith, S. J. (1996). Quantitative analysis of cadherin–catenin–actin reorganization during development of cell–cell adhesion. *J. Cell Biol.* **135**, 1899–911.

Adams, C. L., Chen, Y. T., Smith, S. J., and Nelson, W. J. (1998). Mechanisms of epithelial cell–cell adhesion and cell compaction revealed by high-resolution tracking of E-cadherin-green fluorescent protein. *J. Cell Biol.* **142**, 1105–19.

Afzelius, B. A. (1985). The immotile-cilia syndrome: a microtubule-associated defect. *CRC Crit. Rev. Biochem.* **19**, 63–87.

Agius, E., Oegeschlager, M., Wessely, O., Kemp, C., and De Roberts, E. M. (2000). Endodermal Nodal-related signals and mesoderm induction in *Xenopus*. *Development* **127**, 1173–83.

Agutter, P. S., and Wheatley, D. N. (2000). Random walk and cell size. *BioEssays* **22**, 1018–23.

Akam, M. (1989). Making stripes inelegantly. *Nature* **341**, 282–3.

Akatiya, T., and Bronner-Fraser, M. (1992). Expression of cell adhesion molecules during initiation and cessation of neural crest cell migration. *Dev. Dyn.* **194**, 12–20.

Alber, M. S., Kiskowski, M. A., Glazier, J. A., and Jiang, Y. (2003). On cellular automaton approaches to modeling biological cells. In *Mathematical Systems Theory in Biology, Communication, and Finance* (J. Rosenthal and D. S. Gilliam, eds.), Vol. 134, pp. 1–40. New York, Springer-Verlag.

Alber, M., Hentschel, H. G. E., Kazmierczak, B., and Newman, S. A. (2005). Existence of solutions to a new model of biological pattern formation. *Journal of Mathematical Analysis and Application* **308**, 175–194.

Alberts, B., Johnson, A., Lewis, J., Raff, M., Roberts, K., and Walter, P. (2002). *Molecular Biology of the Cell*. Garland Science, New York.

Allaerts, W. (1991). On the role of gravity and positional information in embryological axis formation and tissue compartmentalization. *Acta Biotheor.* **39**, 47–62.

Amonlirdviman, K., Khare, N. A., Tree, D. R., Chen, W. S., Axelrod, J. D., and Tomlin, C. J. (2005). Mathematical modeling of planar cell polarity to understand domineering nonautonomy. *Science* **307**, 423–6.

Anderson, A. R., and Chaplain, M. A. (1998). Continuous and discrete mathematical models of tumor-induced angiogenesis. *Bull. Math. Biol.* **60**, 857–99.

Anderson, K. V., and Ingham, P. W. (2003). The transformation of the model organism: a decade of developmental genetics. *Nat. Genet.* **33** Suppl., 285–93.

Angres, B., Barth, A., and Nelson, W. J. (1996). Mechanism for transition from initial to stable cell–cell adhesion: kinetic analysis of E-cadherin-mediated adhesion using a quantitative adhesion assay. *J. Cell Biol.* **134**, 549–57.

Armstrong, P. B. (1989). Cell sorting out: the self-assembly of tissues *in vitro*. *Crit. Rev. Biochem. and Mol. Biol.* **24**, 119–49.

Artavanis-Tsakonas, S., Rand, M. D., and Lake, R. J. (1999). Notch signaling: cell fate control and signal integration in development. *Science* **284**, 770–6.

Arthur, W. (1997). *The Origin of Animal Body Plans: A Study in Evolutionary Developmental Biology*. Cambridge, New York, Cambridge University Press.

Atherton-Fessler, S., Hannig, G., and Piwnica-Worms, H. (1993). Reversible tyrosine phosphorylation and cell cycle control. *Semin. Cell Biol.* **4**, 433–42.

Augustin, H. G. (2001). Tubes, branches, and pillars: the many ways of forming a new vasculature. *Circ. Res.* **89**, 645–7.

Aulehla, A., Wehrle, C., Brand-Saberi, B., Kemler, R., Gossler, A., Kanzler, B., and Herrmann, B. G. (2003). Wnt3a plays a major role in the segmentation clock controlling somitogenesis. *Dev. Cell* **4**, 395–406.

Aylsworth, A. S. (2001). Clinical aspects of defects in the determination of laterality. *Am. J. Med. Genet.* **101**, 345–55.

Bachvarova, R. (1985). Gene expression during oogenesis and oocyte development in mammals. *Dev. Biol.* **1**, 453–524.

Ball, W. D. (1974). Development of the rat salivary glands. 3. Mesenchymal specificity in the morphogenesis of the embryonic submaxillary and sublingual glands of the rat. *J. Exp. Zool.* **188**, 277–88.

Ballaro, B., and Reas, P. G. (2000). Chemical and mechanical waves on the cortex of fertilized egg cells: a bioexcitability effect. *Rev. Biol.* **93**, 83–101.

Baoal, D. (2002). *Mechanics of the Cell*. Cambridge University Press, Cambridge.

Barabási, A.-L. (2002). *Linked: The New Science of Networks*. Perseus Publications, Cambridge, MA.

Bard, J. B. (1999). A bioinformatics approach to investigating developmental pathways in the kidney and other tissues. *Int. J. Dev. Biol.* **43**, 397–403.

Barkai, N., and Leibler, S. (2000). Circadian clocks limited by noise. *Nature* **403**, 267–8.

Basu, S., Gerchmann, Y., Collins, C. H., Arnold, F. H., Weiss, R. (2005). A synthetic multicellular system for programmed pattern formation. *Nature* **434**, 1130–4.

Bateman, E. (1998). Autoregulation of eukaryotic transcription factors. *Prog. Nucleic Acid Res. Mol. Biol.* **60**, 133–68.

Bateson, W. (1894). *Materials for the Study of Variation*. Macmillan. London.

Baumgartner, W., Hinterdorfer, P., Ness, W., *et al.* (2000). Cadherin interaction probed by atomic force microscopy. *Proc. Nat. Acad. Sci. USA* **97**, 4005–10.

Bausch, A. R., Moller, W., and Sackmann, E. (1999). Measurement of local viscoelasticity and forces in living cells by magnetic tweezers. *Biophys. J.* **76**, 573–9.

Becker, M., Baumann, C., John, S., *et al.* (2002). Dynamic behavior of transcription factors on a natural promoter in living cells. *EMBO Rep.* **3**, 1188–94.

Beddington, R. S., and Robertson, E. J. (1999). Axis development and early asymmetry in mammals. *Cell* **96**, 195–209.

Bejsovec, A., and Wieschaus, E. (1993). Segment polarity gene interactions modulate epidermal patterning in *Drosophila* embryos. *Development* **119**, 501–17.

Bell, G. I. (1978). Models for the specific adhesion of cells to cells. *Science* **200**, 618–27.

Beloussov, L. (1998). *The Dynamic Architecture of a Developing Organism*. Kluwer Academic Publishers, Dordrecht.

Ben-Avraham, D., and Havlin, S. (2000). *Diffusion and Reactions in Fractals and Disordered Systems*. Cambridge University Press, Cambridge, New York.

Benink, H. A., Mandato, C. A., and Bement, W. M. (2000). Analysis of cortical flow models in vivo. *Mol. Biol. Cell* **11**, 2553–63.

Berg, H. C. (1993). *Random Walks in Biology*. Princeton University Press, Princeton.

Berridge, M. J., Lipp, P., and Bootman, M. D. (2000). The versatility and universality of calcium signalling. *Nat. Rev. Mol. Cell Biol.* **1**, 11–21.

Berry, L. D., and Gould, K. L. (1996). Regulation of Cdc2 activity by phosphorylation at T14/Y15. *Prog. Cell. Cycle Res.* **2**, 99–105.

Bertrand, N., Castro, D. S., and Guillemot, F. (2002). Proneural genes and the specification of neural cell types. *Nat. Rev. Neurosci.* **3**, 517–30.

Bevilacqua, M., Butcher, E., Furie, B., *et al.* (1991). Selectins: a family of adhesion receptors. *Cell* **67**, 233.

Beysens, D. A., Forgacs, G., and Glazier, J. A. (2000). Cell sorting is analogous to phase ordering in fluids. *Proc. Nat. Acad. Sci. USA* **97**, 9467–71.

Bhalla, U. S., and Iyengar, R. (1999). Emergent properties of networks of biological signaling pathways. *Science* **283**, 381–7.

Bissell, M. J., and Barcellos-Hoff, M. H. (1987). The influence of extracellular matrix on gene expression: is structure the message? *J. Cell Sci. Suppl.* **8**, 327–43.

Blair, S. S. (2003). Developmental biology: boundary lines. *Nature* **424**, 379–81.

Boal, D. H. (2002). *Mechanics of the Cell*. Cambridge University Press, Cambridge, New York.

Boggon, T. J., Murray, J., Chappuis-Flament, S., *et al.* (2002). C-cadherin ectodomain structure and implications for cell adhesion mechanisms. *Science* **296**, 1308–13.

Boissonade, J., Dulos, E., and DeKepper, P. (1994). Turing patterns: from myth to reality. In *Chemical Waves and Patterns* (R. Kapral and K. Showalter, eds.), pp. 221–68. Kluwer, Boston.

Bolouri, H., and Davidson, E. H. (2003). Transcriptional regulatory cascades in development: initial rates, not steady state, determine network kinetics. *Proc. Nat. Acad. Sci. USA* **100**, 9371–6.

Bonner, J. T. (1998). The origins of multicellularity. *Integrative Biology* **1**, 27–36.

Boring, L. (1989). Cell–cell interactions determine the dorsoventral axis in embryos of an equally cleaving opisthobranch mollusc. *Dev. Biol.* **136**, 239–53.

Borisuk, M. T., and Tyson, J. J. (1998). Bifurcation analysis of a model of mitotic control in frog eggs. *J. Theor. Biol.* **195**, 69–85.

Borkhvardt, V. G. (2000). The growth and form development of the limb buds in vertebrate animals. *Ontogenez* **31**, 192–200.

Bouligand, Y. (1972). Twisted fibrous arrangements in biological materials and cholesteric mesophases. *Tissue Cell* **4**, 189–217.

Braat, A. K., Zandbergen, T., van de Water, S., Goos, H. J., and Zivkovic, D. (1999). Characterization of zebrafish primordial germ cells: morphology and early distribution of vasa RNA. *Dev. Dyn.* **216**, 153–67.

Braga, V. M. (2002). Cell–cell adhesion and signalling. *Curr. Opin. Cell. Biol.* **14**, 546–56.

Branford, W. W., and Yost, H. J. (2002). Lefty-dependent inhibition of Nodal- and Wnt-responsive organizer gene expression is essential for normal gastrulation. *Curr. Biol.* **12**, 2136–41.

Brasier, M., and Antcliffe, J. (2004). Paleobiology. Decoding the Ediacaran enigma. *Science* **305**, 1115–7.

Breckenridge, R. A., Mohun, T. J., and Amaya, E. (2001). A role for BMP signalling in heart looping morphogenesis in *Xenopus*. *Dev. Biol.* **232**, 191–203.

Brinker, C. J., and Scherer, G. W. (1990). *Sol–Gel Science*. Academic Press, New York.

Brock, H. W., and Fisher, C. L. (2005). Maintenance of gene expression patterns. *Dev. Dyn.* **232**, 633–55.

Bronner-Fraser, M. (1982). Distribution of latex beads and retinal pigment epithelial cells along the ventral neural crest pathways. *Dev. Biol.* **91**, 50–63.

Bronner-Fraser, M. (1984). Latex beads as probes of a neural crest pathway: effects of laminin, collagen, and surface charge on bead translocation. *J. Cell Biol.* **98**, 1947–60.

Bronner-Fraser, M. (1985). Effects of different fragments of the fibronectin molecule on latex bead translocation along neural crest migratory pathways. *Dev. Biol.* **108**, 131–45.

Bronner-Fraser, M., Wolf, J. J., and Murray, B. A. (1992). Effects of antibodies against N-cadherin and N-CAM on the cranial neural crest and neural tube. *Dev. Biol.* **153**, 291–301.

Brown, S. J., Hilgenfeld, R. B., and Denell, R. E. (1994a). The beetle *Tribolium castaneum* has a *fushi tarazu* homolog expressed in stripes during segmentation. *Proc. Nat. Acad. Sci. USA* **91**, 12922–6.

Brown, S. J., Patel, N. H., and Denell, R. E. (1994b). Embryonic expression of the single *Tribolium engrailed* homolog. *Dev. Genet.* **15**, 7–18.

Browne, E. N. (1909). The production of new hydranths in *Hydra* by insertion of small grafts. *J. Exp. Zool.* **7**, 1–23.

Brunet, J. F., and Ghysen, A. (1999). Deconstructing cell determination: proneural genes and neuronal identity. *Bioessays* **21**, 313–8.

Bryant, P. J. (1999). Filopodia: fickle fingers of cell fate? *Curr. Biol.* **9**, R655–7.

Buckley, C. D., Rainger, G. E., Bradfield, P. F., Nash, G. B., and Simmons, D. L. (1998). Cell adhesion: more than just glue. *Mol. Membr. Biol.* **15**, 167–76.

Bugrim, A., Fontanilla, R., Eutenier, B. B., Keizer, J., and Nuccitelli, R. (2003). Sperm initiate a Ca^{2+} wave in frog eggs that is more similar to Ca^{2+} waves initiated by IP_3 than by Ca^{2+}. *Biophys. J.* **84**, 1580–90.

Bugrim, A. E., Zhabotinsky, A. M., and Epstein, I. R. (1997). Calcium waves in a model with a random spatially discrete distribution of Ca^{2+} release sites. *Biophys. J.* **73**, 2897–906.

Callamaras, N., Marchant, J. S., Sun, X. P., and Parker, I. (1998). Activation and co-ordination of InsP3-mediated elementary Ca^{2+} events during global Ca^{2+} signals in *Xenopus* oocytes. *J. Physiol.* **509**, 81–91.

Campochiaro, P. A. (2000). Retinal and choroidal neovascularization. *J. Cell Physiol.* **184**, 301–10.

Canman, J. C., and Bement, W. M. (1997). Microtubules suppress actomyosin-based cortical flow in *Xenopus* oocytes. *J. Cell Sci.* **110**, 1907–17.

Capco, D. G., and McGaughey, R. W. (1986). Cytoskeletal reorganization during early mammalian development: analysis using embedment-free sections. *Dev. Biol.* **115**, 446–58.

Carnac, G., and Gurdon, J. B. (1997). The community effect in *Xenopus* myogenesis is promoted by dorsalizing factors. *Int. J. Dev. Biol.* **41**, 521–4.

Castets, V., Dulos, E., Boissonade, J., and DeKepper, P. (1990). Experimental evidence of a sustained standing Turing-type nonequilibrium chemical pattern. *Phys. Rev. Lett.* **64**, 2953–6.

Cayan, S., Conaghan, J., Schriock, E. D., Ryan, I. P., Black, L. D., and Turek, P. J. (2001). Birth after intracytoplasmic sperm injection with use of testicular sperm from men with Kartagener/immotile cilia syndrome. *Fertil. Steril.* **76**, 612–4.

Chambon, F., and Winter, H. H. (1987). Linear viscoelasticity at the gel point of a crosslinking PDMS with imbalanced stoichiometry. *J. Rheol.* **31**, 683–97.

Chan, A. P., and Etkin, L. D. (2001). Patterning and lineage specification in the amphibian embryo. *Curr. Top. Dev. Biol.* **51**, 1–67.

Chaturvedi, R., Huang, C., Kazmierczak, B., Schneider, T., Izaguirre, J. A., Glimm, T., Hentschel, H. G. E., Newman, S. A., Glazier, J. A., and Alber, M. (2005). On multiscale approaches to three-dimensional modeling of morphogenesis. *J. Roy. Soc. London Interface* **2**, 237–53.

Cheer, A., Vincent, J. P., Nuccitelli, R., and Oster, G. (1987). Cortical activity in vertebrate eggs. I: The activation waves. *J. Theor. Biol.* **124**, 377–404.

Chen, J. N., van Eeden, F. J., Warren, K. S., *et al.* (1997). Left–right pattern of cardiac BMP4 may drive asymmetry of the heart in zebrafish. *Development* **124**, 4373–82.

Chen, J.-Y., Bottjer, D. J., Oliveri, P., *et al.* (2004). Small bilaterian fossils from 40 to 55 million years before the Cambrian. *Science* **305**, 218–22.

Chen, Y., and Schier, A. F. (2002). Lefty proteins are long-range inhibitors of squint-mediated nodal signaling. *Curr. Biol.* **12**, 2124–8.

Cheng, A., Ross, K. E., Kaldis, P., and Solomon, M. J. (1999). Dephosphorylation of cyclin-dependent kinases by type 2C protein phosphatases. *Genes Dev.* **13**, 2946–57.

Christen, B., and Slack, J. (1999). Spatial response to fibroblast growth factor signalling in *Xenopus* embryos. *Development* **126**, 119–25.

Chuong, C. M. (1993). The making of a feather: homeoproteins, retinoids and adhesion molecules. *BioEssays* **15**, 513–21.

Cickovski, T., Huang, C., Chaturvedi, R., *et al.* (2005). A framework for three-dimensional simulation of morphogenesis. *IEEE/ACM Trans. Computat. Biol. Bioinformatics*, in press.

Cinquin, O., and Demongeof, J. (2005). High-dimensional switches and the modelling of cellular differentiation. *J. Theor. Biol.* **233**, 391–411.

Clements, D., Friday, R. V., and Woodland, H. R. (1999). Mode of action of VegT in mesoderm and endoderm formation. *Development* **126**, 4903–11.

Clerk, J. P., Giraud, G., Laugier, J. M., and Luck, J. M. (1990). The AC electrical conductance of binary disordered systems, percolation custers, fractals and related models. *Adv. Phys.* **39**, 191–309.

Cline, C. A., Schatten, H., Balczon, R., and Schatten, G. (1983). Actin-mediated surface motility during sea urchin fertilization. *Cell Motil.* **3**, 513–24.

Clyde, D. E., Corado, M. S., Wu, X., Pare, A., Papatsenko, D., and Small, S. (2003). A self-organizing system of repressor gradients establishes segmental complexity in *Drosophila*. *Nature* **426**, 849–53.

Colas, J. F., and Schoenwolf, G. C. (2001). Towards a cellular and molecular understanding of neurulation. *Dev. Dyn.* **221**, 117–45.

Collier, J. R., Monk, N. A., Maini, P. K., and Lewis, J. H. (1996). Pattern formation by lateral inhibition with feedback: a mathematical model of Delta–Notch intercellular signalling. *J. Theor. Biol.* **183**, 429–46.

Comper, W. D. (1996). *Extracellular Matrix*. Vol. I. *Tissue Function*; Vol. II. *Molecular Components and Interactions*. Harwood Academic Publishers, Amsterdam.

Comper, W. D., Pratt, L., Handley, C. J., and Harper, G. S. (1987). Cell transport in model extracellular matrices. *Arch. Biochem. Biophys.* **252**, 60–70.

Conway Morris, S. (2003). The Cambrian "explosion" of metazoans. In *Origination of Organismal Form: Beyond the Gene in Developmental and*

Evolutionary Biology (G. B. Müller and S. A. Newman, eds.), pp. 13–32. MIT Press, Cambridge, MA.

Cooke, J., Nowak, M. A., Boerlijst, M., and Maynard-Smith, J. (1997). Evolutionary origins and maintenance of redundant gene expression during metazoan development. *Trends Genet.* **13**, 360–4.

Cooke, J., and Zeeman, E. C. (1976). A clock and wavefront model for control of the number of repeated structures during animal morphogenesis. *J. Theor. Biol.* **58**, 455–76.

Cormack, D. H. (1987). *Ham's Histology*, ninth edn. Lippincott, Philadelphia.

Cornish-Bowden, A. (1995). *Fundamentals of Enzyme Kinetics*. Ashgate Publ. Co., London.

Coulombe, J. N., and Bronner-Fraser, M. (1984). Translocation of latex beads after laser ablation of the avian neural crest. *Dev. Biol.* **106**, 121–34.

Crawford, K., and Stocum, D. L. (1988). Retinoic acid coordinately proximalizes regenerate pattern and blastema differential affinity in axolotl limbs. *Development* **102**, 687–98.

Crick, F. H. C. (1970). Diffusion in embryogenesis. *Nature* **225**, 420–2.

Crick, F. H. C., and Lawrence, P. A. (1975). Compartments and polyclones in insect development. *Science* **189**, 340–7.

Cross, N. L., and Elinson, R. P. (1980). A fast block to polyspermy in frogs mediated by changes in the membrane potential. *Dev. Biol.* **75**, 187–98.

Cunliffe, V. T. (2003). Memory by modification: the influence of chromatin structure on gene expression during vertebrate development. *Gene* **305**, 141–50.

Czirok, A., Rupp, P. A., Rongish, B. J., and Little, C. D. (2002). Multi-field 3D scanning light microscopy of early embryogenesis. *J. Microsc.* **206**, 209–17.

Dale, J. K., Maroto, M., Dequeant, M. L., Malapert, P., McGrew, M., and Pourquie, O. (2003). Periodic notch inhibition by *Lunatic fringe* underlies the chick segmentation clock. *Nature* **421**, 275–8.

Danos, M. C., and Yost, H. J. (1996). Role of notochord in specification of cardiac left–right orientation in zebrafish and *Xenopus*. *Dev. Biol.* **177**, 96–103.

Dathe, V., Gamel, A., Manner, J., Brand-Saberi, B., and Christ, B. (2002). Morphological left–right asymmetry of Hensen's node precedes the asymmetric expression of Shh and Fgf8 in the chick embryo. *Anat. Embryol. (Berlin)* **205**, 343–54.

Davidson, E. H. (2001). *Genomic Regulatory Systems: Development and Evolution*. Academic Press, San Diego.

Davidson, E. H., Rast, J. P., Oliveri, P., *et al.* (2002). A genomic regulatory network for development. *Science* **295**, 1669–78.

Davidson, L. A., Koehl, M. A., Keller, R., and Oster, G. F. (1995). How do sea urchins invaginate? Using biomechanics to distinguish between mechanisms of primary invagination. *Development* **121**, 2005–18.

Davidson, L. A., Oster, G. F., Keller, R. E., and Koehl, M. A. (1999). Measurements of mechanical properties of the blastula wall reveal which hypothesized mechanisms of primary invagination are physically plausible in the sea urchin *Strongylocentrotus purpuratus*. *Dev. Biol.* **209**, 221–38.

Dawes, R., Dawson, I., Falciani, F., Tear, G., and Akam, M. (1994). *Dax*, a locust *Hox* gene related to *fushi-tarazu* but showing no pair-rule expression. *Development* **120**, 1561–72.

Dawson, S. P., Keizer, J., and Pearson, J. E. (1999). Fire–diffuse–fire model of dynamics of intracellular calcium waves. *Proc. Nat. Acad. Sci. USA* **96**, 6060–3.

De Felici, M. (2000). Regulation of primordial germ cell development in the mouse. *Int. J. Dev. Biol.* **44**, 575–80.

de Gennes, P. G. (1976a). Critical dimensionality for a special percolation problem (relevant to the gelation in polymers). *J. Physique (Paris)* **30**, 1049–54.

de Gennes, P. G. (1976b). On the relation between percolation and the elasticity of gels. *J. Physique (Paris)* **37**, L1–2.

de Gennes, P. G. (1992). Soft matter. *Science* **256**, 495–7.

de Gennes, P. G., and Prost, J. (1993). *The Physics of Liquid Crystals*. Clarendon Press, Oxford.

Deguchi, R., Shirakawa, H., Oda, S., Mohri, T., and Miyazaki, S. (2000). Spatiotemporal analysis of Ca^{2+} waves in relation to the sperm entry site and animal–vegetal axis during Ca^{2+} oscillations in fertilized mouse eggs. *Dev. Biol.* **218**, 299–313.

DeMarais, A. A., and Moon, R. T. (1992). The armadillo homologs beta-catenin and plakoglobin are differentially expressed during early development of *Xenopus laevis*. *Dev. Biol.* **153**, 337–46.

De Smedt, V., Poulhe, R., Cayla, X., *et al.* (2002). Thr-161 phosphorylation of monomeric Cdc2. Regulation by protein phosphatase 2C in *Xenopus* oocytes. *J. Biol. Chem.* **277**, 28592–600.

Dembo, M., Glushko, V., Aberlin, M. E., and Sonenberg, M. (1979). A method for measuring membrane microviscosity using pyrene excimer formation. Application to human erythrocyte ghosts. *Biochim. Biophys. Acta* **552**, 201–11.

Derganc, J., Bozic, B., Svetina, S., and Zeks, B. (2000). Stability analysis of micropipette aspiration of neutrophils. *Biophys. J.* **79**, 153–62.

Dewar, H., Tanaka, K., Nasmyth, K., and Tanaka, T. U. (2004). Tension between two kinetochores suffices for their bi-orientation on the mitotic spindle. *Nature* **428**, 93–7.

Dickinson, R. B., and Tranquillo, R. T. (1993). A stochastic model for adhesion-mediated cell random motility and haptotaxis. *J. Math. Biol.* **31**, 563–600.

Djabourov, M., Leblond, J., and Papon, P. (1988). Gelation of aqueous gelatin solutions. II. Rheology of the sol–gel transition. *J. Phys. (Paris)* **49**, 333–343.

Djabourov, M., Lechaire, J. P., and Gaill, F. (1993). Structure and rheology of gelatin and collagen gels. *Biorheology* **30**, 191–205.

Doedel, E. J., and Wang, X. J. (1995). AUTO94: Software for continuation and bifurcation problems in ordinary differential equations. Center for Research on Parallel Computing, California Institute of Technology, Pasadena, CA.

Dolmetsch, R. E., Xu, K., and Lewis, R. S. (1998). Calcium oscillations increase the efficiency and specificity of gene expression. *Nature* **392**, 933–6.

Dosch, R., Gawantka, V., Delius, H., Blumenstock, C., and Niehrs, C. (1997). Bmp-4 acts as a morphogen in dorsoventral mesoderm patterning in *Xenopus*. *Development* **124**, 2325–34.

Downie, S. A., and Newman, S. A. (1994). Morphogenetic differences between fore and hind limb precartilage mesenchyme: relation to mechanisms of skeletal pattern formation. *Dev. Biol.* **162**, 195–208.

Downie, S. A., and Newman, S. A. (1995). Different roles for fibronectin in the generation of fore and hind limb precartilage condensations. *Dev. Biol.* **172**, 519–30.

Drasdo, D., and Forgacs, G. (2000). Modeling the interplay of generic and genetic mechanisms in cleavage, blastulation, and gastrulation. *Dev. Dyn.* **219**, 182–91.

Drasdo, D., Kree, R., and McCaskill, J. S. (1995). Monte Carlo approach to tissue-cell populations. *Phys. Rev. E Statistical Physics, Plasmas, Fluids, and Related Interdisciplinary Topics* **52**, 6635–57.

Duband, J. L., Monier, F., Delannet, M., and Newgreen, D. (1995). Epithelium–mesenchyme transition during neural crest development. *Acta Anat. (Basel)* **154**, 63–78.

Dubrulle, J., McGrew, M. J., and Pourquié, O. (2001). FGF signaling controls somite boundary position and regulates segmentation clock control of spatiotemporal *Hox* gene activation. *Cell* **106**, 219–32.

Ducibella, T., Huneau, D., Angelichio, E., Xu, Z., Schultz, R. M., Kopf, G. S., Fissore, R., Madoux, S., and Ozil, J. P. (2002). Egg-to-embryo transition is driven by differential responses to Ca^{2+} oscillation number. *Dev. Biol.* **250**, 280–91.

Duguay, D., Foty, R. A., and Steinberg, M. S. (2003). Cadherin-mediated cell adhesion and tissue segregation: qualitative and quantitative determinants. *Dev. Biol.* **253**, 309–23.

Dumollard, R., Carroll, J., Dupont, G., and Sardet, C. (2002). Calcium wave pacemakers in eggs. *J. Cell Sci.* **115**, 3557–64.

Dumont, J. N., and Brummett, A. R. (1985). Egg envelopes in vertebrates. *Dev. Biol.* **1**, 235–88.

Durand, D., Delsanti, M., Adam, M., and Luck, J. M. (1987). Frequency dependence of viscoelastic properties of branched polymers near gelation threshold. *Europhys. Lett.* **3**, 297–301.

Edelman, G. M. (1992). Morphoregulation. *Dev. Dyn.* **193**, 2–10.

Eichmann, A., Pardanaud, L., Yuan, L., and Moyon, D. (2002). Vasculogenesis and the search for the hemangioblast. *J. Hematother. Stem Cell Res.* **11**, 207–14.

Eidne, K. A., Zabavnik, J., Allan, W. T., Trewavas, A. J., Read, N. D., and Anderson, L. (1994). Calcium waves and dynamics visualized by confocal microscopy in *Xenopus* oocytes expressing cloned TRH receptors. *J. Neuroendocrinol.* **6**, 173–8.

Ekblom, P. (1992). *Renal Development*. Raven Press, New York.

Elinson, R. P., and Rowning, B. (1988). A transient array of parallel microtubules in frog eggs: potential tracks for a cytoplasmic rotation that specifies the dorso-ventral axis. *Dev. Biol.* **128**, 185–97.

Ellis, R. J. (2001). Macromolecular crowding: an important but neglected aspect of the intracellular environment. *Curr. Opin. Struct. Biol.* **11**, 114–9.

Ellis, R. J., and Minton, A. P. (2003). Cell biology: join the crowd. *Nature* **425**, 27–8.

Elowitz, M. B., and Leibler, S. (2000). A synthetic oscillatory network of transcriptional regulators. *Nature* **403**, 335–8.

Entchev, E. V., Schwabedissen, A., and Gonzalez-Gaitan, M. (2000). Gradient formation of the TGF-β homolog Dpp. *Cell* **103**, 981–91.

Erickson, C. A. (1985). Control of neural crest cell dispersion in the trunk of the avian embryo. *Dev. Biol.* **111**, 138–57.

Erickson, C. A. (1988). Control of pathfinding by the avian trunk neural crest. *Development* **103**, 63–80.

Erickson, C. A., and Isseroff, R. R. (1989). Plasminogen activator activity is associated with neural crest cell motility in tissue culture. *J. Exp. Zool.* **251**, 123–33.

Erickson, C. A., and Perris, R. (1993). The role of cell–cell and cell–matrix interactions in the morphogenesis of the neural crest. *Dev. Biol.* **159**, 60–74.

Eshkind, L., Tian, Q., Schmidt, A., *et al.* (2002). Loss of desmoglein 2 suggests essential functions for early embryonic development and proliferation of embryonal stem cells of mice and models: improved animal models for biomedical research. Synaptic vesicle alterations in rod photoreceptors of synaptophysin-deficient mice. *Eur. J. Cell. Biol.* **81**, 592–8.

Espeseth, A., Johnson, E., and Kintner, C. (1995). *Xenopus* F-cadherin, a novel member of the cadherin family of cell adhesion molecules, is expressed at boundaries in the neural tube. *Mol. Cell. Neurosci.* **6**, 199–211.

Essner, J. J., Vogan, K. J., Wagner, M. K., Tabin, C. J., Yost, H. J., and Brueckner, M. (2002). Conserved function for embryonic nodal cilia. *Nature* **418**, 37–8.

Ettensohn, C. A. (1999). Cell movements in the sea urchin embryo. *Curr. Opin. Genet. Dev.* **9**, 461–5.

Ettinger, L., and Doljanski, F. (1992). On the generation of form by the continuous interactions between cells and their extracellular matrix. *Biol. Rev. Camb. Philos. Soc.* **67**, 459–89.

Evans, E., and Yeoung, A. (1989). Apparent viscosity and cortical tension of blood granulocytes determined by micropipet aspiration. *Biophys. J.* **56**, 151–60.

Fagotto, F., Guger, K., and Gumbiner, B. M. (1997). Induction of the primary dorsalizing center in *Xenopus* by the Wnt/GSK/beta-catenin signaling pathway, but not by Vg1, Activin or Noggin. *Development* **124**, 453–60.

Ferrell, J. E., Jr,Wu, M., Gerhart, J. C., and Martin, G. S. (1991). Cell cycle tyrosine phosphorylation of p34cdc2 and a microtubule-associated protein kinase homolog in *Xenopus* oocytes and eggs. *Mol. Cell. Biol.* **11**, 1965–71.

Fitch, J., Fini, M. E., Beebe, D. C., and Linsenmayer, T. F. (1998). Collagen type IX and developmentally regulated swelling of the avian primary corneal stroma. *Dev. Dyn.* **212**, 27–37.

Fleming, T. P., and Goodall, H. (1986). Endocytic traffic in trophectoderm and polarised blastomeres of the mouse preimplantation embryo. *Anat. Rec.* **216**, 490–503.

Folkman, J. (2003). Angiogenesis and proteins of the hemostatic system. *J. Thromb. Haemost.* **1**, 1681–2.

Folkman, J., and Moscona, A. (1978). Role of cell shape in growth control. *Nature* **273**, 345–9.

Fontanilla, R. A., and Nuccitelli, R. (1998). Characterization of the sperm-induced calcium wave in *Xenopus* eggs using confocal microscopy. *Biophys. J.* **75**, 2079–87.

Forgacs, G. (1995). On the possible role of cytoskeletal filamentous networks in intracellular signalling: an approach based on percolation. *J. Cell Sci.* **108**, 2131–2143.

Forgacs, G., and Foty, R. A. (2004). Biological implications of tissue viscoelasticity. In *Function and Regulation of Cellular Systems: Experiments and Models* (A. Deutsch, M. Falke, J. Howard, and W. Zimmerman, eds.), pp. 269–77. Biskhauser Basel.

Forgacs, G., and Newman, S. A. (1994). Phase transitions, interfaces, and morphogenesis in a network of protein fibers. *Int. Rev. Cytol.* **150**, 139–48.

Forgacs, G., Jaikaria, N. S., Frisch, H. L., and Newman, S. A. (1989). Wetting, percolation and morphogenesis in a model tissue system. *J. Theor. Biol.* **140**, 417–430.

Forgacs, G., Newman, S. A., Obukhov, S. P., and Birk, D. E. (1991). Phase transition and morphogenesis in a model biological system. *Phys. Rev. Lett.* **67**, 2399–402.

Forgacs, G., Foty, R. A., Shafrir, Y., and Steinberg, M. S. (1998). Viscoelastic properties of living embryonic tissues: a quantitative study. *Biophys. J.* **74**, 2227–34.

Forgacs, G., Newman, S. A., Hinner, B., Maier, C. W., and Sackmann, E. (2003). Assembly of collagen matrices as a phase transition revealed by structural and rheologic studies. *Biophys. J.* **84**, 1272–80.

Forgacs, G., Yook, S. H., Janmey, P. A., Jeong, H., and Burd, C. G. (2004). Role of the cytoskeleton in signaling networks. *J. Cell. Sci.* **117**, 2769–75.

Foty, R. A., and Steinberg, M. S. (1997). Measurement of tumor cell cohesion and suppression of invasion by E- or P-cadherin. *Cancer Res.* **57**, 5033–6.

Foty, R. A., and Steinberg, M. S. (2005). The differential adhesion hypothesis: a direct evaluation. *Dev. Biol.* **278**, 255–63.

Foty, R. A., Forgacs, G., Pfleger, C. M., and Steinberg, M. S. (1994). Liquid properties of embryonic tissues: measurement of interfacial tensions. *Phys. Rev. Lett.* **72**, 2298–301.

Foty, R. A., Pfleger, C. M., Forgacs, G., and Steinberg, M. S. (1996). Surface tensions of embryonic tissues predict their mutual envelopment behavior. *Development* **122**, 1611–20.

Frasch, M., and Levine, M. (1987). Complementary patterns of *even-skipped* and *fushi tarazu* expression involve their differential regulation by a common set of segmentation genes in *Drosophila*. *Genes Dev.* **1**, 981–95.

Freeman, M. (2002). Morphogen gradients, in theory. *Dev. Cell* **2**, 689–90.

Frenz, D. A., Akiyama, S. K., Paulsen, D. F., and Newman, S. A. (1989a). Latex beads as probes of cell surface–extracellular matrix interactions during chondrogenesis: evidence for a role for amino-terminal heparin-binding domain of fibronectin. *Dev. Biol.* **136**, 87–96.

Frenz, D. A., Jaikaria, N. S., and Newman, S. A. (1989b). The mechanism of precartilage mesenchymal condensation: a major role for interaction of the cell surface with the amino-terminal heparin-binding domain of fibronectin. *Dev. Biol.* **136**, 97–103.

Freyman, T. M., Yannas, I. V., Yokoo, R., and Gibson, L. J. (2001). Fibroblast contraction of a collagen-GAG matrix. *Biomaterials* **22**, 2883–91.

Friedlander, D. R., Mege, R.-M., Cunningham, B. A., and Edelman, G. M. (1989). Cell sorting-out is modulated by both the specificity and amount of different cell adhesion molecules (CAMs) expressed on cell surfaces. *Proc. Nat. Acad. Sci. USA* **86**, 7043–7047.

Fristrom, D., and Chihara, C. (1978). The mechanism of evagination of imaginal discs of *Drosophila melanogaster*. V. Evagination of disc fragments. *Dev. Biol.* **66**, 564–570.

Fujiwara, T., Ritchie, K., Murakoshi, H., Jacobson, K., and Kusumi, A. (2002). Phospholipids undergo hop diffusion in compartmentalized cell membrane. *J. Cell Biol.* **157**, 1071–81.

Fung, Y. C. (1993). *Biomechanics: Mechanical Properties of Living Tissues*. Springer-Verlag, New York.

Furusawa, C., and Kaneko, K. (1998). Emergence of rules in cell society: differentiation, hierarchy, and stability. *Bull. Math. Biol.* **60**, 659–87.

Furusawa, C., and Kaneko, K. (2001). Theory of robustness of irreversible differentiation in a stem cell system: chaos hypothesis. *J. Theor. Biol.* **209**, 395–416.

Gaill, F., Lechaire, J. P., and Denefle, J. P. (1991). Fibrillar pattern of self-assembled and cell-assembled collagen: resemblance and analogy. *Biol. Cell* **72**, 149–58.

Gamba, A., Ambrosi, D., Coniglio, A., *et al.* (2003). Percolation, morphogenesis, and Burgers dynamics in blood vessels formation. *Phys. Rev. Lett.* **90**, 118 101.

Garcia-Bellido, A. (1975). Genetic control of wing disc development in *Drosophila*. *Ciba Found. Sym.* **29**, 169–78.

Garcia-Bellido, A., Ripoll, P., and Morata, G. (1976). Developmental compartmentalization in the dorsal mesothoracic disc of *Drosophila*. *Dev. Biol.* **48**, 132–47.

Garcia-Perez, A. I., Lopez-Beltran, E. A., Kluner, P., Luque, J., Ballesteros, P., and Cerdan, S. (1999). Molecular crowding and viscosity as determinants of translational diffusion of metabolites in subcellular organelles. *Arch. Biochem. Biophys.* **362**, 329–38.

Gardner, R. L. (2001). The initial phase of embryonic patterning in mammals. *Int. Rev. Cytol.* **203**, 233–90.

Gerhart, J. (2002). Changing the axis changes the perspective. *Dev. Dyn.* **225**, 380–3.

Gerhart, J., Ubbels, G., Black, S., Hara, K., and Kirschner, M. (1981). A reinvestigation of the role of the grey crescent in axis formation in *Xenopus laevis*. *Nature* **292**, 511–6.

Ghosh, S., and Comper, W. D. (1988). Oriented fibrillogenesis of collagen *in vitro* by ordered convection. *Connect. Tissue. Res.* **17**, 33–41.

Giancotti, F. G., and Ruoslahti, E. (1999). Integrin signaling. *Science* **285**, 1028–32.

Giansanti, M. G., Bonaccorsi, S., Bucciarelli, E., and Gatti, M. (2001). *Drosophila* male meiosis as a model system for the study of cytokinesis in animal cells. *Cell Struct. Funct.* **26**, 609–17.

Gierer, A. (1977). Physical aspects of tissue evagination and biological form. *Quart. Rev. Biophys.* **10**, 529–93.

Gilbert, S. F. (2003). *Developmental Biology*. Sinauer Associates, Sunderland, MA.

Gilkey, J. C., Jaffe, L. F., Ridgway, E. B., and Reynolds, G. T. (1978). A free calcium wave traverses the activating egg of the medaka, *Oryzias latipes*. *J. Cell Biol.* **76**, 448–66.

Gimlich, R. L. (1985). Cytoplasmic localization and chordamesoderm induction in the frog embryo. *J. Embryol. Exp. Morphol.* **89** Suppl., 89–111.

Gimlich, R. L. (1986). Acquisition of developmental autonomy in the equatorial region of the *Xenopus* embryo. *Dev. Biol.* **115**, 340–52.

Ginsburg, M., Snow, M. H., and McLaren, A. (1990). Primordial germ cells in the mouse embryo during gastrulation. *Development* **110**, 521–8.

Giraud-Guille, M. M. (1996). Twisted liquid crystalline supramolecular arrangements in morphogenesis. *Int. Rev. Cytol.* **166**, 59–101.

Glahn, D., and Nuccitelli, R. (2003). Voltage-clamp study of the activation currents and fast block to polyspermy in the egg of *Xenopus laevis*. *Dev. Growth Differ.* **45**, 187–97.

Glazier, J. A., and Graner, F. (1993). A simulation of the differential adhesion driven rearrangement of biological cells. *Phys. Rev. E* **47**, 2128–54.

Godt, D., and Tepass, U. (1998). *Drosophila* oocyte localization is mediated by differential cadherin-based adhesion. *Nature* **395**, 387–91.

Goldbeter, A. (1996). *Biochemical Oscillations and Cellular Rhythms: the Molecular Bases of Periodic and Chaotic Behaviour*. Cambridge University Press, Cambridge.

Gomperts, M., Wylie, C., and Heasman, J. (1994). Primordial germ cell migration. *Ciba Found. Symp.* **182**, 121–34; discussion 134–9.

Gong, Y., Mo, C., and Fraser, S. E. (2004). Planar cell polarity signalling controls cell division orientation during zebrafish gastrulation. *Nature* **430**, 689–93.

Gonzalez-Reyes, A., and St Johnston, D. (1998). The *Drosophila* AP axis is polarised by the cadherin-mediated positioning of the oocyte. *Development* **125**, 3635–44.

Gonze, D., and Goldbeter, A. (2001). A model for a network of phosphorylation–dephosphorylation cycles displaying the dynamics of dominoes and clocks. *J. Theor. Biol.* **210**, 167–86.

Gosden, R., Krapez, J., and Briggs, D. (1997). Growth and development of the mammalian oocyte. *Bioessays* **19**, 875–82.

Gossler, A., and Hrabe de Angelis, M. (1998). Somitogenesis. *Curr. Top. Dev. Biol.* **38**, 225–87.

Gould, S. E., Upholt, W. B., and Kosher, R. A. (1992). Syndecan 3: a member of the syndecan family of membrane-intercalated proteoglycans that is expressed in high amounts at the onset of chicken limb cartilage differentiation. *Proc. Nat. Acad. Sci. USA* **89**, 3271–75.

Gould, S. J. (1977). *Ontogeny and Phylogeny*. Harvard University Press, Cambridge, MA.

Goulian, M., and Simon, S. M. (2000). Tracking single proteins within cells. *Biophys. J.* **79**, 2188–98.

Graner, F., and Glazier, J. A. (1992). Simulation of biological cell sorting using a two-dimensional extended Potts model. *Phys. Rev. Lett.* **69**, 2013–16.

Green, J. (2002). Morphogen gradients, positional information, and *Xenopus*: interplay of theory and experiment. *Dev. Dyn.* **225**, 392–408.

Greenwald, I. (1998). LIN-12/Notch signaling: lessons from worms and flies. *Genes Dev.* **12**, 1751–62.

Greenwald, I., and Rubin, G. M. (1992). Making a difference: the role of cell–cell interactions in establishing separate identities for equivalent cells. *Cell* **68**, 271–81.

Gregory, P. D., Wagner, K., and Horz, W. (2001). Histone acetylation and chromatin remodeling. *Exp. Cell Res.* **265**, 195–202.

Gritsman, K., Talbot, W. S., and Schier, A. F. (2000). Nodal signaling patterns the organizer. *Development* **127**, 921–32.

Gullberg, D. E., and Lundgren-Akerlund, E. (2002). Collagen-binding I domain integrins – what do they do? *Prog. Histochem. Cytochem.* **37**, 3–54.

Gumbiner, B. M. (1996). Cell adhesion: the molecular basis of tissue architecture and morphogenesis. *Cell* **84**, 345–57.

Gumbiner, B. M. (2000). Regulation of cadherin adhesive activity. *J. Cell. Biol.* **148**, 399–404.

Gurdon, J. B. (1988). A community effect in animal development. *Nature* **336**, 772–4.

Guthrie, S., and Lumsden, A. (1991). Formation and regeneration of rhombomere boundaries in the developing chick hindbrain. *Development* **112**, 221–9.

Haddon, C., Smithers, L., Schneider-Maunoury, S., Coche, T., Henrique, D., and Lewis, J. (1998). Multiple delta genes and lateral inhibition in zebrafish primary neurogenesis. *Development* **125**, 359–70.

Hafner, M., Petzelt, C., Nobiling, R., Pawley, J. B., Kramp, D., and Schatten, G. (1988). Wave of free calcium at fertilization in the sea urchin egg visualized with fura-2. *Cell Motil. Cytoskeleton* **9**, 271–7.

Haga, H., Nagayama, M., Kawabata, K., Ito, E., Ushiki, T., and Sambongi, T. (2000). Time-lapse viscoelastic imaging of living fibroblasts using force modulation mode in AFM. *J. Electron Microsc. (Tokyo)* **49**, 473–81.

Hahn, H. S., Ortoleva, P. J., and Ross, J. (1973). Chemical oscillations and multiple steady states due to variable boundary permeability. *J. Theor. Biol.* **41**, 503–21.

Hall, B. K., and Miyake, T. (1995). Divide, accumulate, differentiate: cell condensation in skeletal development revisited. *Int. J. Dev. Biol.* **39**, 881–93.

Hall, B. K., and Miyake, T. (2000). All for one and one for all: condensations and the initiation of skeletal development. *Bioessays* **22**, 138–47.

Hall, D., and Minton, A. P. (2003). Macromolecular crowding: qualitative and semiquantitative successes, quantitative challenges. *Biochim. Biophys. Acta* **1649**, 127–39.

Hardin, J. (1996). The cellular basis of sea urchin gastrulation. *Curr. Top. Dev. Biol.* **33**, 159–262.

Harding, K., Hoey, T., Warrior, R., and Levine, M. (1989). Autoregulatory and gap gene response elements of the *even-skipped* promoter of *Drosophila*. *EMBO J.* **8**, 1205–12.

Hardman, P., and Spooner, B. S. (1992). Salivary epithelium branching morphogenesis. In *Epithelial Organization and Development*. (T. P. Fleming, ed.), pp. 353–75. Chapman and Hall, London.

Harland, R., and Gerhart, J. (1997). Formation and function of Spemann's organizer. *Ann. Rev. Cell Dev. Biol.* **13**, 611–67.

Harper, G. S., Comper, W. D., and Preston, B. N. (1984). Dissipative structures in proteoglycan solutions. *J. Biol. Chem.* **259**, 10 582–9.

Harrison, L. J. (1993). *Kinetic Theory of Living Form*. Cambridge University Press, Cambridge.

Harrisson, F. (1989). The extracellular matrix and cell surface, mediators of cell interactions in chicken gastrulation. *Int. J. Dev. Biol.* **33**, 417–38.

Hayashi, T., and Carthew, R. W. (2004). Surface mechanics mediate pattern formation in the developing retina. *Nature* **431**, 647–52.

He, X., and Dembo, M. (1997). On the mechanics of the first cleavage division of the sea urchin egg. *Exp. Cell Res.* **233**, 252–73.

Heintzelman, K. F., Phillips, H. M., and Davis, G. S. (1978). Liquid-tissue behavior and differential cohesiveness during chick limb budding. *J. Embryol. Exp. Morphol.* **47**, 1–15.

Helfrich, W. (1973). Elastic properties of lipid bilayers: theory and possible experiments. *Z. Naturforsch. C* **28**, 693–703.

Hentschel, H. G., Glimm, T., Glazier, J. A., and Newman, S. A. (2004). Dynamical mechanisms for skeletal pattern formation in the vertebrate limb. *Proc. R. Soc. Lond. B Biol. Sci.* **271**, 1713–22.

Heyman, I., Faissner, A., and Lumsden, A. (1995). Cell and matrix specialisations of rhombomere boundaries. *Dev. Dyn.* **204**, 301–15.

Hieda, Y., and Nakanishi, Y. (1997). Epithelial morphogenesis in mouse embryonic submandibular gland: its relationships to the tissue organization of epithelium and mesenchyme. *Dev. Growth Differ.* **39**, 1–8.

Hiramoto, Y. (1956). Cell division without mitotic apparatus in sea urchin eggs. *Exp. Cell Res.* **11**, 630–636.

Hiramoto, Y. (1958). A quantitative description of protoplasmic movement during cleavage in the sea-urchin egg. *J. Exp. Biol.* **35**, 407–424.

Hiramoto, Y. (1963). Mechanical properties of sea urchin eggs. I. Surface force and elastic modulus of the cell membrane. *Exp. Cell Res.* **32**, 59–75.

Hiramoto, Y. (1968). The mechanics and mechanism of cleavage in the sea-urchin egg. *Symp. Soc. Exp. Biol.* **22**, 311–27.

Hiramoto, Y. (1978). Mechanical properties of the dividing sea urchin egg. In *Cell Motility: Molecules and Organization* (S. Hatano, H. Ishikawa, and H. Sato, eds.), pp. 653–63. University Park Press, Baltimore.

Hobbie, R. K. (1997). *Intermediate Physics for Medicine and Biology.* Springer-Verlag, New York.

Hofmeyr, J. H., and Cornish-Bowden, A. (1997). The reversible Hill equation: how to incorporate cooperative enzymes into metabolic models. *Comput. Appl. Biosci.* **13**, 377–85.

Holley, S. A., Geisler, R., and Nusslein-Volhard, C. (2000). Control of *her1* expression during zebrafish somitogenesis by a Delta-dependent oscillator and an independent wave-front activity. *Genes Dev.* **14**, 1678–90.

Holley, S. A., Julich, D., Rauch, G. J., Geisler, R., and Nusslein-Volhard, C. (2002). *her1* and the notch pathway function within the oscillator mechanism that regulates zebrafish somitogenesis. *Development* **129**, 1175–83.

Holtzendorff, J., Hung, D., Brende, P., *et al.* (2004). Oscillating global regulator control the genetic circuit driving a bacterial cell cycle. *Science* **304**, 983–7.

Hörstadius, S., and Sellman, S. (1946). Experimentelle Untersuchungen uber die Determination des knorpeligen Kopfskelettes bei Urodelen. *Nova Acta R. Soc. Scient. Upsal. Ser.* 4 **13**, 1–170.

Hoshi, M. (1979). Exogastrulation induced by heavy water in sea urchin larvae. *Cell Differ.* **8**, 431–5.

Howard, J. (2001). *Mechanics of Motor Proteins and the Cytoskeleton*. Sinauer Associates, Inc., Sunderland.

Howard, K., and Ingham, P. (1986). Regulatory interactions between the segmentation genes *fushi tarazu*, *hairy*, and *engrailed* in the *Drosophila* blastoderm. *Cell* **44**, 949–57.

Huang, F. Z., Bely, A. E., and Weisblat, D. A. (2001). Stochastic WNT signaling between nonequivalent cells regulates adhesion but not fate in the two-cell leech embryo. *Curr. Biol.* **11**, 1–7.

Hunkapiller, T., and Hood, L. (1989). Diversity of the immunoglobulin gene superfamily. *Adv. Immunol.* **44**, 1–63.

Huppert, S. S., Jacobsen, T. L., and Muskavitch, M. A. (1997). Feedback regulation is central to Delta–Notch signalling required for *Drosophila* wing vein morphogenesis. *Development* **124**, 3283–91.

Hutson, M. S., Tokutake, Y., Chang, M. S., Bloor, J. W., Venakides, S., Kiehart, D. P., and Edwards, G. S. (2003). Forces for morphogenesis investigated with laser microsurgery and quantitative modeling. *Science* **300**, 145–9.

Hynes, R. O. (1987). Integrins: a family of cell surface receptors. *Cell* **48**, 349–54.

Hynes, R. O. (1992). Integrins: versatility, modulation, and signaling in cell adhesion. *Cell* **69**, 11–25.

Hynes, R. O., and Lander, A. D. (1992). Contact and adhesive specificities in the associations, migrations, and targeting of cells and axons. *Cell* **68**, 303–22.

Ingber, D. E. (1991). Extracellular matrix and cell shape: potential control points for inhibition of angiogenesis. *J. Cell. Biochem.* **47**, 236–41.

Ingber, D. E., and Folkman, J. (1989a). How does extracellular matrix control capillary morphogenesis? *Cell* **58**, 803–5.

Ingber, D. E., and Folkman, J. (1989b). Mechanochemical switching between growth and differentiation during fibroblast growth factor-stimulated angiogenesis *in vitro*: role of extracellular matrix. *J. Cell Biol.* **109**, 317–30.

Ingham, P. W. (1988). The molecular genetics of embryonic pattern formation in *Drosophila*. *Nature* **335**, 25–34.

Ingolia, N. T. (2004). Topology and robustness in the *Drosophila* segment polarity network. *PLoS Biol.* **2**, 805–15.

International Human Genome Consortium (2004). Finishing the euchromatic sequence of the human genome. *Nature* **431**, 931–45.

Irvine, K. D., and Wieschaus, E. (1994). Cell intercalation during *Drosophila* germband extension and its regulation by pair-rule segmentation genes. *Development* **120**, 827–41.

Ish-Horowicz, D., Pinchin, S. M., Ingham, P. W., and Gyurkovics, H. G. (1989). Autocatalytic *ftz* activation and instability induced by ectopic *ftz* expression. *Cell* **57**, 223–232.

Israelachvili, J. N. (1991). *Intermolecular and Surface Forces*. Academic Press, London, San Diego.

Itow, T. (1986). Inhibitors of DNA synthesis change the differentiation of body segments and increase the segment number in horseshoe crab embryos. *Roux's Arch. Dev. Biol.* **195**, 323–33.

Jacobson, A. G., Oster, G. F., Odell, G. M., and Cheng, L. Y. (1986). Neurulation and the cortical tractor model for epithelial folding. *J. Embryol. Exp. Morphol.* **96**, 19–49.

Jacobson, K. A., Sheets, E. D., and Simson, R. (1995). Revisiting the fluid mosaic model of membranes. *Science* **268**, 1441–2.

Jacobson, K. A., Moore, S. E., Yang, B., Doherty, P., Gordon, G. W., and Walsh, F. S. (1997). Cellular determinants of the lateral mobility of neural cell adhesion molecules. *Biochim. Biophys. Acta* **1330**, 138–44.

Jaffe, L. A. (1976). Fast block to polyspermy in sea urchin eggs is electrically mediated. *Nature* **261**, 68–71.

Jaffe, L. A., and Cross, N. L. (1986). Electrical regulation of sperm–egg fusion. *Ann. Rev. Physiol.* **48**, 191–200.

Jaffe, L. A., Sharp, A. P., and Wolf, D. P. (1983). Absence of an electrical polyspermy block in the mouse. *Dev. Biol.* **96**, 317–23.

Jaffe, L. A., Giusti, A. F., Carroll, D. J., and Foltz, K. R. (2001). Ca^{2+} signalling during fertilization of echinoderm eggs. *Semin. Cell Dev. Biol.* **12**, 45–51.

Jakab, K., Neagu, A., Mironov, V., Markwald, R. R., and Forgacs, G. (2004). Engineering biological structures of prescribed shape using self-assembling multicellular systems. *Proc. Nat. Acad. Sci. USA* **101**, 2864–9.

Janmey, P. (1998). The cytoskeleton and cell signaling: component localization and mechanical coupling. *Physiol. Rev.* **78**, 763–81.

Janson, L. W., and Taylor, D. L. (1993). *In vitro* models of tail contraction and cytoplasmic streaming in amoeboid cells. *J. Cell Biol.* **123**, 345–56.

Jen, W. C., Gawantka, V., Pollet, N., Niehrs, C., and Kintner, C. (1999). Periodic repression of Notch pathway genes governs the segmentation of *Xenopus* embryos. *Genes Dev.* **13**, 1486–99.

Jiang, T., Jung, H., Widelitz, R. B., and Chuong, C. (1999). Self-organization of periodic patterns by dissociated feather mesenchymal cells and the regulation of size, number and spacing of primordia. *Development* **126**, 4997–5009.

Jiang, Y., Levine, H., and Glazier, J. A. (1998). Possible cooperation of differential adhesion and chemotaxis in mound formation of *Dictyostelium*. *Biophys. J.* **75**, 2615–25.

Jiang, Y. J., Aerne, B. L., Smithers, L., Haddon, C., Ish-Horowicz, D., and Lewis, J. (2000). Notch signalling and the synchronization of the somite segmentation clock. *Nature* **408**, 475–9.

Jimenez, G., Griffiths, S. D., Ford, A. M., Greaves, M. F., and Enver, T. (1992). Activation of the beta-globin locus control region precedes commitment to the erythroid lineage. *Proc. Nat. Acad. Sci. USA* **89**, 10618–22.

Jones, K. T. (1998). Protein kinase C action at fertilization: overstated or undervalued? *Rev. Reprod.* **3**, 7–12.

Joos, T. O., Whittaker, C. A., Meng, F., DeSimone, D. W., Gnau, V., and Hausen, P. (1995). Integrin alpha 5 during early development of *Xenopus laevis*. *Mech. Dev.* **50**, 187–99.

Jouve, C., Palmeirim, I., Henrique, D., *et al.* (2000). Notch signalling is required for cyclic expression of the *hairy*-like gene HES1 in the presomitic mesoderm. *Development* **127**, 1421–9.

Jouve, C., Iimura, T., and Pourquié, O. (2002). Onset of the segmentation clock in the chick embryo: evidence for oscillations in the somite precursors in the primitive streak. *Development* **129**, 1107–17.

Juan, H., and Hamada, H. (2001). Roles of nodal–lefty regulatory loops in embryonic patterning of vertebrates. *Genes Cells* **6**, 923–30.

Kabata, H., Kurosawa, O., Arai, I., *et al.* (1993). Visualization of single molecules of RNA polymerase sliding along DNA. *Science* **262**, 1561–3.

Kadler, K. E., Holmes, D. F., Trotter, J. A., and Chapman, J. A. (1996). Collagen fibril formation. *Biochem. J.* **316**, 1–11.

Kalodimos, C. G., Biris, N., Bonvin, A. M., *et al.* (2004). Structure and flexibility adaptation in nonspecific and specific protein–DNA complexes. *Science* **305**, 386–9.

Kamawaki, Y., Raya, A., Raya, R. M., Rodriquez-Esteban, C., and Belmont, J. C. I. (2005). Retinoic acid signalling links, left–right asymmetric patterning and bilaterally symmetric somitogenesis in the zebrafish embryo. *Nature* **435**, 165–71.

Kamei, N., Swanson, W. J., and Glabe, C. G. (2000). A rapidly diverging EGF protein regulates species-specific signal transduction in early sea urchin development. *Dev. Biol.* **225**, 267–76.

Kaneko, K. (2003). Organization through intra–inter dynamics. In *Origination of Organismal Form: Beyond the Gene in Developmental and Evolutionary Biology* (G. B. Müller and S. A. Newman, eds.), pp. 195–220. MIT Press, Cambridge, MA.

Kaneko, K., and Yomo, T. (1994). Cell division, differentiation and dynamic clustering. *Physica D* **75**, 89–102.

Kaneko, K., and Yomo, T. (1997). Isologous diversification: a theory of cell differentiation. *Bull. Math. Biol.* **59**, 139–96.

Kaneko, K., and Yomo, T. (1999). Isologous diversification for robust development of cell society. *J. Theor. Biol.* **199**, 243–56.

Karr, T. L., Weir, M. P., Ali, Z., and Kornberg, T. (1989). Patterns of engrailed protein in early *Drosophila* embryos. *Development* **105**, 605–612.

Kawaki, Y., Raya, A., Raya, R. M., Rodriguez-Esteban, C., and Belmonte, J. C. I. (2005). Retinoic acid signalling links left–right asymmetric patterning and bilaterally symmetric somitogenesis in the zebrafish embryo. *Nature* **435**, 165–71.

Keller, A. D. (1995). Model genetic circuits encoding autoregulatory transcription factors. *J. Theor. Biol.* **172**, 169–85.

Keller, R. (2000). The origin and morphogenesis of amphibian somites. *Curr. Top. Dev. Biol.* **47**, 183–246.

Keller, R. (2002). Shaping the vertebrate body plan by polarized embryonic cell movements. *Science* **298**, 1950–4.

Keller, R., and Danilchik, M. (1988). Regional expression, pattern and timing of convergence and extension during gastrulation of *Xenopus laevis*. *Development* **103**, 193–209.

Keller, R. E., Danilchik, M., Gimlich, R., and Shih, J. (1985). The function and mechanism of convergent extension during gastrulation of *Xenopus laevis*. *J. Embryol. Exp. Morphol.* **89** Suppl., 185–209.

Keller, R., Cooper, M. S., Danilchik, M., Tibbetts, P., and Wilson, P. A. (1989). Cell intercalation during notochord development in *Xenopus laevis*. *J. Exp. Zool.* **251**, 134–54.

Keller, R., Shih, J., and Sater, A. (1992a). The cellular basis of the convergence and extension of the *Xenopus* neural plate. *Dev. Dyn.* **193**, 199–217.

Keller, R., Shih, J., Sater, A. K., and Moreno, C. (1992b). Planar induction of convergence and extension of the neural plate by the organizer of *Xenopus*. *Dev. Dyn.* **193**, 218–34.

Keller, R., Davidson, L., Edlund, A. *et al.* (2000). Mechanisms of convergence and extension by cell intercalation. *Philos. Trans. R. Soc. Lond. B Biol. Sci.* **355**, 897–922.

Kerszberg, M., and Changeux, J. P. (1998). A simple molecular model of neurulation. *Bioessays* **20**, 758–70.

Kerszberg, M., and Wolpert, L. (1998). Mechanisms for positional signalling by morphogen transport: a theoretical study. *J. Theor. Biol.* **191**, 103–14.

Kimmins, S., and Sassone-Corsi, P. (2005). Chromatin remodeling and epigenetic features of germ cells. *Nature* **434**, 583–9.

Kiskowski, M. A., Alber, M. S., Thomas, G. L., *et al.* (2004). Interplay between activator–inhibitor coupling and cell–matrix adhesion in a cellular automaton model for chondrogenic patterning. *Dev. Biol.* **271**, 372–87.

Klein, C., and Hurlbut, C. S. (2002). *Manual of Mineral Science*. Wiley, New York.

Knoll, A. H. (2003). *Life on a Young Planet: the First Three Billion Years of Evolution on Earth*. Princeton University Press, Princeton, NJ.

Kofron, M., Spagnuolo, A., Klymkowsky, M., Wylie, C., and Heasman, J. (1997). The roles of maternal alpha-catenin and plakoglobin in the early *Xenopus* embryo. *Development* **124**, 1553–60.

Kofron, M., Heasman, J., Lang, S. A., and Wylie, C. C. (2002). Plakoglobin is required for maintenance of the cortical actin skeleton in early *Xenopus* embryos and for cdc42-mediated wound healing. *J. Cell Biol.* **158**, 695–708.

Kohn, K. W. (1999). Molecular interaction map of the mammalian cell cycle control and DNA repair systems. *Mol. Biol. Cell.* **10**, 2703–34.

Kondo, S., and Asai, R. (1995). A reaction–diffusion wave on the skin of the marine angelfish *Pomacanthus*. *Nature* **376**, 765–8.

Kosher, R. A., Savage, M. P., and Chan, S. C. (1979). *In vitro* studies on the morphogenesis and differentiation of the mesoderm subjacent to the apical ectodermal ridge of the embryonic chick limb-bud. *J. Embryol. Exp. Morphol.* **50**, 75–97.

Kosher, R. A., Walker, K. H., and Ledger, P. W. (1982). Temporal and spatial distribution of fibronectin during development of the embryonic chick limb bud. *Cell. Differ.* **11**, 217–228.

Kruse, K., Pantazis, P., Bollenbach, T., Julicher, F., and Gonzalez-Gaitan, M. (2004). Dpp gradient formation by dynamin-dependent endocytosis: receptor trafficking and the diffusion model. *Development* **131**, 4843–56.

Kubota, H. Y., Yoshimoto, Y., Yoneda, M., and Hiramoto, Y. (1987). Free calcium wave upon activation in *Xenopus* eggs. *Dev. Biol.* **119**, 129–36.

Kulesa, P. M., and Fraser, S. E. (2000). *In ovo* time-lapse analysis of chick hindbrain neural crest cell migration shows cell interactions during migration to the branchial arches. *Development* **127**, 1161–72.

Kulesa, P. M., and Fraser, S. E. (2002). Cell dynamics during somite boundary formation revealed by time-lapse analysis. *Science* **298**, 991–5.

Kumano, G., and Smith, W. C. (2002). Revisions to the *Xenopus* gastrula fate map: implications for mesoderm induction and patterning. *Dev. Dyn.* **225**, 409–21.

Lallier, T. E., Whittaker, C. A., and DeSimone, D. W. (1996). Integrin alpha 6 expression is required for early nervous system development in *Xenopus laevis*. *Development* **122**, 2539–54.

Lander, A. D., Nie, Q., and Wan, F. Y. (2002). Do morphogen gradients arise by diffusion? *Dev. Cell* **2**, 785–96.

Lane, M. C., and Sheets, M. D. (2000). Designation of the anterior/posterior axis in pregastrula *Xenopus laevis*. *Dev. Biol.* **225**, 37–58.

Lane, M. C., and Sheets, M. D. (2002). Rethinking axial patterning in amphibians. *Dev. Dyn.* **225**, 434–47.

Lane, M. C., and Smith, W. C. (1999). The origins of primitive blood in *Xenopus*: implications for axial patterning. *Development* **126**, 423–34.

Lane, M. C., Koehl, M. A., Wilt, F., and Keller, R. (1993). A role for regulated secretion of apical extracellular matrix during epithelial invagination in the sea urchin. *Development* **117**, 1049–60.

Langille, R. M., and Hall, B. K. (1993). Pattern formation and the neural crest. In *The Vertebrate Skull* (J. Hanken and B. K. Hall, eds.), Vol. 1, *Development*, pp. 77–111. University of Chicago Press, Chicago.

Langman, J. (1981). *Medical Embryology*. Williams & Wilkins, Baltimore.

Lawrence, P. A. (1992). *The Making of a Fly: the Genetics of Animal Design*. Blackwell Scientific Publications, Oxford, Boston.

Lawson, K. A. (1983). Stage specificity in the mesenchyme requirement of rodent lung epithelium *in vitro:* a matter of growth control? *J. Embryol. Exp. Morphol.* **74**, 183–206.

Lechleiter, J., Girard, S., Clapham, D., and Peralta, E. (1991). Subcellular patterns of calcium release determined by G protein-specific residues of muscarinic receptors. *Nature* **350**, 505–8.

Lechleiter, J. D., John, L. M., and Camacho, P. (1998). Ca^{2+} wave dispersion and spiral wave entrainment in *Xenopus laevis* oocytes overexpressing Ca^{2+} ATPases. *Biophys. Chem.* **72**, 123–9.

Leal, L. G. (1992). *Laminar Flow and Convective Processes: Scaling Principles and Asymptotic Analysis*. Butterworth-Heinemann, Boston.

Lengyel, I., and Epstein, I. R. (1991). Diffusion-induced instability in chemically reacting systems: steady-state multiplicity, oscillation, and chaos. *Chaos* **1**, 69–76.

Lengyel, I., and Epstein, I. R. (1992). A chemical approach to designing Turing patterns in reaction–diffusion systems. *Proc. Nat. Acad. Sci. USA* **89**, 3977–9.

Leonard, C. M., Fuld, H. M., Frenz, D. A., Downie, S. A., Massague, J., and Newman, S. A. (1991). Role of transforming growth factor-β in chondrogenic pattern formation in the embryonic limb: stimulation of mesenchymal condensation and fibronectin gene expression by exogenous

TGF-beta and evidence for endogenous TGF-β-like activity. *Dev. Biol.* **145**, 99–109.

Lercher, M. J., Urrutia, A. O., and Hurst, L. D. (2002). Clustering of housekeeping genes provides a unified model of gene order in the human genome. *Nat. Genet.* **31**, 180–3.

Leslie, P. H. (1948). Some further notes on the use of matrices in population mathematics. *Biometrika* **35**, 213–45.

Levi, G., Ginsberg, D., Girault, J. M., Sabanay, I., Thiery, J. P., and Geiger, B. (1991). EP-cadherin in muscles and epithelia of *Xenopus laevis* embryos. *Development* **113**, 1335–44.

Lewis, J. (2003). Autoinhibition with transcriptional delay: a simple mechanism for the zebrafish somitogenesis oscillator. *Curr. Biol.* **13**, 1398–408.

Li, S., Piotrowicz, R. S., Levin, E. G., Shyy, Y. J., and Chien, S. (1996). Fluid shear stress induces the phosphorylation of small heat shock proteins in vascular endothelial cells. *Am. J. Physiol.* **271**, C994–1000.

Li, S., Zhou, D., Lu, M. M., and Morrisey, E. E. (2004). Advanced cardiac morphogenesis does not require heart tube fusion. *Science* **305**, 1619–22.

Linsenmayer, T. F., Fitch, J. M., Gordon, M. K., Cai, C. X., Igoe, F., Marchant, J. K., and Birk, D. E. (1998). Development and roles of collagenous matrices in the embryonic avian cornea. *Prog. Retin. Eye Res.* **17**, 231–65.

Lowery, L. A., and Sive, H. (2004). Strategies of vertebrate neurulation and a re-evaluation of teleost neural tube formation. *Mech. Dev.* **121**, 1189–97.

Lubarsky, B., and Krasnow, M. A. (2003). Tube morphogenesis: making and shaping biological tubes. *Cell* **112**, 19–28.

Lubkin, S. R., and Li, Z. (2002). Force and deformation on branching rudiments: cleaving between hypotheses. *Biomech. Model Mechanobiol.* **1**, 5–16.

Lucchetta, E. M., Lee, J. H., Fu, L. A., Patel, N. H., and Ismagilov, R. F. (2005). Dynamics of *Drosophila* embryonic patterning network perturbed in space and time using microfluidics. *Nature* **434**, 1134–8.

Luo, Y., Kostetskii, I., and Radiche, G. L. (2005). N-cadherin is not essential for limb mesenchymal chondrogenesis. *Dev. Dyn.* **232**, 336–44.

Mandato, C. A., Benink, H. A., and Bement, W. M. (2000). Microtubule-actomyosin interactions in cortical flow and cytokinesis. *Cell Motil. Cytoskeleton* **45**, 87–92.

Mandelbrot, B. B. (1983). *The Fractal Geometry of Nature*. W. H. Freeman, New York.

Maniatis, T., and Tasic, B. (2002). Alternative pre-mRNA splicing and proteome expansion in metazoans. *Nature* **418**, 236–43.

Manner, J. (2000). Cardiac looping in the chick embryo: a morphological review with special reference to terminological and biomechanical aspects of the looping process. *Anat. Rec.* **259**, 248–62.

Mannervik, M., Nibu, Y., Zhang, H., and Levine, M. (1999). Transcriptional coregulators in development. *Science* **284**, 606–9.

Marom, K., Shapira, E., and Fainsod, A. (1997). The chicken caudal genes establish an anterior–posterior gradient by partially overlapping temporal and spatial patterns of expression. *Mech. Dev.* **64**, 41–52.

Marsden, M., and DeSimone, D. W. (2003). Integrin-ECM interactions regulate cadherin-dependent cell adhesion and are required for convergent extension in *Xenopus*. *Curr. Biol.* **13**, 1182–91.

Marshall, B. T., Long, M., Piper, J. W., Yago, T., McEver, R. P., and Zhu, C. (2003). Direct observation of catch bonds involving cell-adhesion molecules. *Nature* **423**, 190–3.

Martin, G. R. (1998). The roles of FGFs in the early development of vertebrate limbs. *Genes Dev.* **12**, 1571–86.

Martin, J. E., Adolf, D., and Wilcoxon, J. P. (1989). Rheology of the incipient gel: theory and data for epoxies. *Polym. Prepr. Am. Chem. Soc. Div. Polym. Chem.* **30**, 83–84.

Martin, V. J., Littlefield, C. L., Archer, W. E., and Bode, H. R. (1997). Embryogenesis in hydra. *Biol. Bull.* **192**, 345–63.

Maynard Smith, J. (1978). *Models in Ecology*. Cambridge University Press, Cambridge, New York.

Mayr, E. (1982). *The Growth of Biological Thought: Diversity, Evolution, and Inheritance*. Belknap Press, Cambridge, MA.

McCarthy, R. A., and Hay, E. D. (1991). Collagen I, laminin, and tenascin: ultrastructure and correlation with avian neural crest formation. *Int. J. Dev. Biol.* **35**, 437–52.

McDougall, A., Shearer, J., and Whitaker, M. (2000). The initiation and propagation of the fertilization wave in sea urchin eggs. *Biol. Cell.* **92**, 205–14.

McDowell, N., Gurdon, J. B., and Grainger, D. J. (2001). Formation of a functional morphogen gradient by a passive process in tissue from the early *Xenopus* embryo. *Int. J. Dev. Biol.* **45**, 199–207.

McKim, K. S., Jang, J. K., and Manheim, E. A. (2002). Meiotic recombination and chromosome segregation in *Drosophila* females. *Ann. Rev. Genet.* **36**, 205–32.

McLaren, A. (1984). Meiosis and differentiation of mouse germ cells. *Symp. Soc. Exp. Biol.* **38**, 7–23.

Medalia, O., Weber, I., Frangakis, A. S., Nicastro, D., Gerisch, G., and Baumeister, W. (2002). Macromolecular architecture in eukaryotic cells visualized by cryoelectron tomography. *Science* **298**, 1209–13.

Meek, K. M., and Fullwood, N. J. (2001). Corneal and scleral collagens – a microscopist's perspective. *Micron* **32**, 261–72.

Meier, S. (1984). Somite formation and its relationship to metameric patterning of the mesoderm. *Cell Differ.* **14**, 235–43.

Meinhardt, H. (1982). *Models of Biological Pattern Formation*. Academic Press, New York.

Meinhardt, H. (2001). Organizer and axes formation as a self-organizing process. *Int. J. Dev. Biol.* **45**, 177–88.

Meinhardt, H., and Gierer, A. (2000). Pattern formation by local self-activation and lateral inhibition. *Bioessays* **22**, 753–60.

Meir, E., von Dassow, G., Munro, E., and Odell, G. M. (2002). Robustness, flexibility, and the role of lateral inhibition in the neurogenic network. *Curr. Biol.* **12**, 778–86.

Melnick, M., and Jaskoll, T. (2000). Mouse submandibular gland morphogenesis: a paradigm for embryonic signal processing. *Crit. Rev. Oral Biology and Medicine* **11**, 199–215.

Merks, R. M. H., Newman, S. A., and Glazier, J. A. (2004). Cell-oriented modeling of *in vitro* capillary development. In *Cellular Automata: Proc. 6th International Conf. on Cellular Automata for Research and Industry* (P. M. A. Sloot, B. Chopard, and A. G. Hoekstra, eds.), pp. 425–34. Springer-Verlag, Amsterdam, The Netherlands.

Metropolis, N., Rosenbluth, M. N., Rosenbluth, A., Teller, H., and Teller, E. (1953). Equations of state calculations by fast computing machines. *J. Chem. Phys.* **21**, 1087–91.

Minelli, A. (2003). *The Development of Animal Form: Ontogeny, Morphology, and Evolution*. Cambridge University Press, Cambridge, New York.

Minelli, A., and Fusco, G. (2004). Evo-devo perspectives on segmentation: model organisms, and beyond. *Trends Ecol. Evol.* **19**, 423–9.

Miranti, C. K., and Brugge, J. S. (2002). Sensing the environment: a historical perspective on integrin signal transduction. *Nat. Cell. Biol.* **4**, E83–E90.

Misteli, T. (2001). Protein dynamics: implications for nuclear architecture and gene expression. *Science* **291**, 843–7.

Mittenthal, J. E., and Mazo, R. M. (1983). A model for shape generation by strain and cell–cell adhesion in the epithelium of an arthropod leg segment. *J. Theor. Biol.* **100**, 443–83.

Miura, T., and Maini, P. K. (2004). Speed of pattern appearance in reaction–diffusion models: implications in the pattern formation of limb bud mesenchyme cells. *Bull. Math. Biol.* **66**, 627–49.

Miura, T., and Shiota, K. (2000a). Extracellular matrix environment influences chondrogenic pattern formation in limb bud micromass culture: experimental verification of theoretical models. *Anat. Rec.* **258**, 100–7.

Miura, T., and Shiota, K. (2000b). TGFβ2 acts as an "activator" molecule in reaction–diffusion model and is involved in cell sorting phenomenon in mouse limb micromass culture. *Dev. Dyn.* **217**, 241–9.

Miura, T., and Shiota, K. (2000c). Time-lapse observation of branching morphogenesis of the lung bud epithelium in mesenchyme-free culture and its relationship with the localization of actin filaments. *Int. J. Dev. Biol.* **44**, 899–902.

Miura, T., Komori, M., and Shiota, K. (2000). A novel method for analysis of the periodicity of chondrogenic patterns in limb bud cell culture: correlation of *in vitro* pattern formation with theoretical models. *Anat. Embryol. (Berlin)* **201**, 419–28.

Miyazaki, S., Shirakawa, H., Nakada, K., and Honda, Y. (1993). Essential role of the inositol 1,4,5-trisphosphate receptor/Ca^{2+} release channel in Ca^{2+} waves and Ca^{2+} oscillations at fertilization of mammalian eggs. *Dev. Biol.* **158**, 62–78.

Mlodzik, M. (2002). Planar cell polarization: do the same mechanisms regulate *Drosophila* tissue polarity and vertebrate gastrulation? *Trends Genet.* **18**, 564–71.

Moftah, M. Z., Downie, S. A., Bronstein, N. B., Mezentseva, N., Pu, J., Maher, P. A., and Newman, S. A. (2002). Ectodermal FGFs induce perinodular inhibition of limb chondrogenesis *in vitro* and *in vivo* via FGF receptor 2. *Dev. Biol.* **249**, 270–82.

Mombach, J. C., Glazier, J. A., Raphael, R. C., and Zajac, M. (1995). Quantitative comparison between differential adhesion models and cell sorting in the presence and absence of fluctuations. *Phys. Rev. Lett.* **75**, 2244–7.

Monk, N. A. (2003). Oscillatory expression of *Hes1*, *p53*, and *NF-kappaB* driven by transcriptional time delays. *Curr. Biol.* **13**, 1409–13.

Montalta-He, H., and Reichert, H. (2003). Impressive expressions: developing a systematic database of gene-expression patterns in *Drosophila* embryogenesis. *Genome Biol.* **4**, 205.

Montero, J. A., and Heisenberg, C. P. (2003). Adhesive crosstalk in gastrulation. *Dev. Cell* **5**, 190–1.

Morisco, C., Seta, K., Hardt, S. E., Lee, Y., Vatner, S. F., and Sadoshima, J. (2001). Glycogen synthase kinase 3β regulates GATA4 in cardiac myocytes. *J. Biol. Chem.* **276**, 28 586–97.

Morrison, S. J., Perez, S. E., Qiao, Z. *et al.* (2000). Transient Notch activation initiates an irreversible switch from neurogenesis to gliogenesis by neural crest stem cells. *Cell* **101**, 499–510.

Muratov, C. B. (1997). Synchronization, chaos, and the breakdown of the collective domain oscillations in reaction–diffusion systems. *Phys. Rev. E* **55**, 1463–77.

Murray, A. W., and Hunt, T. (1993). *The Cell Cycle: An Introduction*. W. H. Freeman, New York.

Murray, A. W., and Kirschner, M. W. (1989). Dominoes and clocks: the union of two views of the cell cycle. *Science* **246**, 614–21.

Murray, J. D. (2002). *Mathematical biology*. Springer, New York.

Nagafuchi, A., and Takeichi, M. (1988). Cell binding function of E-cadherin is regulated by the cytoplasmic domain. *EMBO J.* **7**, 3679–84.

Nagar, B., Overduin, M., Ikura, M., and Rini, J. M. (1996). Structural basis of calcium-induced E-cadherin rigidification and dimerization. *Nature* **380**, 360–4.

Nagata, W., Harrison, L. G., and Wehner, S. (2003). Reaction–diffusion models of growing plant tips: bifurcations on hemispheres. *Bull. Math. Biol.* **65**, 571–607.

Nakanishi, Y., Morita, T., and Nogawa, H. (1987). Cell proliferation is not required for the initiation of early cleft formation in mouse embryonic submandibular epithelium *in vitro*. *Development* **99**, 429–37.

Nakatsuji, N., Snow, M. H., and Wylie, C. C. (1986). Cinemicrographic study of the cell movement in the primitive-streak-stage mouse embryo. *J. Embryol. Exp. Morphol.* **96**, 99–109.

Nakayama, T., Yakubo, K., and Orbach, R. (1994). Dynamical properties of fractal networks: scaling, numerical simulations and physical realizations. *Rev. Mod. Phys.* **66**, 381–443.

Nanjundiah, V. (2005). Mathematics and biology. *Current Science* **88**, 388–93.

Narbonne, G. M. (2004). Modular construction of early Ediacaran complex life forms. *Science* **305**, 1141–4.

Needham, D., and Hochmuth, R. M. (1992). A sensitive measure of surface stress in the resting neutrophil. *Biophys. J.* **61**, 1664–70.

Neff, A. W., Malacinski, G. M., Wakahara, M., and Jurand, A. (1983). Pattern formation in amphibian embryos prevented from undergoing the classical "rotation response" to egg activation. *Dev. Biol.* **97**, 103–12.

Neff, A. W., Wakahara, M., Jurand, A., and Malacinski, G. M. (1984). Experimental analyses of cytoplasmic rearrangements which follow fertilization and accompany symmetrization of inverted *Xenopus* eggs. *J. Embryol. Exp. Morphol.* **80**, 197–224.

Newgreen, D. F. (1989). Physical influences on neural crest cell migration in avian embryos: contact guidance and spatial restriction. *Dev. Biol.* **131**, 136–48.

Newgreen, D. F., and Minichiello, J. (1995). Control of epitheliomesenchymal transformation. I. Events in the onset of neural crest cell migration are separable and inducible by protein kinase inhibitors. *Dev. Biol.* **170**, 91–101.

Newman, S. A. (1977). Lineage and pattern in the developing wing bud. In *Vertebrate Limb and Somite Morphogenesis* (D. A. Ede, J. R. Hinchliffe, and M. Balls, eds.), pp. 181–97. Cambridge University Press, Cambridge.

Newman, S. A. (1988). Lineage and pattern in the developing vertebrate limb. *Trends Genet.* **4**, 329–32.

Newman, S. A. (1993). Is segmentation generic? *BioEssays* **15**, 277–83.

Newman, S. A. (1994). Generic physical mechanisms of tissue morphogenesis: a common basis for development and evolution. *J. Evol. Biol.* **7**, 467–88.

Newman, S. A. (1995). Interplay of genetics and physical processes of tissue morphogenesis in development and evolution: the biological fifth dimension. In "*Interplay of Genetic and Physical Processes in the Development of Biological Form* (D. Beysens, G. Forgacs, and F. Gaill, eds.), pp. 3–12. World Scientific, Singapore.

Newman, S. A. (1998a). Epithelial morphogenesis: a physico-evolutionary interpretation. In *Molecular Basis of Epithelial Appendage Morphogenesis* (C.-M. Chuong, ed.), pp. 341–58. R. G. Landes, Austin, TX.

Newman, S. A. (1998b). Networks of extracellular fibers and the generation of morphogenetic forces. In *Dynamical Networks in Physics and Biology* (D. Beysens and G. Forgacs, eds.), pp. 139–48. Springer-Verlag, Berlin.

Newman, S. A. (2003a). The fall and rise of systems biology. *GeneWatch* **16**, 8–12.

Newman, S. A. (2003b). From physics to development: the evolution of morphogenetic mechanisms. In *Origination of Organismal Form: Beyond the Gene in Developmental and Evolutionary Biology*. (G. B. Müller and S. A. Newman, eds.), pp. 221–39. MIT Press, Cambridge, MA.

Newman, S. A., and Comper, W. D. (1990). "Generic" physical mechanisms of morphogenesis and pattern formation. *Development* **110**, 1–18.

Newman, S. A., and Frisch, H. L. (1979). Dynamics of skeletal pattern formation in developing chick limb. *Science* **205**, 662–8.

Newman, S. A., and Müller, G. B. (2000). Epigenetic mechanisms of character origination. *J. Exp. Zool. (Mol. Evol. Dev.)* **288**, 304–17.

Newman, S. A., and Müller, G. B. (eds.) (2003). *Origination of Organismal Form: Beyond the Gene in Developmental and Evolutionary Biology*. MIT Press, Cambridge, MA.

Newman, S. A., and Tomasek, J. J. (1996). Morphogenesis of connective tissues. In *Extracellular Matrix* (W. D. Comper, ed.), Vol. 2, *Molecular Components and Interactions*, pp. 335–69. Harwood Academic Publishers, Amsterdam.

Newman, S. A., Frisch, H. L., Perle, M. A., and Tomasek, J. J. (1981). Limb development: aspects of differentiation, pattern formation and morphogenesis. In *Morphogenesis and Pattern Formation* (T. G. Connolly, L. L. Brinkley, and B. M. Carlson, eds.), pp. 163–78. Raven Press, New York.

Newman, S. A., Frenz, D. A., Tomasek, J. J., and Rabuzzi, D. D. (1985). Matrix-driven translocation of cells and nonliving particles. *Science* **228**, 885–9.

Newman, S. A., Frenz, D. A., Hasegawa, E., and Akiyama, S. K. (1987). Matrix-driven translocation: dependence on interaction of amino-terminal domain of fibronectin with heparin-like surface components of cells or particles. *Proc. Nat. Acad. Sci. USA* **84**, 4791–5.

Newman, S. A., Frisch, H. L., and Percus, J. K. (1988). On the stationary state analysis of reaction–diffusion mechanisms for biological pattern formation *J. Theor. Biol.* **134**, 183–197 (published erratum appears in *J. Theor. Biol.* **135**, 137 (1988)).

Newman, S. A., Cloitre, M., Allain, C., Forgacs, G., and Beysens, D. (1997). Viscosity and elasticity during collagen assembly *in vitro*: relevance to matrix-driven translocation. *Biopolymers* **41**, 337–47.

Newman, S. A., Forgacs, G., Hinner, B., Maier, C. W., and Sackmann, E. (2004). Phase transformations in a model mesenchymal tissue. *Phys. Biol.* **1**, 100–9.

Newport, J., and Kirschner, M. (1982a). A major developmental transition in early *Xenopus* embryos: I. Characterization and timing of cellular changes at the midblastula stage. *Cell* **30**, 675–86.

Newport, J., and Kirschner, M. (1982b). A major developmental transition in early *Xenopus* embryos: II. Control of the onset of transcription. *Cell* **30**, 687–96.

Nicklas, R. B., and Koch, C. A. (1969). Chromosome micromanipulation. 3. Spindle fiber tension and the reorientation of mal-oriented chromosomes. *J. Cell Biol.* **43**, 40–50.

Nieuwkoop, P. D. (1969). The formation of mesoderm in *Urodelean* amphibians. I. Induction by the endoderm. *Wilhelm Roux' Arch. Entw. Mech. Org.* **162**, 341–73.

Nieuwkoop, P. D. (1973). The "organization center" of the amphibian embryo: its origin, spatial organization and morphogenetic action. *Adv. Morphogen.* **10**, 1–39.

Nieuwkoop, P. D. (1992). The formation of the mesoderm in *Urodelean* amphibians VI. The self-organizing capacity of the induced meso-endoderm. *Roux's Arch. Dev. Biol.* **201**, 18–29.

Noden, D. M. (1984). Craniofacial development: new views on old problems. *Anat. Rec.* **208**, 1–13.

Noden, D. M. (1988). Interactions and fates of avian craniofacial mesenchyme. *Development Suppl.* **103**, 121–40.

Nogawa, H., and Ito, T. (1995). Branching morphogenesis of embryonic mouse lung epithelium in mesenchyme-free culture. *Development* **121**, 1015–22.

Nogawa, H., and Takahashi, Y. (1991). Substitution for mesenchyme by basement-membrane-like substratum and epidermal growth factor in inducing branching morphogenesis of mouse salivary epithelium. *Development* **112**, 855–61.

Nonaka, S., Tanaka, Y., Okada, Y., *et al.* (1998). Randomization of left–right asymmetry due to loss of nodal cilia generating leftward flow of extraembryonic fluid in mice lacking KIF3B motor protein. *Cell* **95**, 829–37.

Nonaka, S., Shiratori, H., Saijoh, Y., and Hamada, H. (2002). Determination of left–right patterning of the mouse embryo by artificial nodal flow. *Nature* **418**, 96–9.

Norel, R., and Agur, Z. (1991). A model for the adjustment of the mitotic clock by cyclin and MPF levels. *Science* **251**, 1076–8.

Novak, B., and Tyson, J. J. (1993). Numerical analysis of a comprehensive model of M-phase control in *Xenopus* oocyte extracts and intact embryos. *J. Cell Sci.* **106**, 1153–68.

Novak, B., Csikasz-Nagy, A., Gyorffy, B., Nasmyth, K., and Tyson, J. J. (1998). Model scenarios for evolution of the eukaryotic cell cycle. *Philos. Trans. R. Soc. Lond. B Biol. Sci.* **353**, 2063–76.

Novak, B., Toth, A., Csikasz-Nagy, A., Gyorffy, B., Tyson, J. J., and Nasmyth, K. (1999). Finishing the cell cycle. *J. Theor. Biol.* **199**, 223–33.

Oates, A. C., and Ho, R. K. (2002). *Hairy/E(spl)*-related (*Her*) genes are central components of the segmentation oscillator and display redundancy with the Delta/Notch signaling pathway in the formation of anterior segmental boundaries in the zebrafish. *Development* **129**, 2929–46.

Oberlender, S. A., and Tuan, R. S. (1994). Expression and functional involvement of N-cadherin in embryonic limb chondrogenesis. *Development* **120**, 177–87.

Oda, S., Deguchi, R., Mohri, T., Shikano, T., Nakanishi, S., and Miyazaki, S. (1999). Spatiotemporal dynamics of the $[Ca^{2+}]i$ rise induced by microinjection of sperm extract into mouse eggs: preferential induction of a Ca^{2+} wave from the cortex mediated by the inositol 1,4,5-trisphosphate receptor. *Dev. Biol.* **209**, 172–85.

Odell, G. M., Oster, G., Alberch, P., and Burnside, B. (1981). The mechanical basis of morphogenesis. I. Epithelial folding and invagination. *Dev. Biol.* **85**, 446–62.

Oheim, M., and Stuhmer, W. (2000). Tracking chromaffin granules on their way through the actin cortex. *Eur. Biophys. J.* **29**, 67–89.

Oliveri, P., and Davidson, E. H. (2004). Gene regulatory network controlling embryonic specification in the sea urchin. *Curr. Opin. Genet. Dev.* **14**, 351–60.

Oliveri, P., Carrick, D. M., and Davidson, E. H. (2002). A regulatory gene network that directs micromere specification in the sea urchin embryo. *Dev. Biol.* **246**, 209–28.

Opas, M., Davies, J. R., Zhou, Y., and Dziak, E. (2001). Formation of retinal pigment epithelium *in vitro* by transdifferentiation of neural retina cells. *Int. J. Dev. Biol.* **45**, 633–42.

Ornitz, D. M., and Marie, P. J. (2002). FGF signaling pathways in endochondral and intramembranous bone development and human genetic disease. *Genes Dev.* **16**, 1446–65.

Ouyang, Q., and Swinney, H. (1991). Transition from a uniform state to hexagonal and striped Turing patterns. *Nature* **352**, 610–12.

Owen, M. R., and Sherratt, J. A. (1998). Mathematical modelling of juxtacrine cell signalling. *Math. Biosci.* **153**, 125–50.

Owen, M. R., Sherratt, J. A., and Myers, S. R. (1999). How far can a juxtacrine signal travel? *Proc. R. Soc. Lond. B Biol. Sci.* **266**, 579–85.

Owen, M. R., Sherratt, J. A., and Wearing, H. J. (2000). Lateral induction by juxtacrine signaling is a new mechanism for pattern formation. *Dev. Biol.* **217**, 54–61.

Ozdamar, B., Bose, R., Barrios-Rodiles, M., Wang, H.-R., Zhang, Y., and Wrana, J. L. (2005). Regulation of the polarity protein Par6 by TGFβ receptors controls epithelial cell plasticity. *Science* **307**, 1603–9.

Pagan-Westphal, S. M., and Tabin, C. J. (1998). The transfer of left–right positional information during chick embryogenesis. *Cell* **93**, 25–35.

Palmeirim, I., Henrique, D., Ish-Horowicz, D., and Pourquié, O. (1997). Avian *hairy* gene expression identifies a molecular clock linked to vertebrate segmentation and somitogenesis. *Cell* ***91***, 639–48.

Pardanaud, L., and Dieterlen-Lievre, F. (2000). Ontogeny of the endothelial system in the avian model. *Adv. Exp. Med. Biol.* **476**, 67–78.

Pardanaud, L., Luton, D., Prigent, M., Bourcheix, L. M., Catala, M., and Dieterlen-Lievre, F. (1996). Two distinct endothelial lineages in ontogeny, one of them related to hemopoiesis. *Development* **122**, 1363–71.

Parkinson, J., Kadler, K. E., and Brass, A. (1995). Simple physical model of collagen fibrillogenesis based on diffusion limited aggregation. *J. Mol. Biol.* **247**, 823–31.

Patan, S. (2000). Vasculogenesis and angiogenesis as mechanisms of vascular network formation, growth and remodeling. *J. Neurooncol.* **50**, 1–15.

Patel, N. H. (1994). Developmental evolution: insights from studies of insect segmentation. *Science* **266**, 581–90.

Patel, N. H., Kornberg, T. B., and Goodman, C. S. (1989). Expression of *engrailed* during segmentation in grasshopper and crayfish. *Development* **107**, 201–12.

Patel, N. H., Ball, E. E., and Goodman, C. S. (1992). Changing role of *even-skipped* during the evolution of insect pattern formation. *Nature* **357**, 339–42.

Patel, N. H., Condron, B. G., and Zinn, K. (1994). Pair-rule expression patterns of *even-skipped* are found in both short- and long-germ beetles. *Nature* **367**, 429–34.

Pearson, J. E., and Ponce-Dawson, S. (1998). Crisis on skid row. *Physica A* **257**, 141–48.

Pecht, I., and Lancet, D. (1976). In *Chemical Relaxation in Molecular Biology* (R. Riegler and I. Pecht, eds.) Springer-Verlag, Heidelberg.

Perrin, F. (1934). Mouvement Brownien d'un ellipsoide (I). Dispersion dielectrique pour des molecules ellipsoidales. *J. Physique et la Radium* Serie 7, **5**, 497–511.

Perrin, F. (1936). Mouvement Brownien d'un ellipsoide (II). Rotation libre et depolarisation des fluorescences. Translation et diffusion des molecules ellipsoidales. *J. Physique et la Radium* Serie 7, **7**, 1–11.

Perris, R., and Perissonotto, D. (2000). Role of the extracellular matrix during neural crest migration. *Mech. Dev.* **95**, 3–21.

Peters, K. G., Werner, S., Chen, G., and Williams, L. T. (1992). Two FGF receptor genes are differentially expressed in epithelial and mesenchymal tissues during limb formation and organogenesis in the mouse. *Development* **114**, 233–43.

Phillips, H. M. (1969). Equilibrium measurements of embryonic cell adhesiveness: physical formulation and testing of the differential adhesion hypothesis. Ph.D. thesis, Johns Hopkins University, Baltimore.

Picton, H., Briggs, D., and Gosden, R. (1998). The molecular basis of oocyte growth and development. *Mol. Cell. Endocrinol.* **145**, 27–37.

Pollack, G. H. (2001). Is the gel a gel – and why does it matter? *Jpn J. Physiol.* **51**, 649–60.

Pourquié, O. (2000). Skin development: delta laid bare. *Curr. Biol.* **10**, R425–8.

Pourquié, O. (2001). Vertebrate somitogenesis. *Ann. Rev. Cell Dev. Biol.* **17**, 311–50.

Pourquié, O. (2003). The segmentation clock: converting embryonic time into spatial pattern. *Science* **301**, 328–30.

Pourquié, O., and Goldbeter, A. (2003). Segmentation clock: insights from computational models. *Curr. Biol.* **13**, R632–4.

Preston, B. N., Laurent, T. C., Comper, W. D., and Checkley, G. J. (1980). Rapid polymer transport in concentrated solutions through the formation of ordered structures. *Nature* **287**, 499–503.

Primmett, D. R., Norris, W. E., Carlson, G. J., Keynes, R. J., and Stern, C. D. (1989). Periodic segmental anomalies induced by heat shock in the chick embryo are associated with the cell cycle. *Development* **105**, 119–30.

Purcell, E. M. (1977). Life at low Reynolds number. *Am. J. Phys.* **45**, 3–11.

Quaas, J., and Wylie, C. (2002). Surface contraction waves (SCWs) in the *Xenopus* egg are required for the localization of the germ plasm and are dependent upon maternal stores of the kinesin-like protein Xklp1. *Dev. Biol.* **243**, 272–80.

Raff, R. A. (1996). *The Shape of Life: Genes, Development, and the Evolution of Animal Form*. University of Chicago Press, Chicago.

Ramirez-Weber, F. A., and Kornberg, T. B. (1999). Cytonemes: cellular processes that project to the principal signaling center in *Drosophila* imaginal discs. *Cell* **97**, 599–607.

Rappaport, R. (1966). Experiments concerning the cleavage furrow in invertebrate eggs. *J. Exp. Zool.* **161**, 1–8.

Rappaport, R. (1986). Establishment of the mechanism of cytokinesis in animal cells. *Int. Rev. Cytol.* **105**, 245–81.

Rappaport, R. (1996). *Cytokinesis in Animal Cells*. Cambridge University Press, Cambridge.

Rashevsky, N. (1960). *Mathematical Biophysics: Physico-mathematical Foundations of Biology*, Vol. 1, Dover, New York.

Raya, A., Kawakami, Y., Rodriguez-Esteban, C., *et al.* (2003). Notch activity induces Nodal expression and mediates the establishment of left–right asymmetry in vertebrate embryos. *Genes Dev.* **17**, 1213–8.

Raya, A., Kawakami, Y., Rodriguez-Esteban, C., *et al.* (2004). Notch activity acts as a sensor for extracellular calcium during vertebrate left–right determination. *Nature* **427**, 121–8.

Redner, S. (2001). *A Guide to First Passage Processes.* Cambridge University Press, Cambridge.

Reilly, K. M., and Melton, D. A. (1996). Short-range signaling by candidate morphogens of the TGF beta family and evidence for a relay mechanism of induction. *Cell* **86**, 743–54.

Reinitz, J., Mjolsness, E., and Sharp, D. H. (1995). Model for cooperative control of positional information in *Drosophila* by *bicoid* and maternal *hunchback*. *J. Exp. Zool.* **271**, 47–56.

Resnick, N., Collins, T., Atkinson, W., Bonthron, D. T., Dewey, C. F., Jr, and Gimbrone, M. A., Jr (1993). Platelet-derived growth factor B chain promoter contains a cis-acting fluid shear-stress-responsive element. *Proc. Nat. Acad. Sci. USA* **90**, 4591–5 (published erratum appears in *Proc. Nat. Acad. Sci. USA* **90**, 7908 (1993)).

Riedl, R. (1977). A systems-analytical approach to macro-evolutionary phenomena. *Q. Rev. Biol.* **52**, 351–70.

Rieu, J. P., Kataoka, N., and Sawada, Y. (1998). Quantitative analysis of cell motion during sorting in 2D aggregates of dissociated hydra cells. *Phys. Rev. E* **57**, 924–31.

Rieu, J. P., Upadhyaya, A., Glazier, J. A., Ouchi, N. B., and Sawada, Y. (2000). Diffusion and deformations of single hydra cells in cellular aggregates. *Biophys. J.* **79**, 1903–14.

Robinson, E. E., Zazzali, K. M., Corbett, S. A., and Foty, R. A. (2003). Alpha5beta1 integrin mediates strong tissue cohesion. *J. Cell. Sci.* **116**, 377–86.

Rodriguez, I., and Basler, K. (1997). Control of compartmental affinity boundaries by *hedgehog*. *Nature* **389**, 614–8.

Roegiers, F., McDougall, A., and Sardet, C. (1995). The sperm entry point defines the orientation of the calcium-induced contraction wave that directs the first phase of cytoplasmic reorganization in the ascidian egg. *Development* **121**, 3457–66.

Roegiers, F., Djediat, C., Dumollard, R., Rouviere, C., and Sardet, C. (1999). Phases of cytoplasmic and cortical reorganizations of the ascidian zygote between fertilization and first division. *Development* **126**, 3101–17.

Rooke, J. E., and Xu, T. (1998). Positive and negative signals between interacting cells for establishing neural fate. *Bioessays* **20**, 209–14.

Ross, M. H., Kaye, G. I., and Pawlina, W. (2003). *Histology: A Text and Atlas*. Lippincott Williams & Wilkins, Philadelphia, PA.

Roy, P., Petroll, W. M., Cavanagh, H. D., and Jester, J. V. (1999). Exertion of tractional force requires the coordinated up-regulation of cell contractility and adhesion. *Cell Motil. Cytoskeleton* **43**, 23–34.

Rubinstein, M., Colby, R. H., and Gillmor, J. R. (1989). Dynamic scaling for polymer gelation. *Polym. Prepr. Am. Chem. Soc. Div. Polym. Chem.* **30**, 81–2.

Rudnick, J., and Gaspari, G. (2004). *Elements of Random Walk*. Cambridge University Press, Cambridge.

Runft, L. L., Jaffe, L. A., and Mehlmann, L. M. (2002). Egg activation at fertilization: where it all begins. *Dev. Biol.* **245**, 237–54.

Rupp, R. A., Singhal, N., and Veenstra, G. J. (2002). When the embryonic genome flexes its muscles. *Eur. J. Biochem* **269**, 2294–9.

Rupp, P. A., Czirok, A., and Little, C. D. (2003). Novel approaches for the study of vascular assembly and morphogenesis in avian embryos. *Trends Cardiovasc Med.* **13**, 283–8.

Rustom, A., Saffrich, R., Markovic, I., Walther, P., and Gerdes, H. H. (2004). Nanotubular highways for intercellular organelle transport. *Science* **303**, 1007–10.

Ryan, P. L., Foty, R. A., Kohn, J., and Steinberg, M. S. (2001). Tissue spreading on implantable substrates is a competitive outcome of cell–cell vs. cell–substratum adhesivity. *Proc. Nat. Acad. Sci. USA* **98**, 4323–7.

Sahimi, M. (1994). *Applications of Percolation Theory*. Taylor & Francis, London.

Sakai, T., Larsen, M., and Yamada, K. M. (2003). Fibronectin requirement in branching morphogenesis. *Nature* **423**, 876–81.

Sakuma, R., Ohnishi Yi, Y., Meno, C. *et al.* (2002). Inhibition of Nodal signalling by Lefty mediated through interaction with common receptors and efficient diffusion. *Genes Cells* **7**, 401–12.

Salazar-Ciudad, I., and Jernvall, J. (2002). A gene network model accounting for development and evolution of mammalian teeth. *Proc. Nat. Acad. Sci. USA* **99**, 8116–20.

Salazar-Ciudad, I., Garcia-Fernandez, J., and Solé, R. V. (2000). Gene networks capable of pattern formation: from induction to reaction–diffusion. *J. Theor. Biol.* **205**, 587–603.

Salazar-Ciudad, I., Newman, S. A., and Solé, R. (2001a). Phenotypic and dynamical transitions in model genetic networks. I. Emergence of patterns and genotype–phenotype relationships. *Evolution & Development* **3**, 84–94.

Salazar-Ciudad, I., Solé, R., and Newman, S. A. (2001b). Phenotypic and dynamical transitions in model genetic networks. II. Application to the evolution of segmentation mechanisms. *Evolution & Development* **3**, 95–103.

Salazar-Ciudad, I., Jernvall, J., and Newman, S. A. (2003). Mechanisms of pattern formation in development and evolution. *Development* **130**, 2027–37.

Salthe, S. N. (1993). *Development and Evolution: Complexity and Change in Biology*. MIT Press, Cambridge, MA.

Sanchez, L., and Thieffry, D. (2001). A logical analysis of the *Drosophila* gap-gene system. *J. Theor. Biol.* **211**, 115–41.

Sanders, E. J. (1991). Embryonic cell invasiveness: an *in vitro* study of chick gastrulation. *J. Cell Sci.* **98**, 403–7.
Sardet, C., Roegiers, F., Dumollard, R., Rouviere, C., and McDougall, A. (1998). Calcium waves and oscillations in eggs. *Biophys. Chem.* **72**, 131–40.
Sardet, C., Prodon, F., Dumollard, R., Chang, P., and Chenevert, J. (2002). Structure and function of the egg cortex from oogenesis through fertilization. *Dev. Biol.* **241**, 1–23.
Sato, Y., Yasuda, K., and Takahashi, Y. (2002). Morphological boundary forms by a novel inductive event mediated by Lunatic fringe and Notch during somitic segmentation. *Development* **129**, 3633–44.
Saunders, J. W., Jr (1948). The proximo-distal sequence of origin of the parts of the chick wing and the role of the ectoderm. *J. Exp. Zool.* **108**, 363–402.
Savill, N. J., and Sherratt, J. A. (2003). Control of epidermal stem cell clusters by Notch-mediated lateral induction. *Dev. Biol.* **258**, 141–53.
Saxton, M. J., and Jacobson, K. (1997). Single-particle tracking: applications to membrane dynamics. *Ann. Rev. Biophys. Biomol. Struct.* **26**, 373–99.
Schier, A. F., and Gehring, W. J. (1993). Analysis of a *fushi tarazu* autoregulatory element: multiple sequence elements contribute to enhancer activity. *EMBO J.* **12**, 1111–9.
Schlosser, G. and Wagner, G. P. (eds.) (2004). *Modularity in Development and Evolution.* University of Chicago Press, Chicago.
Schmalhausen, I. I. (1949). *Factors of Evolution.* Blakiston, Philadelphia.
Schroeder, T. E. (1975). Dynamics of the contractile ring. *Soc. Gen. Physiol. Ser.* **30**, 305–34.
Schulte-Merker, S., and Smith, J. C. (1995). Mesoderm formation in response to Brachyury requires FGF signalling. *Curr. Biol.* **5**, 62–7.
Schuster, S., Marhl, M., and Hofer, T. (2002). Modelling of simple and complex calcium oscillations. From single-cell responses to intercellular signalling. *Eur. J. Biochem.* **269**, 1333–55.
Seifert, U. (1997). Configurations of fluid membranes and vesicles. *Adv. Phys.* **46**, 13–137.
Seilacher, A. (1992). *Vendobionta* and *Psammocorallia* – lost constructions of precambrian evolution. *J. Geolog. Soc. Lond.* **149**, 607–13.
Serini, G., Ambrosi, D., Giraudo, E., Gamba, A., Preziosi, L., and Bussolino, F. (2003). Modeling the early stages of vascular network assembly. *EMBO J.* **22**, 1771–9.
Shafrir, Y., and Forgacs, G. (2002). Mechanotransduction through the cytoskeleton. *Am. J. Physiol. Cell Physiol.* **282**, C479–86.
Shafrir, Y., ben-Avraham, D., and Forgacs, G. (2000). Trafficking and signaling through the cytoskeleton: a specific mechanism. *J. Cell Sci.* **113**, 2747–57.
Shapiro, L., Fannon, A. M., Kwong, P. D., *et al.* (1995). Structural basis of cell–cell adhesion by cadherins. *Nature* **374**, 327–37.
Shav-Tal, Y., Darzacq, X., Shenoy, S. M., *et al.* (2004). Dynamics of single mRNPs in nuclei of living cells. *Science* **304**, 1797–800.
Sheetz, M. P. (2001). Cell control by membrane-cytoskeleton adhesion. *Nat. Rev. Mol. Cell Biol.* **2**, 392–6.
Sherwood, D. R., and McClay, D. R. (2001). LvNotch signaling plays a dual role in regulating the position of the ectoderm–endoderm boundary in the sea urchin embryo. *Development* **128**, 2221–32.
Shih, J., and Keller, R. (1992). Cell motility driving mediolateral intercalation in explants of *Xenopus laevis*. *Development* **116**, 901–14.

Shyy, J. Y., Li, Y. S., Lin, M. C., Chen, W., Yuan, S., Usami, S., and Chien, S. (1995a). Multiple cis-elements mediate shear stress-induced gene expression. *J. Biomech.* **28**, 1451–7.

Shyy, J. Y., Lin, M. C., Han, J., Lu, Y., Petrime, M., and Chien, S. (1995b). The cis-acting phorbol ester "12-O-tetradecanoylphorbol 13-acetate"- responsive element is involved in shear stress-induced monocyte chemotactic protein 1 gene expression. *Proc. Nat. Acad. Sci. USA* **92**, 8069–73.

Simonneau, L., Kitagawa, M., Suzuki, S., and Thiery, J. P. (1995). *Cadherin 11* expression marks the mesenchymal phenotype: towards new functions for cadherins? *Cell Adhes. Commun.* **3**, 115–30.

Singer, S. J., and Nicolson, G. L. (1972). The fluid mosaic model of the structure of cell membranes. *Science* **175**, 720–31.

Sivasankar, S., Brieher, W., Lavrik, N., Gumbiner, B., and Leckband, D. (1999). Direct molecular force measurements of multiple adhesive interactions between cadherin ectodomains. *Proc. Nat. Acad. Sci. USA* **96**, 11820–4.

Sivasankar, S., Gumbiner, B., and Leckband, D. (2001). Direct measurements of multiple adhesive alignments and unbinding trajectories between cadherin extracellular domains. *Biophys. J.* **80**, 1758–68.

Small, S., Kraut, R., Hoey, T., Warrior, R., and Levine, M. (1991). Transcriptional regulation of a pair-rule stripe in *Drosophila*. *Genes Dev.* **5**, 827–39.

Small, S., Blair, A., and Levine, M. (1992). Regulation of *even-skipped* stripe 2 in the *Drosophila* embryo. *EMBO J.* **11**, 4047–57.

Small, S., Blair, A., and Levine, M. (1996). Regulation of two pair-rule stripes by a single enhancer in the *Drosophila* embryo. *Dev. Biol.* **175**, 314–24.

Solnica-Krezel, L. (2003). Vertebrate development: taming the nodal waves. *Curr. Biol.* **13**, R7–R9.

Spehr, M., Gisselmann, G., Poplawski, A., *et al.* (2003). Identification of a testicular odorant receptor mediating human sperm chemotaxis. *Science* **299**, 2054–8.

Spemann, H., and Mangold, H. (1924). Über Induktion von Embryonalanlagen durch Implantation artfremder Organisatoren. *Wilhelm Roux' Arch. Entw. Mech. Org.* **100**, 599–638.

Spooner, B. S., and Wessells, N. K. (1972). An analysis of salivary gland morphogenesis: role of cytoplasmic microfilaments and microtubules. *Dev. Biol.* **27**, 38–54.

St Johnston, D., and Nusslein-Volhard, C. (1992). The origin of pattern and polarity in the *Drosophila* embryo. *Cell* **68**, 201–19.

Standley, H. J., Zorn, A. M., and Gurdon, J. B. (2001). eFGF and its mode of action in the community effect during *Xenopus* myogenesis. *Development* **128**, 1347–57.

Starz-Gaiano, M., and Lehmann, R. (2001). Moving towards the next generation. *Mech. Dev.* **105**, 5–18.

Stauber, M., Jackle, H., and Schmidt-Ott, U. (1999). The anterior determinant bicoid of *Drosophila* is a derived *Hox* class 3 gene. *Proc. Nat. Acad. Sci. USA* **96**, 3786–9.

Steinberg, M. S. (1963). Reconstruction of tissues by dissociated cells. Some morphogenetic tissue movements and the sorting out of embryonic cells may have a common explanation. *Science* **141**, 401–8.

Steinberg, M. S. (1978). Specific cell ligands and the differential adhesion hypothesis: how do they fit together? In *Specificity of Embryological Interactions* (D. R. Garrod, ed.), pp. 97–130. Chapman and Hall, London.

Steinberg, M. S. (1998). Goal-directedness in embryonic development. *Integrative Biology* **1**, 49–59.

Steinberg, M. S., and Foty, R. A. (1997). Intercellular adhesions as determinants of tissue assembly and malignant invasion. *J. Cell Physiol.* **173**, 135–9.

Steinberg, M. S., and Poole, T. J. (1982). Liquid behavior of embryonic tissues. In *Cell Behavior* (R. Bellairs and A. S. G. Curtis, eds.), pp. 583–607. Cambridge University Press, Cambridge.

Steinberg, M. S., and Takeichi, M. (1994). Experimental specification of cell sorting, tissue spreading, and specific spatial patterning by quantitative differences in cadherin expression. *Proc. Nat. Acad. Sci. USA* **91**, 206–9.

Stern, C. D., and Bellairs, R. (1984). Mitotic activity during somite segmentation in the early chick embryo. *Anat. Embryol. (Berlin)* **169**, 97–102.

Stollewerk, A., Schoppmeier, M., and Damen, W. G. (2003). Involvement of *Notch* and *Delta* genes in spider segmentation. *Nature* **423**, 863–5.

Stopak, D., and Harris, A. K. (1982). Connective tissue morphogenesis by fibroblast traction. I. Tissue culture observations. *Dev. Biol.* **90**, 383–398.

Stossel, T. P. (2001). Manifesto for a cytoplasmic revolution. *Science* **293**, 611.

Stricker, S. A. (1999). Comparative biology of calcium signaling during fertilization and egg activation in animals. *Dev. Biol.* **211**, 157–76.

Strogatz, S. H. (1994). *Nonlinear Dynamics and Chaos: With Applications to Physics, Biology, Chemistry, and Engineering*. Perseus, Cambridge, MA.

Strogatz, S. H. (2003). *Sync: The Emerging Science of Spontaneous Order*. Theia, New York.

Subtelny, S., and Penkala, J. E. (1984). Experimental evidence for a morphogenetic role in the emergence of primordial germ cells from the endoderm in *Rana pipiens*. *Differentiation* **26**, 211–19.

Sun, B., Bush, S., Collins-Racie, L., *et al.* (1999). *derriere*: a TGF-beta family member required for posterior development in *Xenopus*. *Development* **126**, 1467–82.

Sun, Q. Y. (2003). Cellular and molecular mechanisms leading to cortical reaction and polyspermy block in mammalian eggs. *Microsc. Res. Tech.* **61**, 342–8.

Supp, D. M., Witte, D. P., Potter, S. S., and Brueckner, M. (1997). Mutation of an axonemal dynein affects left–right asymmetry in inversus viscerum mice. *Nature* **389**, 963–6.

Supp, D. M., Potter, S. S., and Brueckner, M. (2000). Molecular motors: the driving force behind mammalian left–right development. *Trends Cell Biol.* **10**, 41–5.

Suzuki, K., Tanaka, Y., Nakajima, Y., *et al.* (1995). Spatiotemporal relationships among early events of fertilization in sea urchin eggs revealed by multiview microscopy. *Biophys. J.* **68**, 739–48.

Sveiczer, A., Csikasz-Nagy, A., Gyorffy, B., Tyson, J. J., and Novak, B. (2000). Modeling the fission yeast cell cycle: quantized cycle times in wee1–cdc25Delta mutant cells. *Proc. Nat. Acad. Sci. USA* **97**, 7865–70.

Szebenyi, G., Savage, M. P., Olwin, B. B., and Fallon, J. F. (1995). Changes in the expression of fibroblast growth factor receptors mark distinct stages of chondrogenesis *in vitro* and during chick limb skeletal patterning. *Dev. Dyn.* **204**, 446–56.

Takahashi, Y., and Nogawa, H. (1991). Branching morphogenesis of mouse salivary epithelium in basement membrane-like substratum separated from mesenchyme by the membrane filter. *Development* **111**, 327–35.

Vernon, R. B., Lara, S. L., Drake, C. J., *et al.* (1995). Organized type I collagen influences endothelial patterns during "spontaneous angiogenesis *in vitro*": planar cultures as models of vascular development. *In Vitro Cell Dev. Biol. Anim.* **31**, 120–31.

Vilar, J. M., Kueh, H. Y., Barkai, N., and Leibler, S. (2002). Mechanisms of noise resistance in genetic oscillators. *Proc. Nat. Acad. Sci. USA* **99**, 5988–92.

von Dassow, G., and Munro, E. (1999). Modularity in animal development and evolution: elements of a conceptual framework for EvoDevo. *J. Exp. Zool. (Mol. Dev. Evol.)* **285**, 307–25.

von Dassow, G., Meir, E., Munro, E. M., and Odell, G. M. (2000). The segment polarity network is a robust developmental module. *Nature* **406**, 188–92.

von Hippel, P. H., and Berg, O. G. (1989). Facilitated target location in biological systems. *J. Biol. Chem.* **264**, 675–8.

Voronov, D. A., and Taber, L. A. (2002). Cardiac looping in experimental conditions: effects of extraembryonic forces. *Dev. Dyn.* **224**, 413–21.

Voronov, D. A., Alford, P. W., Xu, G., and Taber, L. A. (2004). The role of mechanical forces in dextral rotation during cardiac looping in the chick embryo. *Dev. Biol.* **272**, 339–50.

Waddington, C. H. (1942). Canalization of development and the inheritance of acquired characters. *Nature* **150**, 563–5.

Wagner, G. P., and Altenberg, L. (1996). Complex adaptations and the evolution of evolvability. *Evolution* **50**, 967–6.

Wakamatsu, Y., Maynard, T. M., and Weston, J. A. (2000). Fate determination of neural crest cells by NOTCH-mediated lateral inhibition and asymmetrical cell division during gangliogenesis. *Development* **127**, 2811–21.

Wakely, J., and England, M. A. (1977). Scanning electron microscopy (SEM) of the chick embryo primitive streak. *Differentiation* **7**, 181–6.

Wallingford, J. B., and Harland, R. M. (2002). Neural tube closure requires Dishevelled-dependent convergent extension of the midline. *Development* **129**, 5815–25.

Wang, N., and Stamenovic, D. (2000). Contribution of intermediate filaments to cell stiffness, stiffening, and growth. *Am. J. Physiol. Cell Physiol.* **279**, C188–94.

Wang, N., Butler, J. P., and Ingber, D. E. (1993). Mechanotransduction across the cell surface and through the cytoskeleton. *Science* **260**, 1124–7.

Wassarman, P., Chen, J., Cohen, N., Litscher, E., Liu, C., Qi, H., and Williams, Z. (1999). Structure and function of the mammalian egg zona pellucida. *J. Exp. Zool.* **285**, 251–8.

Wassarman, P. M. (1999). Mammalian fertilization: molecular aspects of gamete adhesion, exocytosis, and fusion. *Cell* **96**, 175–83.

Waters, C. M., Oberg, K. C., Carpenter, G., and Overholser, K. A. (1990). Rate constants for binding, dissociation, and internalization of EGF: effect of receptor occupancy and ligand concentration. *Biochemistry* **29**, 3563–9.

Wearing, H. J., Owen, M. R., and Sherratt, J. A. (2000). Mathematical modelling of juxtacrine patterning. *Bull. Math. Biol.* **62**, 293–320.

Webb, S. D., and Owen, M. R. (2004). Oscillations and patterns in spatially discrete models for developmental intercellular signalling. *J. Math. Biol.* **48**, 444–76.

Weidinger, G., Wolke, U., Koprunner, M., Thisse, C., Thisse, B., and Raz, E. (2002). Regulation of zebrafish primordial germ cell migration by attraction towards an intermediate target. *Development* **129**, 25–36.

Weng, W., and Stemple, D. L. (2003). Nodal signaling and vertebrate germ layer formation. *Birth Defects Res. Part C Embryo Today* **69**, 325–32.

Wessel, G. M., and McClay, D. R. (1987). Gastrulation in the sea urchin embryo requires the deposition of crosslinked collagen within the extracellular matrix. *Dev. Biol.* **121**, 149–65.

West-Eberhard, M. J. (2003). *Developmental Plasticity and Evolution*. Oxford University Press, Oxford, New York.

Wheelock, M. J., and Johnson, K. R. (2003). Cadherins as modulators of cellular phenotype. *Ann. Rev. Cell Dev. Biol.* **19**, 207–35.

White, J. G., and Borisy, G. G. (1983). On the mechanisms of cytokinesis in animal cells. *J. Theor. Biol.* **101**, 289–316.

Wikramanayake, A. H., Huang, L., and Klein, W. H. (1998). beta-Catenin is essential for patterning the maternally specified animal–vegetal axis in the sea urchin embryo. *Proc. Nat. Acad. Sci. USA* **95**, 9343–8.

Wilkins, A. S. (1992). *Genetic analysis of animal development*. Wiley-Liss, New York.

Wilkins, A. S. (1997). Canalization: a molecular genetic perspective. *BioEssays* **19**, 257–62.

Wilkins, A. S. (2002). *The Evolution of Developmental Pathways*. Sinauer Associates, Sunderland, MA.

Williams, A. F., and Barclay, A. N. (1988). The immunoglobulin superfamily – domains for cell surface recognition. *Ann. Rev. Immunol.* **6**, 381–405.

Wilson, P. D. (2004). Polycystic kidney disease: new understanding in the pathogenesis. *Int. J. Biochem. Cell Biol.* **36**, 1868–73.

Wimmer, E. A., Carleton, A., Harjes, P., Turner, T., and Desplan, C. (2000). Bicoid-independent formation of thoracic segments in *Drosophila*. *Science* **287**, 2476–9.

Winther, R. G. (2001). Varieties of modules: kinds, levels, origins, and behaviors. *J. Exp. Zool.* **291**, 116–29.

Wolf, D. M., and Eeckman, F. H. (1998). On the relationship between genomic regulatory element organization and gene regulatory dynamics. *J. Theor. Biol.* **195**, 167–86.

Wolf, J., and Heinrich, R. (1997). Dynamics of two-component biochemical systems in interacting cells; synchronization and desynchronization of oscillations and multiple steady states. *Biosystems* **43**, 1–24.

Wolfram, S. (2002). *A New Kind of Science*. Wolfram Media, Champaign, IL.

Wolpert, L. (1969). Positional information and the spatial pattern of cellular differentiation. *J. Theor. Biol.* **25**, 1–47.

Wolpert, L. (2002). *Principles of Development*. Oxford University Press, Oxford, New York.

Wong, G. K., Allen, P. G., and Begg, D. A. (1997). Dynamics of filamentous actin organization in the sea urchin egg cortex during early cleavage divisions: implications for the mechanism of cytokinesis. *Cell Motil. Cytoskeleton* **36**, 30–42.

Woolley, D. M., and Vernon, G. G. (2002). Functional state of the axonemal dyneins during flagellar bend propagation. *Biophys. J.* **83**, 2162–9.

Wylie, C. C., and Heasman, J. (1993). Migration, proliferation, and potency of primordial germ cells. *Seminars in Dev. Biol.* **4**, 161–170.

Xiao, S., and Knoll, A. H. (2000). Phosphatized animal embryos from the Neoproterozoic Doushantuo Formation in Weng'an, Guizhou, South China. *Paleontology* **74**, 767–88.

Xiao, S., Yuan, X., and Knoll, A. H. (2000). Eumetazoan fossils in terminal Proterozoic phosphorites? *Proc. Nat. Acad. Sci. USA* **97**, 13 684–9.

Xu, Z., and Tung, V. W. (2001). Temporal and spatial variations in slow axonal transport velocity along peripheral motoneuron axons. *Neuroscience* **102**, 193–200.

Yamada, S., Wirtz, D., and Kuo, S. C. (2000). Mechanics of living cells measured by laser tracking microrheology. *Biophys. J.* **78**, 1736–47.

Yamamoto, K., and Yoneda, M. (1983). Cytoplasmic cycle in meiotic division of starfish oocytes. *Dev. Biol.* **96**, 166–72.

Yamamoto, M., Saijoh, Y., Perea-Gomez, A., *et al.* (2004). Nodal antagonists regulate formation of the anteroposterior axis of the mouse embryo. *Nature* **428**, 387–92.

Yanagimachi, R., and Noda, Y. D. (1970). Electron microscope studies of sperm incorporation into the golden hamster egg. *Am. J. Anat.* **128**, 429–62.

Yoneda, M. (1973). Tension at the surface of sea urchin eggs on the basis of "liquid-drop" concept. *Adv. Biophys.* **4**, 153–90.

Yoshida, M., Inaba, K., and Morisawa, M. (1993). Sperm chemotaxis during the process of fertilization in the ascidians *Ciona savignyi* and *Ciona intestinalis*. *Dev. Biol.* **157**, 497–506.

Yurchenco, P. D., and O'Rear, J. J. (1994). Basal lamina assembly. *Curr. Opin. Cell. Biol.* **6**, 674–681.

Zachariae, W., and Nasmyth, K. (1999). Whose end is destruction: cell division and the anaphase-promoting complex. *Genes Dev.* **13**, 2039–58.

Zahalak, G. I., Wagenseil, J. E., Wakatsuki, T., and Elson, E. L. (2000). A cell-based constitutive relation for bio-artificial tissues. *Biophys. J.* **79**, 2369–81.

Zajac, M., Jones, G. L., and Glazier, J. A. (2000). Model of convergent extension in animal morphogenesis. *Phys. Rev. Lett.* **85**, 2022–5.

Zajac, M., Jones, G. L., and Glazier, J. A. (2003). Simulating convergent extension by way of anisotropic differential adhesion. *J. Theor. Biol.* **222**, 247–59.

Zeng, W., Thomas, G. L., Newman, S. A., and Glazier, J. A. (2003). A novel mechanism for mesenchymal condensation during limb chondrogenesis *in vitro*. In *Mathematical Modelling and Computing in Biology and Medicine, Proc. Fifth ESMTB Conf 2002* (V. Capasso, ed.), pp. 80–6. Società Editrice Esculapio, Bologna, Italy.

Zhang, J., Houston, D. W., King, M. L., Payne, C., Wylie, C., and Heasman, J. (1998). The role of maternal VegT in establishing the primary germ layers in *Xenopus* embryos. *Cell* **94**, 515–24.

Index